This book is based on expository lectures by six internationally known experts presented at the 2002 MSRI introductory workshop on commutative algebra. They focus on the interaction of commutative algebra with other areas of mathematics, including algebraic geometry, group cohomology and representation theory, and combinatorics, with all necessary background provided. Short complementary papers describing work at the research frontier are also included. The unusual scope and format make the book invaluable reading for graduate students and researchers interested in commutative algebra and its various uses.

Mathematical Sciences Research Institute
Publications

51

Trends in Commutative Algebra

Mathematical Sciences Research Institute Publications

Volumes 1–4 and 6–27 are published by Springer-Verlag

Trends in
Commutative Algebra

Edited by

Luchezar L. Avramov
University of Nebraska

Mark Green
University of California, Los Angeles

Craig Huneke
University of Kansas

Karen E. Smith
University of Michigan

Bernd Sturmfels
University of California, Berkeley

CAMBRIDGE
UNIVERSITY PRESS

CAMBRIDGE UNIVERSITY PRESS
Cambridge, New York, Melbourne, Madrid, Cape Town, Singapore,
São Paulo, Delhi, Dubai, Tokyo, Mexico City

Cambridge University Press
The Edinburgh Building, Cambridge CB2 8RU, UK

Published in the United States of America by Cambridge University Press, New York

www.cambridge.org
Information on this title: www.cambridge.org/9780521168724

First published 2004
First paperback edition 2010

A catalogue record for this publication is available from the British Library

Library of Congress Cataloguing in Publication data

Trends in commutative algebra / edited by Luchezar L. Avramov . . . [et al.].
p. cm.
(Mathematical Sciences Research Institute publications ; 51)
Includes bibliographical references and index.
ISBN 0-521-83195-4 (hb)
1. Commutative algebra. I. Avramov, L. L. (Luchezar L.), 1948–
II. Title. III. Series.
QA251.3.T74 2005
512′.44–dc22   2004022712

ISBN 978-0-521-83195-6 Hardback
ISBN 978-0-521-16872-4 Paperback

Trends in Commutative Algebra
MSRI Publications
Volume 51, 2004

Contents

Trends in Commutative Algebra
MSRI Publications
Volume **51**, 2004

Preface

Over the last fifteen years, commutative algebra has experienced a striking evolution. During this period the outlook of the subject has been altered, new connections to other areas have been established, and powerful techniques have been developed. To foster further development a year-long program on commutative algebra was held at MSRI during the 2002–03 academic year, starting with an introductory workshop on September 9–13, 2002. This workshop concentrated on the interplay and growing connections between commutative algebra and other areas, such as algebraic geometry, the cohomology of groups, and combinatorics.

Six main speakers each gave a series of three talks during the week: David Benson, David Eisenbud, Mark Haiman, Melvin Hochster, Rob Lazarsfeld, and Bernard Teissier. The workshop was very well attended, with more than 120 participants. Every series of main talks was supplemented by a discussion/talk session presented by a young researcher: Manuel Blickle, Ana Bravo, Srikanth Iyengar, Graham Leuschke, Ezra Miller, and Jessica Sidman. Each of these speakers has contributed a paper, or in some cases a combined paper, in this volume.

David Benson spoke on the cohomology of groups, presenting some of the many questions which are unanswered and which have a close relationship to modern commutative algebra. He gave us many convincing reasons for working in the "graded" commutative case, where signs are introduced when commuting elements of odd degree. Srikanth Iyengar gives background information for Benson's notes.

David Eisenbud spoke on a classical subject in commutative algebra: free resolutions. In his paper with a chapter by Jessica Sidman, he visits this classic territory with a different perspective, by drawing close ties between graded free resolutions and the geometry of projective varieties. He leads us through recent developments, including Mark Green's proof of the linear syzygy conjecture.

Mark Haiman lectured on the commutative algebra of n points in the plane. This leads quite rapidly to the geometry of the Hilbert scheme, and to substantial combinatorial questions (and answers) which can be phrased in terms of common questions in commutative algebra such as asking about the Cohen–Macaulay property for certain Rees algebras. Ezra Miller writes an appendix about the Hilbert scheme of n points in the plane.

Mel Hochster gave three lectures on tight closure, telling eleven reasons to study tight closure. Hochster presents tight closure as a test for ideal membership which is necessary, but not sufficient, except for certain rings such as regular rings. Graham Leuschke's appendix gives examples of computation of tight closure.

The theory of multiplier ideals has been expanding rapidly in the last few years and bears a close relationship to commutative algebra, particularly tight closure. Rob Lazarsfeld and Manuel Blickle present a gentle introduction to this theory, with emphasis on the important theorems and concepts, applications, and examples.

Resolution of singularities has long played a crucial role in algebraic geometry and commutative algebra. Bernard Teissier talked about new ideas for understanding resolution coming from the simplest of all polynomials: monomials and binomials. Toric geometry of course enters into this story in a crucial way. Ana Bravo provides a summary of results on SAGBI bases which enter into this story.

The editors of this volume, who formed the organizing committee for the year program, would like to thank the many people who made the year possible, and thank the speakers for their wonderful contributions. A special thanks to David Eisenbud, the director of MSRI, without whom none of this would have been possible. We thank Michael Singer, the acting director of MSRI during the academic year when the program took place, for his generous help, and for the loan of Eisenbud to participate in our program. The great staff at MSRI were unfailingly helpful, friendly and professional. We thank the MSRI editor, Silvio Levy, for all his work on this volume. Finally, we thank the National Science Foundation for its support of institutes of mathematics in general, and of MSRI in particular.

We hope the papers in this volume will be a springboard for further learning and research for both experts and beginners.

Luchezar Avramov
Mark Green
Craig Huneke
Karen Smith
Bernd Sturmfels

Note: The lectures this volume is based on were videotaped. They are available on streaming video and for downloading at www.msri.org/publications/video or www.cambridge.org/0521831954.

Trends in Commutative Algebra
MSRI Publications
Volume **51**, 2004

Commutative Algebra
in the Cohomology of Groups

DAVE BENSON

ABSTRACT. Commutative algebra is used extensively in the cohomology
of groups. In this series of lectures, I concentrate on finite groups, but I
also discuss the cohomology of finite group schemes, compact Lie groups,
p-compact groups, infinite discrete groups and profinite groups. I describe
the role of various concepts from commutative algebra, including finite generation, Krull dimension, depth, associated primes, the Cohen–Macaulay
and Gorenstein conditions, local cohomology, Grothendieck's local duality,
and Castelnuovo–Mumford regularity.

CONTENTS

The author is partly supported by NSF grant DMS-9988110.

1. Introduction

The purpose of these lectures is to explain how commutative algebra is used in the cohomology of groups. My interpretation of the word "group" is catholic: the kinds of groups in which I shall be interested include finite groups, finite group schemes, compact Lie groups, p-compact groups, infinite discrete groups, and profinite groups, although later in the lectures I shall concentrate more on the case of finite groups, where representation theoretic methods are most effective. In each case, there are finite generation theorems which state that under suitable conditions, the cohomology ring is a graded commutative Noetherian ring; over a field k, this means that it is a finitely generated graded commutative k-algebra.

Although graded commutative is not quite the same as commutative, the usual concepts from commutative algebra apply. These include the maximal/prime ideal spectrum, Krull dimension, depth, associated primes, the Cohen–Macaulay and Gorenstein conditions, local cohomology, Grothendieck's local duality, and so on. One of the themes of these lectures is that the rings appearing in group cohomology theory are quite special. Most finitely generated graded commutative k-algebras are not candidates for the cohomology ring of a finite (or compact Lie, or virtual duality, or p-adic Lie, or ...) group. The most powerful restrictions come from local cohomology spectral sequences such as the Greenlees spectral sequence $H_{\mathfrak{m}}^{s,t} H^*(G,k) \implies H_{-s-t}(G,k)$, which can be viewed as a sort of duality theorem. We describe how to construct such spectral sequences and obtain information from them.

The companion article to this one, [Iyengar 2004], explains some of the background material that may not be familiar to commutative algebraists. A number of references are made to that article, and for distinctiveness, I write [Sri].

2. Some Examples

For motivation, let us begin with some examples. We defer until the next section the definition of group cohomology

$$H^*(G,k) = \mathrm{Ext}^*_{kG}(k,k)$$

(or see § 6 of [Sri]). All the examples in this section are for finite groups G over a field of coefficients k.

(2.1)   The first comment is that in the case where k is a field of characteristic zero or characteristic not dividing the order of G, Maschke's theorem in representation theory shows that all kG-modules are projective (see Theorem 3.1 of [Sri]). So for any kG-modules M and N, and all $i > 0$, we have $\mathrm{Ext}^i_{kG}(M,N) = 0$. In particular, $H^*(G,k)$ is just k, situated in degree zero. Given this fact, it makes sense to look at examples where k has characteristic p dividing $|G|$.

(2.2)   Next, we discuss finite abelian groups. See also § 7.4 of [Sri]. The Künneth theorem implies that

(2.2.1) $H^*(G_1 \times G_2, k) \cong H^*(G_1, k) \otimes_k H^*(G_2, k).$

So we decompose our finite abelian group as a direct product of cyclic groups of prime power order. The factors of order coprime to the characteristic may be thrown away, using (2.1). For a cyclic p-group in characteristic p, there are two possibilities (Proposition 7.3 of [Sri]). If $p = 2$ and $|G| = 2$, then $H^*(G, k) = k[x]$ where x has degree one. In all other cases (i.e., p odd, or $p = 2$ and $|G| \geq 4$), we have $H^*(G, k) = k[x, y]/(x^2)$ where x has degree one and y has degree two. It follows that if G is any finite abelian group then $H^*(G, k)$ is a tensor product of a polynomial ring and a (possibly trivial) exterior algebra.

(2.2.2)   In particular, if G is a finite elementary abelian p-group of rank r (i.e., a product of r copies of $\mathbb{Z}/p$) and k is a field of characteristic p, then the cohomology ring is as follows. For $p = 2$, we have

$$H^*((\mathbb{Z}/2)^r, k) = k[x_1, \ldots, x_r]$$

with $|x_i| = 1$, while for p odd, we have

$$H^*((\mathbb{Z}/p)^r, k) = \Lambda(x_1, \ldots, x_r) \otimes k[y_1, \ldots, y_r]$$

with $|x_i| = 1$ and $|y_i| = 2$. In the latter case, the nil radical is generated by $x_1, \ldots, x_r$, and in both cases the quotient by the nil radical is a polynomial ring in r generators.

(2.3)   The next comment is that if S is a Sylow p-subgroup of G then a transfer argument shows that the restriction map from $H^*(G, k)$ to $H^*(S, k)$ is injective. What's more, the stable element method of Cartan and Eilenberg [1956] identifies the image of this restriction map. For example, if $S \trianglelefteq G$ then $H^*(G, k) = H^*(S, k)^{G/S}$, the invariants of G/S acting on the cohomology of S (see § 7.6 of [Sri]). It follows that really important case is where G is a p-group and k has characteristic p. Abelian p-groups are discussed in (2.2), so let's look at some nonabelian p-groups.

(2.4)   Consider the quaternion group of order eight,

(2.4.1) $Q_8 = \langle g, h \mid gh = h^{-1}g = hg^{-1} \rangle.$

There is an embedding

$$g \mapsto i, \quad h \mapsto j, \quad gh \mapsto k, \quad g^2 = h^2 = (gh)^2 \mapsto -1$$

of Q_8 into the unit quaternions (i.e., $SU(2)$), which form a three dimensional sphere S^3. So left multiplication gives a free action of Q_8 on S^3; in other words, each nonidentity element of the group has no fixed points on the sphere. The quotient S^3/Q_8 is an orientable three dimensional manifold, whose cohomology therefore satisfies Poincaré duality. The freeness of the action implies that we can choose a CW decomposition of S^3 into cells permuted freely by Q_8. Taking cellular chains with coefficients in $\mathbb{F}_2$, we obtain a complex of free $\mathbb{F}_2Q_8$-modules of length four, whose homology consists of one copy of $\mathbb{F}_2$ at the beginning and another copy at the end. Making suitable choices for the cells, this looks as follows.

$$0 \to \mathbb{F}_2Q_8 \xrightarrow{\binom{g-1}{h-1}} (\mathbb{F}_2Q_8)^2 \xrightarrow{\left(\begin{smallmatrix} h-1 & hg+1 \\ gh+1 & g-1 \end{smallmatrix}\right)} (\mathbb{F}_2Q_8)^2 \xrightarrow{(g-1 \ h-1)} \mathbb{F}_2Q_8 \to 0$$

So we can form a Yoneda splice of an infinite number of copies of this sequence to obtain a free resolution of $\mathbb{F}_2$ as an $\mathbb{F}_2Q_8$-module. The upshot of this is that we obtain a decomposition for the cohomology ring

$$(2.4.2) \qquad H^*(Q_8, \mathbb{F}_2) = \mathbb{F}_2[z] \otimes_{\mathbb{F}_2} H^*(S^3/Q_8; \mathbb{F}_2)$$
$$= \mathbb{F}_2[x, y, z]/(x^2 + xy + y^2, x^2y + xy^2),$$

where z is a polynomial generator of degree four and x and y have degree one. This structure is reflected in the Poincaré series

$$\sum_{i=0}^{\infty} t^i \dim H^i(Q_8, \mathbb{F}_2) = (1 + 2t + 2t^2 + t^3)/(1 - t^4).$$

The decomposition (2.4.2) into a polynomial piece and a finite Poincaré duality piece can be expressed as follows (cf. § 11):

$$\boxed{H^*(Q_8, \mathbb{F}_2) \text{ IS A GORENSTEIN RING.}}$$

(2.5)   We recall that the meanings of Cohen–Macaulay and Gorenstein in this context are as follows. Let R be a finitely generated graded commutative k-algebra with $R_0 = k$ and $R_i = 0$ for $i < 0$. Then Noether's normalization lemma guarantees the existence of a homogeneous polynomial subring $k[x_1, \ldots, x_r]$ over which R is finitely generated as a module.

PROPOSITION 2.5.1. *If R is of the type described in the previous paragraph, then the following are equivalent.*

(a) *There exists a homogeneous polynomial subring $k[x_1, \ldots, x_r] \subseteq R$ such that R is finitely generated and free as a module over $k[x_1, \ldots, x_r]$.*

(b) *If $k[x_1, \ldots, x_r] \subseteq R$ is a homogeneous polynomial subring such that R is finitely generated as a $k[x_1, \ldots, x_r]$-module then R a free $k[x_1, \ldots, x_r]$-module.*

(c) *There exist homogeneous elements of positive degree $x_1, \ldots, x_r$ forming a regular sequence, and $R/(x_1, \ldots, x_r)$ has finite rank as a k-vector space.*

We say that R is *Cohen–Macaulay* of dimension r if the equivalent conditions of the above proposition hold.

(2.6) If R is Cohen–Macaulay, and the quotient ring $R/(x_1, \ldots, x_r)$ has a simple socle, then we say that R is *Gorenstein*. Whether this condition holds is independent of the choice of the polynomial subring. Another way to phrase the condition is that $R/(x_1, \ldots, x_r)$ is injective as a module over itself. This quotient satisfies Poincaré duality, in the sense that if the socle lies in degree d (d is called the *dualizing degree*) and we write

$$p(t) = \sum_{i=0}^{\infty} t^i \dim_k (R/(x_1, \ldots, x_r))_i$$

then

(2.6.1) $t^d p(1/t) = p(t).$

Setting

$$P(t) = \sum_{i=0}^{\infty} t^i \dim_k R_i,$$

the freeness of R over $k[x_1, \ldots, x_r]$ implies that $P(t)$ is the power series expansion of the rational function $p(t)/\prod_{i=1}^{r}(1 - t^{|x_i|})$. So plugging in equation (2.6.1), we obtain the functional equation

(2.6.2) $P(1/t) = (-t)^r t^{-a} P(t),$

where $a = d - \sum_{i=1}^{r}(|x_i| - 1)$. We say that R is Gorenstein with *a-invariant* a.

Another way of expressing the Gorenstein condition is as follows. If R (as above) is Cohen–Macaulay, then the local cohomology $H_{\mathfrak{m}}^{s,t} R$ is only nonzero for $s = r$. The graded dual of $H_{\mathfrak{m}}^{r,*} R$ is called the *canonical module*, and written Ω_R. To say that R is Gorenstein with *a-invariant* a is the same as saying that Ω_R is a copy of R shifted so that the identity element lies in degree $r - a$.

In the case of $H^*(Q_8, \mathbb{F}_2)$, we can choose the polynomial subring to be $k[z]$. The ring $H^*(Q_8, \mathbb{F}_2)$ is a free module over $k[z]$ on six generators, corresponding to a basis for the graded vector space $H^*(S^3/Q_8; \mathbb{F}_2) \cong H^*(Q_8, \mathbb{F}_2)/(z)$, which satisfies Poincaré duality with $d = 3$. So in this case the a-invariant is $3 - (4-1) = 0$. We have $p(t) = 1 + 2t + 2t^2 + t^3$ and $P(t) = p(t)/(1 - t^4)$.

(2.7) A similar pattern to the one seen above for Q_8 holds for other groups. Take for example the group $\mathrm{GL}(3,2)$ of 3×3 invertible matrices over $\mathbb{F}_2$. This is a finite simple group of order 168. Its cohomology is given by

$$H^*(\mathrm{GL}(3,2), \mathbb{F}_2) = \mathbb{F}_2[x, y, z]/(x^3 + yz)$$

where $\deg x = 2$, $\deg y = \deg z = 3$. A homogeneous system of parameters for this ring is given by y and z, and these elements form a regular sequence. Modulo the ideal generated by y and z, we get $\mathbb{F}_2(x)/(x^3)$. This is a finite Poincaré duality ring whose dualizing degree is 4. Again, this means that the cohomology is a Gorenstein ring with a-invariant $4 - (3-1) - (3-1) = 0$, but it

does not decompose as a tensor product the way it did for the quaternion group
(2.4.2).

(2.8) It is not true that the cohomology ring of a finite group is always Gorenstein. For example, the semidihedral group of order 2^n $(n \geq 4)$,

(2.8.1) $G = SD_{2^n} = \langle g, h \mid g^{2^{n-1}} = 1, \ h^2 = 1, \ h^{-1}gh = g^{2^{n-2}-1} \rangle$

has cohomology ring

$$H^*(SD_{2^n}, \mathbb{F}_2) = \mathbb{F}_2[x, y, z, w]/(xy, y^3, yz, z^2 + x^2 w)$$

with $\deg x = \deg y = 1$, $\deg z = 3$ and $\deg w = 4$. This ring is not even Cohen–
Macaulay. But what is true is that whenever the ring is Cohen–Macaulay, it is
Gorenstein with a-invariant zero. See §11 for further details.

Even if the cohomology ring is not Cohen–Macaulay, there is still a certain
kind of duality, but it is expressed in terms of a spectral sequence of Greenlees,
$H_{\mathfrak{m}}^{s,t} H^*(G, k) \implies H_{-s-t}(G, k)$. Let us see in the case above of the semidihedral
group, what this spectral sequence looks like. And let's do it in pictures. We'll
draw the cohomology ring as follows.

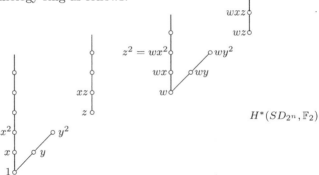

The vertical coordinate indicates cohomological degree, and the horizontal coordinate is just for separating elements of the same degree. To visualize the
homology, just turn this picture upside down by rotating the page, as follows.

(2.8.2)

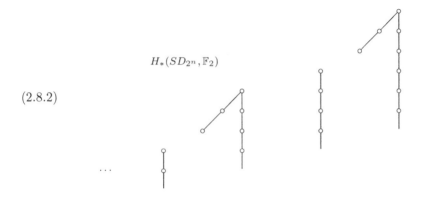

We compute local cohomology using the stable Koszul complex for the homogeneous system of parameters w, x,

$$0 \to H^*(G, \mathbb{F}_2) \to H^*(G, \mathbb{F}_2)[w^{-1}] \oplus H^*(G, \mathbb{F}_2)[x^{-1}] \to H^*(G, \mathbb{F}_2)[w^{-1}x^{-1}] \to 0$$

where the subscripts denote localization by inverting the named element. A picture of this stable Koszul complex is as follows.

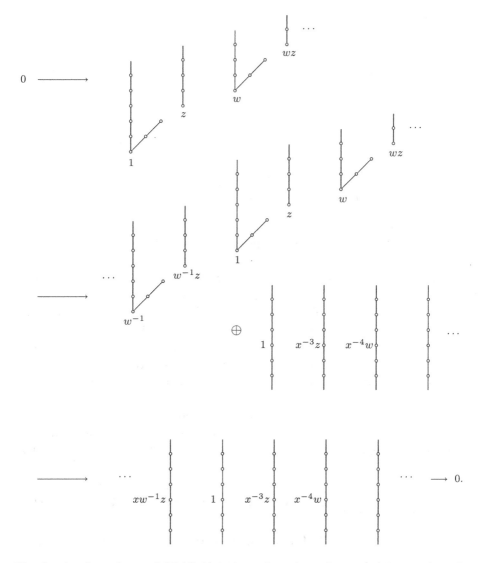

The local cohomology of $H^*(G, k)$ is just the cohomology of this complex. In degree zero there is no cohomology. In degree one there is some cohomology, namely the hooks that got introduced when w was inverted,

$$H^1_{\mathfrak{m}} H^*(SD_{2^n}, \mathbb{F}_2) =$$

In degree two, we get the part of the plane not hit by either of the two degree one pieces,

$$H^2_{\mathfrak{m}} H^*(SD_{2^n}, \mathbb{F}_2) =$$

Now the differential d_2 in this spectral sequence increases local cohomological degree by two and decreases internal degree by one, and the higher differentials are only longer. So there is no room in this example for nonzero differentials. It follows that the spectral sequence takes the form of a short exact sequence

$$0 \to H^{1,t-1}_{\mathfrak{m}} H^*(SD_{2^n}, \mathbb{F}_2) \to H_{-t}(SD_{2^n}, \mathbb{F}_2) \to H^{2,t-2}_{\mathfrak{m}} H^*(SD_{2^n}, \mathbb{F}_2) \to 0.$$

This works fine, because $H_*(SD_{2^n}, \mathbb{F}_2)$ is the graded dual of $H^*(SD_{2^n}, \mathbb{F}_2)$, as shown in (2.8.2). So the short exact sequence places the hooks of $H^1_{\mathfrak{m}}$ underneath every second nonzero column in $H^2_{\mathfrak{m}}$ to build $H_*(SD_{2^n}, \mathbb{F}_2)$. Notice that the hooks appear inverted, so that there is a separate Poincaré duality for a hook.

The same happens as in this case whenever the depth and the Krull dimension differ by one. The kernel of multiplication by the last parameter, modulo the previous parameters, satisfies Poincaré duality with dualizing degree determined by the degrees of the parameters; in particular, the top degree of this kernel is determined. In the language of commutative algebra, this can be viewed in terms of the Castelnuovo–Mumford regularity of the cohomology ring. See § 14 for more details.

The reader who wishes to understand these examples better can skip directly to § 14, and refer back to previous sections as necessary to catch up on definitions. Conjecture 14.6.1 says that for a finite group G, $\operatorname{Reg} H^*(G, k)$ is always zero. This conjecture is true when the depth and the Krull dimension differ by at most one, as in the above example. It is even true when the difference is two, by a more subtle transfer argument sketched in § 14 and described in detail in [Benson 2004].

3. Group Cohomology

For general background material on cohomology of groups, the textbooks I recommend are [Adem and Milgram 1994; Benson 1991b; Brown 1982; Cartan and Eilenberg 1956; Evens 1991]. The commutative algebra texts most relevant to these lectures are [Bruns and Herzog 1993; Eisenbud 1995; Grothendieck 1965; 1967; Matsumura 1989].

(3.1)   For a *discrete group*, the easiest way to think of group cohomology is as the Ext ring (see §5 of [Sri]). If G is a group and k is a commutative ring of coefficients, we define group cohomology via

$$H^*(G, k) = \text{Ext}^*_{\mathbb{Z}G}(\mathbb{Z}, k) \cong \text{Ext}^*_{kG}(k, k).$$

Here, the group ring kG consists of formal linear combinations $\sum \lambda_i g_i$ of elements of the group G with coefficients in k. The cup product in cohomology comes from the fact that kG is a Hopf algebra (see §1.8 of [Sri]), with comultiplication $\Delta(g) = g \otimes g$. Another part of the Hopf structure on kG is the augmentation map $kG \to k$, $\sum \lambda_i g_i \mapsto \sum \lambda_i$, which is what allows us to regard k as a kG-module.

Cup product and Yoneda product define the same multiplicative structure, and this makes cohomology into a *graded commutative* ring, in the sense that

$$ab = (-1)^{|a||b|} ba,$$

where $|a|$ denotes the degree of an element a (see Prop. 5.5 of [Sri]). In contrast, the Ext ring of a commutative local ring is seldom graded commutative; this happens only for a restricted class of complete intersections. The group ring of an abelian group is an example of a complete intersection (see §1.4 of [Sri]).

More generally, if M is a left kG-module then

$$H^*(G, M) = \text{Ext}^*_{\mathbb{Z}G}(\mathbb{Z}, M) \cong \text{Ext}^*_{kG}(k, M)$$

is a graded *right* $H^*(G, k)$-module.

It is a nuisance that most texts on commutative algebra are written for *strictly commutative* graded rings, where $ab = ba$ with no sign. I do not know of an instance where the signs make a theorem from commutative algebra fail. It is worth pointing out that if a is an element of odd degree in a graded commutative ring then $2a^2 = 0$. So $2a$ is nilpotent, and it follows that modulo the nil radical the ring is strictly commutative. On the other hand, it is *more than a nuisance* that commutative algebraists often assume that their graded rings are generated by elements of degree one, because this is not at all true for cohomology rings. Nor, for that matter, is it true for rings of invariants.

(3.2)   A homomorphism of groups $\rho \colon H \to G$ gives rise to a map the other way

$$\rho^* \colon H^*(G, M) \to H^*(H, M)$$

for any kG-module M. If $\rho \colon H \to G$ is an inclusion, this is called the *restriction map*, and denoted $\text{res}_{G,H}$. If G is a quotient group of H and $\rho \colon H \to G$ is the quotient map, then it is called the *inflation map*, and denoted $\inf_{G,H}$.

(3.3)   For a *topological group* (this includes compact Lie groups as well as discrete groups), a theorem of Milnor [1956] says that the infinite join

$$EG = G \star G \star \cdots$$

is weakly contractible, G acts freely on it, and the quotient $BG = EG/G$ together with the principal G-bundle $p: EG \to BG$ forms a classifying space for principal G-bundles over a paracompact base. A topologist refers to $H^*(BG; k)$ as the classifying space cohomology of G. Again, it is a graded commutative ring. For example, for the compact unitary group $U(n)$, the cohomology ring

(3.3.1) $$H^*(BU(n); k) \cong k[c_1, \ldots, c_n]$$

is a polynomial ring over k on n generators $c_1, \ldots, c_n$ with $|c_i| = 2i$, called the *Chern classes*. Similarly, for the orthogonal group $O(2n)$, if k is a field of characteristic not equal to two, then we have

(3.3.2) $$H^*(BO(2n); k) \cong k[p_1, \ldots, p_n]$$

is a polynomial ring over k on n generators $p_1, \ldots, p_n$ with $|p_i| = 4i$, called the *Pontrjagin classes*. For $SO(2n)$ we have

(3.3.3) $$H^*(BSO(2n); k) \cong k[p_1, \ldots, p_{n-1}, e].$$

where $e \in H^{2n}(BSO(2n); k)$ is called the *Euler class*, and satisfies $e^2 = p_n$. We shall discuss these examples further in § 12.

If G is a discrete group then BG is an *Eilenberg–Mac Lane space* for G; in other words, $\pi_1(BG) \cong G$ and $\pi_i(BG) = 0$ for $i > 1$. The relationship between group cohomology and classifying space cohomology for G discrete is that the singular chains $C_*(EG)$ form a free resolution of $\mathbb{Z}$ as a $\mathbb{Z}G$-module. Then there are isomorphisms

$$H^*(BG; k) = H^* \operatorname{Hom}_{\mathbb{Z}}(C_*(BG), k) \cong H^* \operatorname{Hom}_{\mathbb{Z}G}(C_*(EG), k) \cong H^*(G, k),$$

and the topologically defined product on the left agrees with the algebraically defined product on the right.

(3.4)   Another case of interest is *profinite groups*. A profinite group is defined to be an inverse limit of a system of finite groups, which makes it a compact, Hausdorff, totally disconnected topological group. For example, writing $\hat{\mathbb{Z}}_p$ for the ring of p-adic integers, $\mathrm{SL}_n(\hat{\mathbb{Z}}_p)$ is a profinite group. The open subgroups of a profinite group are the subgroups of finite index.

Classifying space cohomology turns out to be the wrong concept for a profinite group. A better concept is continuous cohomology, which is defined as follows [Serre 1965a]. Let $G = \varprojlim_{U \in \mathscr{U}} G/U$ be a profinite group, where $\mathscr{U}$ is a system of open normal subgroups with $\bigcap_{U \in \mathscr{U}} U = \{1\}$. We restrict attention to modules M such that $M = \bigcup_{U \in \mathscr{U}} M^U$, and continuous cohomology is then defined as

$$H^i_c(G, M) = \varinjlim_{U \in \mathscr{U}} H^i(G/U, M^U).$$

Again, if k is a commutative ring of coefficients then $H^*_c(G, k)$ is a graded commutative ring.

(3.5)   In all of the above situations, if p is a prime number and k is the finite field $\mathbb{F}_p$, then there are *Steenrod operations*

(3.5.1)
$$\begin{cases} \mathrm{Sq}^i \colon H^n(G, \mathbb{F}_2) \to H^{n+i}(G, \mathbb{F}_2) & (p = 2), \\ P^i \colon H^n(G, \mathbb{F}_p) \to H^{n+2i(p-1)}(G, \mathbb{F}_p) & (p \text{ odd}) \end{cases}$$

$(i \geq 0)$ acting on the cohomology of any group (Sq^0 and P^0 act as the identity operation).[1] These operations satisfy some identities called the *Adem relations*, and the *Steenrod algebra* is the graded algebra generated by the Steenrod operations subject to the Adem relations. The action of the Steenrod operations is related to the multiplicative structure of cohomology by the *Cartan formula*

(3.5.2)
$$\begin{cases} \mathrm{Sq}^n(xy) = \sum_{i+j=n} \mathrm{Sq}^i(x)\mathrm{Sq}^j(y) & (p = 2), \\ P^n(xy) = \sum_{i+j=n} P^i(x)P^j(y) & (p \text{ odd}). \end{cases}$$

Finally, the action of the Steenrod operations on group cohomology satisfies the *unstable axiom*, which states that

(3.5.3)
$$\begin{cases} \mathrm{Sq}^i(x) = 0 \quad \text{for} \quad i > |x| \quad \text{and} \quad \mathrm{Sq}^{|x|}(x) = x^2 & (p = 2) \\ P^i(x) = 0 \quad \text{for} \quad i > 2(p-1)|x| \quad \text{and} \quad P^{2(p-1)|x|}(x) = x^p & (p \text{ odd}). \end{cases}$$

The Cartan formula and the unstable axiom make the cohomology ring of a group (or more generally, the cohomology ring of any space) with $\mathbb{F}_p$ coefficients into an *unstable algebra* over the Steenrod algebra. For more details, see [Steenrod 1962; Schwartz 1994].

(3.6)   There are some important variations on the definitions of group cohomology. For example, for a finite group, *Tate cohomology* is defined using complete resolutions, and gives a $\mathbb{Z}$-graded ring $\hat{H}^*(G, k)$. More precisely, if G is a finite group, k is a field[2] and N is a kG-module, then we splice together an injective

---

[1] For p odd, there is also a separate Bockstein operation $\beta \colon H^n(G, \mathbb{F}_p) \to H^{n+1}(G, \mathbb{F}_p)$ which we shall systematically ignore. For $p = 2$, the Bockstein operation is equal to Sq^1, so it is not a separate operation.

[2] Tate cohomology is defined over an arbitrary commutative ring of coefficients, but the definition is slightly different to the one given here. See [Cartan and Eilenberg 1956, Chapter XII].

resolution and a projective resolution of N to give an exact sequence

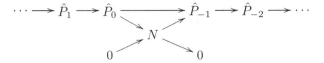

The fact that injective kG-modules are the same as projective kG-modules (see Theorem 3.6 as well as the paragraph following Corollary 3.7 in [Sri]) means that this is an exact sequence of projective modules such that the image of the middle map is equal to N, which is the definition of a complete resolution. If M is another kG-module, we define $\widehat{\mathrm{Ext}}^*_{kG}(N, M)$ to be the cohomology of the cochain complex obtained by taking homomorphisms from the complete resolution $\hat{P}_*$ to M. In the case where $N = k$, we define

$$\hat{H}^*(G, M) = \widehat{\mathrm{Ext}}^*_{kG}(k, M).$$

If M is also equal to k, then $\hat{H}^*(G, k)$ is a graded commutative ring. If M is a left kG-module then $\hat{H}^*(G, M)$ is a *right* module over $\hat{H}^*(G, k)$.

There is a map from a complete resolution of N to the projective resolution of N used to make it

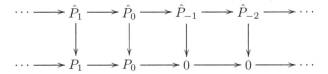

which is an isomorphism in nonnegative degrees and the zero map in negative degrees. This induces a map from $\mathrm{Ext}^*_{kG}(N, M)$ to $\widehat{\mathrm{Ext}}^*_{kG}(N, M)$ which is an isomorphism in positive degrees, surjective in degree zero, and the zero map in negative degrees. In particular, we obtain this way a map from ordinary cohomology to Tate cohomology, $H^*(G, M) \to \hat{H}^*(G, M)$. This is a homomorphism of graded k-algebras.

Tate duality says that for any kG-module M and every $n \in \mathbb{Z}$, the k-vector space $\widehat{\mathrm{Ext}}^{n-1}_{kG}(M, k)$ is the dual of $\hat{H}^{-n}(G, M)$,

(3.6.1) $\widehat{\mathrm{Ext}}^{n-1}_{kG}(M, k) \cong \mathrm{Hom}_k(\hat{H}^{-n}(G, M), k).$

The case $M = k$ of this statement can be interpreted as saying that the Tate cohomology is isomorphic to its graded dual, shifted in degree by one. This implies that it is selfinjective as a ring.

4. Finite Generation

There are various finite generation theorems, which provide the Noetherian hypothesis as starting point for the application of commutative algebra.

(4.1)   We begin with finite groups, where we have the following algebraic theorem of Evens [1961] (see also [Golod 1959] for the case of a finite p-group).

THEOREM 4.1.1. *Let G be a finite group, k a commutative ring of coefficients and M a kG-module. If M is Noetherian as a k-module then $H^*(G, M)$ is Noetherian as an $H^*(G, k)$-module. In particular, if k is Noetherian then $H^*(G, k)$ is a finitely generated k-algebra.*

This can be contrasted with the situation in commutative algebra, where the Ext ring of a commutative local ring is Noetherian if and only if the ring is a complete intersection [Bøgvad and Halperin 1986]. The following extension of the theorem above also appears in [Evens 1961].

THEOREM 4.1.2. *Let H be a subgroup of a finite group G, let k be a commutative ring of coefficients, and let N be a Noetherian kH-module. Then $H^*(H, N)$ is a finitely generated module over $H^*(G, k)$ via the restriction map (3.2) from G to H.*

In contrast, Tate cohomology is almost never finitely generated. In fact, if k is a field of characteristic p, then there is a dichotomy [Benson and Krause 2002]. Either

(4.1.3) G has no subgroups isomorphic to $\mathbb{Z}/p \times \mathbb{Z}/p$ (i.e., the Sylow p-subgroups of G are cyclic, or $p = 2$ and the Sylow 2-subgroups of G are generalized quaternion) and $\hat{H}^*(G, k)$ is periodic and Noetherian, of the form $k[x, x^{-1}]$ tensored with a finite dimensional part, or

(4.1.4) G has a subgroup isomorphic to $\mathbb{Z}/p \times \mathbb{Z}/p$ and $\hat{H}^*(G, k)$ is not Noetherian. In this case, the negative degree cohomology $\hat{H}^-(G, k)$ is nilpotent, in the sense that there is some integer n such that every product of n or more elements of $\hat{H}^-(G, k)$ gives zero. In fact, if the depth of $H^*(G, k)$ is bigger than one then all products in $\hat{H}^-(G, k)$ vanish [Benson and Carlson 1992].

(4.2)   Evens' theorem generalizes in a number of directions. The following is a theorem of Friedlander and Suslin [1997].

THEOREM 4.2.1. *Let G be a finite group scheme over a field k (i.e., kG is a finite dimensional cocommutative Hopf algebra), and let M be a finitely generated kG-module. Then $H^*(G, k) = \mathrm{Ext}_{kG}^*(k, k)$ is a finitely generated k-algebra and $H^*(G, M) = \mathrm{Ext}_{kG}^*(k, M)$ is a finitely generated $H^*(G, k)$-module.*

(4.3)   For compact Lie groups, the following is a theorem of Venkov [1959].

THEOREM 4.3.1. *Let G be a compact Lie group and k a Noetherian ring of coefficients. If $G \to \mathrm{U}(n)$ is a faithful unitary representation[3] of G then $H^*(BG; k)$*

---

[3]The Peter–Weyl theorem implies that every compact Lie group has a faithful unitary representation, so that this is not a restriction on G.

is finitely generated as a module over the image of the Chern classes (3.3.1):

$$H^*(\mathrm{BU}(n); k) = k[c_1, \ldots, c_n] \to H^*(BG; k).$$

In particular, $H^*(BG; k)$ is a finitely generated k-algebra. If H is a closed subgroup of G then $H^*(BH; k)$ is a finitely generated module over the image of the restriction map (3.2) from G to H.

(4.4)   There is an interesting generalization of compact Lie groups which has been extensively investigated by Dwyer and Wilkerson, among others. A *loop space* is by definition a space X together with another space Y and a homotopy equivalence $X \simeq \Omega Y$ between X and the space of pointed maps from S^1 to Y. If, furthermore, $H_*(X; \mathbb{Z})$ is finitely generated as an abelian group (in other words, if each $H_i(X; \mathbb{Z})$ is finitely generated, and only nonzero for a finite number of different values of i), so that X looks homologically like a finite CW-complex, then we say that X is a *finite loop space*. For example, if G is a compact Lie group then $G \simeq \Omega BG$ and G is a finite loop space. For this reason, in general, the notation for the space Y is BX, and it is called the *classifying space* of the loop space X. But in spite of the notation, the space Y cannot be recovered from the space X, so naming $Y = BX$ is regarded as part of the structure of the finite loop space X. The following theorem of Dwyer and Wilkerson [1994] generalizes Venkov's Theorem 4.3.1.

THEOREM 4.4.1. *If X is a finite loop space, then for any field k, the algebra $H^*(BX; k)$ is finitely generated.*

A closely related concept is that of a *p-compact group*, which is by definition a loop space X which is $\mathbb{F}_p$-complete in the sense of [Bousfield and Kan 1972][4], $\mathbb{F}_p$-finite in the sense that $H^*(X; \mathbb{F}_p)$ is finite, and such that $\pi_0 X$ is a finite p-group. The $\mathbb{F}_p$-completion of a finite loop space is an example of a p-compact group. The following theorem is also proved by Dwyer and Wilkerson [1994].

THEOREM 4.4.2. *If X is a p-compact group then $H^*(BX; \mathbb{F}_p)$ is a finitely generated $\mathbb{F}_p$-algebra.*

(4.5)   For infinite discrete groups, the question of finite generation is more delicate, and there are various theorems for some special classes of infinite groups. For example, the cohomology of an arithmetic group with coefficients in $\mathbb{Z}$ or a field is finitely generated. More generally, we have the following theorem.

THEOREM 4.5.1. *If a discrete group G has a subgroup H of finite index, such that there is a classifying space BH which is a finite CW complex, and k is a Noetherian commutative ring of coefficients, then $H^*(G, k)$ is Noetherian.*

---

[4]Bousfield–Kan $\mathbb{F}_p$-completion of a space is analogous to completion with respect to a prime ideal, inasmuch as it isolates the homotopy theoretic information at the prime p and kills torsion coprime to p. It is used in order to do homotopy theory one prime at a time.

PROOF. We can take H to be normal, and then the spectral sequence

$$H^*(G/H, H^*(H, k)) \implies H^*(G, k)$$

has Noetherian E_2 page, and so $H^*(G, k)$ is Noetherian.   □

(4.6)   A *pro p-group* is defined to be in inverse limit of finite p-groups. For pro p-groups, we have the following finite generation theorem of Minh and Symonds [2004].

THEOREM 4.6.1. *Let G be a pro p-group.*

(i) *The cohomology ring $H_c^*(G, \mathbb{F}_p)$ is finitely generated over $\mathbb{F}_p$ if and only if G contains an open normal torsion-free subgroup U such that $H_c^*(U, \mathbb{F}_p)$ is finite.*

(ii) *If $H_c^*(G, \mathbb{F}_p)$ is finitely generated then the number of conjugacy classes of finite subgroups of G is finite.*

(iii) *$H_c^*(G, \mathbb{F}_p)$ modulo its nil radical is a finitely generated $\mathbb{F}_p$-algebra if and only if G has only a finite number of conjugacy classes of finite elementary abelian p-subgroups.*

5. Krull Dimension

(5.1)   For a finitely generated graded commutative algebra R over a field k, there are several ways to define Krull dimension, which all give the same answer.

(5.1.1)   Noether normalization (2.5) guarantees the existence of a homogeneous polynomial ring $k[x_1, \ldots, x_r]$ over which R is finitely generated as a module. The integer r is the Krull dimension of R.

(5.1.2)   If $\mathfrak{m} = \mathfrak{p}_0 \supset \mathfrak{p}_1 \supset \mathfrak{p}_2 \supset \cdots \supset \mathfrak{p}_r$ is the longest chain of homogeneous prime ideals in R and proper inclusions then r is the Krull dimension of R.

(5.1.3)   Set $p_R(t) = \sum_{i=0}^{\infty} t^i \dim_k R_i$. Then $p_R(t)$ is a rational function of t, and the order of the pole at $t = 1$ is the Krull dimension of R.

(5.2)   The first results on Krull dimension for cohomology rings are due to Quillen [1971b; 1971c], and are expressed in terms of the *p-rank* $r_p(G)$, where $p \geq 0$ is the characteristic of k. If p is a prime, this is defined to be the largest r such that G has an elementary abelian subgroup of rank r; in other words, such that $(\mathbb{Z}/p)^r \subseteq G$. If G is a compact Lie group, this is at least as big as the *Lie rank* $r_0(G)$, which is defined to be the rank r_0 of a maximal torus $(S^1)^{r_0} \subseteq G$. Quillen's theorem is as follows.

THEOREM 5.2.1. *Let G be a compact Lie group, and k be a field of characteristic $p \geq 0$. Then the Krull dimension of $H^*(BG; k)$ is equal to $r_p(G)$.*

(5.3)   In the special case where G is finite, this means that the Krull dimension of $H^*(G, k)$ is equal to $r_p(G)$. Quillen [1971c] proved that the same holds more generally for groups of *finite virtual cohomological dimension*; in other words, for discrete groups G containing a normal subgroup H of finite index such that $H^n(H, M) = 0$ for all large enough values of n and all $\mathbb{Z}H$-modules M. The cohomology ring of G is not necessarily finitely generated in this situation, but it is finitely generated modulo its nil radical.

In fact, in these cases, Quillen obtained much more than just the Krull dimension. He obtained a complete description of the maximal ideal spectrum of $H^*(BG; k)$ up to inseparable isogeny, in terms of the *Quillen category* $\mathscr{A}_p(G)$ whose objects are the finite elementary abelian p-subgroups of G, and whose arrows are the monomorphisms induced by conjugation in G.

THEOREM 5.3.1. *The restriction map (3.2)*

$$(5.3.2) \qquad\qquad H^*(BG; k) \to \varprojlim_{E \in \mathscr{A}_p(G)} H^*(BE; k)$$

is an inseparable isogeny. In other words, the kernel of this map consists of nilpotent elements, and if x is an element of the right hand side then x^{p^a} is in the image for some $a \geq 0$.

We interpret Theorem 5.3.1 in terms of varieties in § 9. But for now, notice that the cohomology of a finite elementary abelian p-group of rank r is described in (2.2.2). In particular, modulo its nil radical it is always a polynomial ring in r generators.

COROLLARY 5.3.3. *The minimal primes in $H^*(BG; k)$ are in one-one correspondence with the conjugacy classes of maximal elementary abelian p-subgroups of G, with respect to inclusion. If E is a maximal elementary abelian p-subgroup, then the corresponding minimal prime is $\sqrt{\ker(\mathrm{res}_{G,E})}$, the radical of the kernel of the restriction map (3.2) from G to E. The Krull dimension of the quotient by this minimal prime is equal to the rank $r_p(E)$.*

(5.4)   The analog of the inseparable isogeny (5.3.2) was also proved by Quillen [1971c, Proposition 13.4] in the case of a profinite group with a finite number of conjugacy classes of elementary abelian p-subgroups.

6. Depth

(6.1)   In contrast with Krull dimension, the depth of a cohomology ring is harder to compute. There are many interesting classes of groups for which the cohomology is known to be Cohen–Macaulay, even though this is less common for general finite groups, let alone for more general classes of groups.

(6.1.1)   Groups with abelian Sylow p-subgroups have Cohen–Macaulay cohomology [Duflot 1981].

(6.1.2) $GL_n(\mathbb{F}_q)$ in characteristic not dividing q [Quillen 1972], as well as various other finite groups of Lie type away from their defining characteristic, have

Cohen–Macaulay cohomology (see [Fiedorowicz and Priddy 1978] for the classical groups and [Kleinerman 1982] for the groups of exceptional Lie type).

(6.1.3) ($p = 2$) Groups with almost extraspecial Sylow 2-subgroups[5] have Cohen–Macaulay cohomology [Quillen 1971a].

(6.1.4) ($p = 2$) Finite simple groups of 2-rank at most three have Cohen–Macaulay cohomology [Adem and Milgram 1995].

(6.1.5) Finite groups of Lie type in the defining characteristic and finite symmetric groups almost never have Cohen–Macaulay cohomology, because they have maximal elementary abelian p-subgroups of different ranks, and hence minimal primes with quotients of different dimensions by Corollary 5.3.3.

(6.1.6) ($p = 2$) By computations of Carlson [≥ 2004], of the 267 isomorphism classes of 2-groups of order 64, 119 have Cohen–Macaulay cohomology rings. The depth differs from the Krull dimension by one in 126 cases and by two in the remaining 22 cases. See the Appendix at the end of these notes for more detailed information.

(6.2) As far as group theoretic characterizations of depth are concerned, the best theorems to date only give bounds on the depth. For example, Duflot's theorem [1981] gives the following lower bound.

THEOREM 6.2.1. *Let G be a finite group and k a field of characteristic p. Then the depth of $H^*(G, k)$ is greater than or equal to the p-rank of the center of a Sylow p-subgroup S of G. In particular, if $|G|$ is divisible by p then $H^*(G, k)$ has strictly positive depth.*

The bound of Theorem 6.2.1 gives the exact value for the depth for 193 of the 267 groups of order 64.

(6.3) Theorem 6.2.1 generalizes to compact Lie groups as follows. If G is a compact Lie group and T is a maximal torus, then the inverse image $S \subseteq G$ of a Sylow p-subgroup of $N_G(T)/T$ is called a *Sylow p-toral subgroup* of G. The Sylow p-toral subgroups play the same role for compact Lie groups that Sylow p-subgroups do for finite groups. The crucial property as far as cohomology is concerned is that the restriction map (3.2) $H^*(BG; k) \to H^*(BS; k)$ followed by the transfer map $H^*(BS; k) \to H^*(BG; k)$ of [Becker and Gottlieb 1975] is the identity.[6] Since the transfer map is a map of $H^*(BG; k)$-modules, it follows that $H^*(BG; k)$ is a direct summand of $H^*(BS; k)$ as an $H^*(BG; k)$-module. In particular, the depth of $H^*(BG; k)$ is at least as big as the depth of $H^*(BS; k)$.

---

[5]A finite p-group P is said to be *almost extraspecial* if P has a central subgroup $Z \cong \mathbb{Z}/p$ such that P/Z is elementary abelian.

[6]For any closed subgroup, the composite of the restriction and the transfer is multiplication by the Euler characteristic of the quotient. The analog of the third Sylow theorem says that $\chi(G/S)$ is congruent to 1 modulo p.

Stong (unpublished) showed how to generalize Duflot's proof to this situation, and showed that the depth of $H^*(BG; k)$ is at least as big as $r_p(Z(S))$. Broto and Henn [1993] gave another proof which is conceptually easier, and goes as follows. We begin by establishing the notation.

If C is a central elementary abelian p-subgroup of S, then the multiplication maps $C \times C \to C$ and $\mu\colon C \times S \to S$ are group homomorphisms. This means that $H^*(BC; k)$ is a graded commutative and cocommutative Hopf algebra, $H^*(BS; k)$ is an $H^*(BC; k)$-comodule algebra via

$$\mu^*\colon H^*(BS; k) \to H^*(BC; k) \otimes_k H^*(BS; k),$$

and $H^*(BG; k)$ is a sub-comodule algebra.

Since $H^*(BC; k)$ is Cohen–Macaulay (2.2.2), we can find elements $\zeta_1, \ldots, \zeta_r \in H^*(BG; k)$ whose restriction to C form a homogeneous system of parameters and hence a regular sequence $x_1, \ldots, x_r \in H^*(BC; k)$. One way to do this is to use Theorem 4.3.1 and throw away redundant Chern classes. We claim that $\zeta_1, \ldots, \zeta_r$ form a regular sequence in $H^*(BG; k)$. Setting $\xi_i = \mu^*(\zeta_i)$, the elements $\xi_1, \ldots, \xi_r$ are elements of $H^*(B(C \times G); k) \cong H^*(BC; k) \otimes_k H^*(BG; k)$ (see (2.2.1)) whose restrictions to the first factor are $x_1, \ldots, x_r$ and whose restrictions to the second factor are $\zeta_1, \ldots, \zeta_r$. So

$$\mu^*(\zeta_i) = x_i \otimes 1 + \cdots + 1 \otimes \zeta_i.$$

We begin with ζ_1. If y is a nonzero element in $H^d(BG; k)$ such that $\zeta_1 y = 0$, then $\mu^*(\zeta_1 y) = 0$. Then $\mu^*(y)$ is a sum of tensors, and we separate out the terms whose degree in $H^*(BC; k)$ are highest, say

$$\mu^*(y) = \sum_j u_j \otimes v_j + \text{terms of lower first degree}.$$

Then

$$\mu^*(\zeta_1 y) = \sum_j x_1 u_j \otimes v_j + \text{terms of lower first degree}.$$

But multiplication by $x_1 \otimes 1$ is injective on $H^*(BC; k) \otimes H^*(BG; k)$, so the only way this can be zero is for $\sum_j u_j \otimes v_j$ to be zero. This means that $\mu^*(y) = 0$ and so $y = 0$.

The same argument works inductively, because the map μ^* passes down to a well defined map

$$\mu^*\colon H^*(BG; k)/(\zeta_1, \ldots, \zeta_i) \to H^*(BC; k)/(x_1, \ldots, x_i) \otimes H^*(BG; k)/(\zeta_1, \ldots, \zeta_i).$$

Applying the same argument to ζ_{i+1} using this map, we see that multiplication by ζ_{i+1} is injective on $H^*(BG; k)/(\zeta_1, \ldots, \zeta_i)$. This inductive argument proves that $\zeta_1, \ldots, \zeta_r$ is a regular sequence in $H^*(BG; k)$, and completes the Broto–Henn proof of the following version of Duflot's theorem.

THEOREM 6.3.1. *Let G be a compact Lie group and k a field of characteristic p. Then the depth of $H^*(BG; k)$ is greater than or equal to the p-rank of the center of a Sylow p-toral subgroup S of G.* □

(6.4) Green [2003] extended the above idea to prove a stronger result. An element $x \in H^*(BS; k)$ is said to be primitive if its image under μ^* is equal to $1 \otimes x$. Since μ^* is a ring homomorphism, the primitives form a subring of $H^*(BS; k)$. Since both μ and the projection onto the second factor of $C \times S$ have the same composite with the quotient map $S \to S/C$, it follows that the image of the inflation map (see (3.2)) $H^*(B(S/C); k) \to H^*(BS; k)$ consists of primitive elements. If I is an ideal in $H^*(BG; k)$ generated by a regular sequence of primitive elements, then we can replace $H^*(BG; k)$ by $H^*(BG; k)/I$ in the above argument for Duflot's theorem, to obtain the following.

THEOREM 6.4.1. *Let G be a compact Lie group with Sylow p-toral subgroup S, and set $r = r_p(Z(S))$. If there is a regular sequence of length s in $H^*(BG; k)$ which consists of primitive elements, then the depth of $H^*(BG; k)$ is at least $r + s$.*

7. Associated Primes and Steenrod Operations

(7.1) Depth of the cohomology ring is closely linked with the action of the Steenrod operations (3.5). For example, an analysis of unstable algebras over the Steenrod algebra gives rise to a way of computing the depth with a single test sequence.

The test sequence takes the form of Dickson invariants. If $\mathbb{F}_p[x_1, \ldots, x_r]$ is a polynomial ring, then the general linear group $\mathrm{GL}_r(\mathbb{F}_p)$ acts by linear substitutions, and Dickson [1911] proved that the invariants form a polynomial ring

$$\mathbb{F}_p[x_1, \ldots, x_r]^{\mathrm{GL}_r(\mathbb{F}_p)} = \mathbb{F}_p[c_{r,r-1}, \ldots, c_{r,0}].$$

The Dickson invariant $c_{r,i}$ has degree $p^r - p^i$ in the variables $x_1, \ldots, x_r$. The Dickson invariants are further studied in [Wilkerson 1983], where the action of the Steenrod operations on them is also described.

If G is a compact Lie group of p-rank r, then it is shown in [Benson and Wilkerson 1995] that it follows from Quillen's Theorem 5.3.1 that as an algebra over the Steenrod algebra, $H^*(BG; \mathbb{F}_p)$ always contains a copy of the Dickson invariants as a homogeneous system of parameters, where, for a suitable integer $a \geq 0$ depending on G, $x_1, \ldots, x_r$ are taken to have degree 2^a if $p = 2$ and $2p^a$ if p is odd. So

$$|c_{r,i}| = \begin{cases} 2^{a+r} - 2^{a+i} & (p = 2) \\ 2(p^{a+r} - p^{a+i}) & (p \text{ odd}). \end{cases}$$

If E is an elementary abelian p-subgroup of G of rank $s \leq r$, then the restriction to $H^*(BE; k)$ of $c_{r,i} \in H^*(BG; k)$ is equal to $c_{s,i-r+s}^{p^{a+r-s}}$; thus it is a power of the Dickson invariant in the polynomial generators (2.2.2) in degree one ($p = 2$) or two (p odd) of $H^*(BE; k)$.

The Landweber–Stong conjecture, proved by Bourguiba and Zarati [1997], implies the following.

THEOREM 7.1.1. *The depth of $H^*(BG; \mathbb{F}_p)$ is equal to the maximum value of d for which $c_{r,r-1}, \ldots, c_{r,r-d}$ is a regular sequence.*

Theorem 8.1.8 of [Neusel 2000] proves the much stronger statement that such a copy of the Dickson algebra can be found in any Noetherian unstable algebra over the Steenrod algebra. The conclusion of Theorem 7.1.1 holds in this more general context. The proof given by Bourguiba and Zarati makes heavy use of the machinery of unstable algebras over the Steenrod algebra. A proof without this machinery, but which only works in the context of the cohomology of a finite group, can be found in [Benson 2004].

The Dickson invariants have further desirable properties among all homogeneous systems of parameters in cohomology. For example, if H is a closed subgroup of G of p-rank s then the restrictions of $c_{r,r-1}, \ldots, c_{r,r-s}$ are Dickson invariants forming a homogeneous system of parameters in $H^*(BH; k)$. Furthermore, $c_{r,i}$ is a sum of transfers from centralizers of elementary abelian subgroups of rank i.

(7.2)   To get further with depth, it is necessary to get some understanding of the associated primes in group cohomology. In general, the dimension of the quotient by an associated prime is an upper bound for the depth. For general commutative rings, the depth cannot be computed from the dimensions of the quotients by associated primes, but in the case of cohomology of a finite group, we have the following conjecture of Carlson [1999]. Partial results on this conjecture have been obtained by Green [2003].

CONJECTURE 7.2.1. *Let G be a finite group and k a field. Then $H^*(G, k)$ has an associated prime $\mathfrak{p}$ such that the Krull dimension of $H^*(G, k)/\mathfrak{p}$ is equal to the depth of the $H^*(G, k)$.*

(7.3)   The following theorem of Wilkerson [1982] shows that associated primes are invariant under the action of the Steenrod operations. Since it is not easy to find an explicit reference, we include a complete proof here.

THEOREM 7.3.1. *Let H be a graded commutative unstable algebra over the Steenrod algebra. For example, these hypotheses are satisfied if H is the mod p cohomology ring of a space, see (3.5). Then the radical of the annihilator of any element is invariant under the action of the Steenrod operations.*

More explicitly, for $p = 2$, if y annihilates x and $2^n > |x|$ then $(\mathrm{Sq}^k y)^{2^n} x = 0$ for all $k > 0$. For p odd, if y annihilates x and $p^n > 2(p-1)|x|$ then $(P^k y)^{p^n} = 0$ for all $k > 0$.

PROOF. We give the proof for $p = 2$; the proof for p odd is the same, with P^i instead of Sq^i and p^n instead of 2^n.

Let $x \in H$, let I be the annihilator of x, and let $y \in I$. We have $y^{2^n} x = 0$, and so for any $k > 0$ we have $\mathrm{Sq}^{2^n k}(y^{2^n} x) = 0$. Using the Cartan formula (3.5.2) we obtain

$$\sum_{i=0}^{2^n k} \mathrm{Sq}^{2^n k - i}(y^{2^n}) \mathrm{Sq}^i x = 0.$$

The Cartan formula and divisibility properties of binomial coefficients imply that $\mathrm{Sq}^{2^n k - i}(y^{2^n}) = 0$ unless i is of the form $2^n j$, and in that case we have $\mathrm{Sq}^{2^n (k-j)}(y^{2^n}) = (\mathrm{Sq}^{k-j} y)^{2^n}$. So the above equation becomes

$$\sum_{j=0}^{k} (\mathrm{Sq}^{k-j} y)^{2^n} \mathrm{Sq}^{2^n j} x = 0.$$

Since $2^n > |x|$, the unstable condition implies that the only term which survives in this sum is the term with $j = 0$. So we have $(\mathrm{Sq}^k y)^{2^n} x = 0$.   □

Since associated primes are annihilators, we get the following.

COROLLARY 7.3.2. *The associated primes in a mod p cohomology ring of a space are invariant under the Steenrod operations.*   □

(7.4) The importance of the Steenrod invariance of the associated primes in the cohomology of groups comes from the following theorem of Serre [1965b, Proposition 1].

THEOREM 7.4.1. *Let E be an elementary abelian p-group. The Steenrod invariant prime ideals in $H^*(E, \mathbb{F}_p)$ are in one-one correspondence with the subgroups of E. If E' is a subgroup of E then the corresponding Steenrod invariant prime ideal is $\sqrt{\ker(\mathrm{res}_{E,E'})}$, the radical of the kernel of restriction from E to E'.*

(7.5) Combining Theorem 7.4.1 with Quillen's Theorem 5.3.1, it follows that for any of the classes of groups for which that theorem holds, the Steenrod invariant primes in the mod p cohomology ring are the radicals of the kernels of restriction to elementary abelian subgroups. So using Corollary 7.3.2 we have the following.

THEOREM 7.5.1. *Let G be a compact Lie group. Then the Steenrod invariant prime ideals in $H^*(BG; \mathbb{F}_p)$ are the ideals of the form $\sqrt{\ker(\mathrm{res}_{G,E})}$, where E is an elementary abelian p-subgroup of G. In particular, the associated primes are of this form.*

(7.6) The corresponding result holds for the cohomology of groups of finite virtual cohomological dimension, and continuous cohomology of profinite groups with a finite number of conjugacy classes of elementary abelian p-subgroups.

(7.7) The question of exactly which elementary abelian subgroups give the associated primes is difficult. In the next section, we relate this to the question of finding upper bounds for the depth.

8. Associated Primes and Transfer

(8.1)   Upper bounds on the depth of the cohomology ring of a finite group come from a careful analysis of transfer and its relationship to the associated primes. If H is a subgroup of a finite group G and M is a kG-module, then the transfer is a map

$$\mathrm{Tr}_{H,G} \colon H^n(H, M) \to H^n(G, M).$$

It is defined by choosing a set of left coset representatives g_i of H in G. Let P_* be a projective resolution of k as a kG-module. Given a representative cocycle, which is a kH-module homomorphism $\hat{\zeta} \colon P_n \to M$, the transfer $\mathrm{Tr}_{H,G}(\zeta)$ is represented by $\sum_i g_i(\hat{\zeta})$, which is a kG-module homomorphism.

The reason why the transfer map is relevant is that if H is a subgroup of G and M is a kG-module, then $\mathrm{Tr}_{H,G}$ is $H^*(G, k)$-linear, when $H^*(H, M)$ is viewed as a right module over $H^*(G, k)$ via the restriction map $H^*(G, k) \to H^*(H, k)$. In other words, the following identity holds. If $\zeta \in H^*(G, k)$ and $\eta \in H^*(H, M)$ then

(8.1.1) $$\mathrm{Tr}_{H,G}(\eta.\mathrm{res}_{G,H}(\zeta)) = \mathrm{Tr}_{H,G}(\eta).\zeta.$$

In particular, if η annihilates $\mathrm{res}_{G,H}(\zeta)$ then $\mathrm{Tr}_{H,G}(\eta)$ annihilates ζ. For example, if ζ restricts to zero on some set of subgroups, then all transfers from those subgroups annihilate ζ.

(8.2)   One way to exploit the above observation is to use the following transfer theorem from [Benson 1993]. This generalizes a theorem of Carlson [1987] relating transfers from all proper subgroups of a p-group to the kernel of restriction to the center.

THEOREM 8.2.1. *Suppose that G is a finite group, and k is a field of characteristic p. Let $\mathscr{H}$ be a collection of subgroups of G. Let $\mathscr{K}$ denote the collection of all elementary abelian p-subgroups K of G with the property that the Sylow p-subgroups of the centralizer $C_G(K)$ are not conjugate to a subgroup of any of the groups in $\mathscr{H}$.*

Let J be the sum of the images of transfer from subgroups in $\mathscr{H}$, which is an ideal in $H^*(G, k)$ by (8.1.1). Let J' be the intersection of the kernels of restriction to subgroups in $\mathscr{K}$, which is again an ideal in $H^*(G, k)$ (in case $\mathscr{K}$ is empty, this intersection is taken to be the ideal of all elements of positive degree). Then J and J' have the same radical, $\sqrt{J} = \sqrt{J'}$.

(8.3)   Theorem 8.2.1 is the main ingredient in the proof of the following theorem of Carlson [1995] relating the associated primes with detection on centralizers.

THEOREM 8.3.1. *Let G be a finite group. Suppose that $H^*(G, k)$ has a nonzero element ζ which restricts to zero on $C_G(E)$ for each elementary abelian p-subgroup $E \leq G$ of rank s. Then $H^*(G, k)$ has an associated prime $\mathfrak{p}$ such* *that the Krull dimension of $H^*(G,k)/\mathfrak{p}$ is strictly less than s. In particular, the depth of $H^*(G,k)$ is strictly less than s.*

PROOF. In Theorem 8.2.1, we take $\mathcal{H}$ to be the set of centralizers of elementary abelian p-subgroups of rank s. Then the elementary abelian p-subgroups in $\mathcal{K}$ have rank strictly less than s. So the theorem implies that the ideal J has dimension strictly less than s.

If ζ is an element of $H^*(G,k)$ which restricts to zero on every element of $\mathcal{H}$, then by (8.1.1), ζ is annihilated by all transfers from $\mathcal{H}$. In other words, the annihilator of ζ contains J. Since the associated primes are the maximal annihilators, there is an associated prime containing J, and such an associated prime has dimension strictly less than s.   □

(8.4)   Another way of stating the conclusion to the theorem above is that if $H^*(G,k)$ has depth at least s then cohomology is detected on centralizers of rank s elementary abelian p-sugroups of G.

9. Idempotent Modules and Varieties

(9.1)   There is a method for systematically exploiting the connections between representation theory and cohomology, which was first introduced by Carlson [1981a; 1981b; 1983] for finitely generated modules, and by Benson, Carlson and Rickard [Benson et al. 1995; 1996] for infinitely generated modules.

Let G be a finite group, and let k be an algebraically closed field of characteristic p. We write V_G for the maximal ideal spectrum of $H^*(G,k)$. This is a closed homogeneous affine variety. For example, if $G \cong (\mathbb{Z}/p)^r$ then $V_G = \mathbb{A}^r(k)$, affine r-space over k. Quillen's Theorem 5.3.1 can be interpreted as saying that for any finite group G, the natural map

$$\varinjlim_{E \in \mathscr{A}_G} V_E \to V_G$$

is bijective at the level of sets of points. However, it is usually not invertible in the category of varieties.

If M is a finitely generated kG-module, then the kernel of the natural map

$$H^*(G,k) = \mathrm{Ext}^*_{kG}(k,k) \xrightarrow{M \otimes_k -} \mathrm{Ext}^*_{kG}(M,M)$$

is an ideal in $H^*(G,k)$, which defines a closed homogeneous subvariety $V_G(M)$ of V_G. The same subvariety can be obtained by taking the intersection of the annihilators of $\mathrm{Ext}^*_{kG}(S,M)$ as S runs over the simple kG-modules. Properties of varieties for modules include (9.1.1)–(9.1.5) below.

(9.1.1) $V_G(M) = \{0\}$ if and only if M is projective,

(9.1.2) $V_G(M \oplus N) = V_G(M) \cup V_G(N)$,

(9.1.3) $V_G(M \otimes N) = V_G(M) \cap V_G(N)$,[7]

(9.1.4)   If $0 \neq \zeta \in H^n(G, k)$ is represented by a cocycle $\hat{\zeta} \colon \Omega^n(k) \to k$, let L_ζ be the kernel of $\hat{\zeta}$. Then $V_G(L_\zeta)$ is the hypersurface in V_G determined by regarding ζ as an element of the coordinate ring of V_G.

(9.1.5)   If V is any closed homogeneous subvariety of V_G, then we can choose homogeneous elements $\zeta_1, \ldots, \zeta_t \in H^*(G, k)$ so that the intersection of the hypersurfaces they define is equal to V. Properties (9.1.3) and (9.1.4) then imply that

$$V_G(L_{\zeta_1} \otimes \cdots \otimes L_{\zeta_t}) = V.$$

So every closed homogeneous subvariety is the variety of some module.

(9.2)   For infinite dimensional modules,[8] the definitions are more difficult. A tentative definition was given in [Benson et al. 1995], and the definition was modified in [Benson et al. 1996] to remedy some defects. We begin with some background on the stable module category. We write $\mathsf{Mod}(kG)$ for the category of kG-modules and module homomorphisms. The stable module category $\mathsf{StMod}(kG)$ has the same objects as $\mathsf{Mod}(kG)$, but the morphisms are

$$\underline{\mathrm{Hom}}_{kG}(M, N) = \mathrm{Hom}_{kG}(M, N)/\mathrm{PHom}_{kG}(M, N),$$

where $\mathrm{PHom}_{kG}(M, N)$ is the subspace consisting of maps which factor through some projective kG-module. One of the advantages of $\mathsf{StMod}(kG)$ over $\mathsf{Mod}(kG)$ is that if we define ΩM to be the kernel of a surjection from a projective module P onto M, then Ω is a functor on $\mathsf{StMod}(kG)$. Since, over kG, projective modules are the same as injective modules, Ω is a self-equivalence of $\mathsf{StMod}(kG)$. Its inverse Ω^{-1} is defined by embedding M into an injective kG-module I and writing $\Omega^{-1}M$ for the quotient I/M.

The category $\mathsf{Mod}(kG)$ is abelian, but $\mathsf{StMod}(kG)$ is not. Instead it is a triangulated category. The triangles are of the form

$$A \to B \to C \to \Omega^{-1}A$$

where $0 \to A \to B \to C \to 0$ is a short exact sequence in $\mathsf{Mod}(kG)$. We write $\mathsf{mod}(kG)$ and $\mathsf{stmod}(kG)$ for the full subcategories of finitely generated modules.

Write $\mathsf{Proj}\, H^*(G, k)$ for the set of closed homogeneous irreducible subvarieties of V_G, and let $\mathscr{V}$ be a subset of $\mathsf{Proj}\, H^*(G, k)$ which is closed under specialization,

---

[7]When we write $M \otimes N$ for kG-modules M and N, we mean $M \otimes_k N$ with diagonal G-action. So an element $g \in G$ acts via $g(m \otimes n) = gm \otimes gn$. But the general element of kG does not act in this fashion; rather, we extend linearly from the action of the group elements. See § 2.11 of [Sri].

[8]The reason for interest in infinite dimensional modules in this context is similar to the reason for the interest in infinite CW complexes in algebraic topology. Namely, the representing objects for functors often turn out to be infinite. For example in algebraic topology, Eilenberg–Mac Lane spaces are the representing objects for cohomology, BU for K-theory, MU for cobordism, and so on.

in the sense that if $V \in \mathscr{V}$ and $W \subseteq V$ then $W \in \mathscr{V}$. Let $\mathscr{M}$ be the full subcategory of $\mathsf{stmod}(kG)$ consisting of finitely generated modules M such that $V_G(M)$ is a finite union of elements of $\mathscr{V}$. Then $\mathscr{M}$ is a *thick subcategory* of $\mathsf{stmod}(kG)$; in other words it is a full triangulated subcategory of $\mathsf{stmod}(kG)$ with the same definition of triangles, and is closed under taking direct summands. Furthermore, a tensor product of any module with a module in $\mathscr{M}$ gives an answer in $\mathscr{M}$, so we say that $\mathscr{M}$ is a *tensor closed* thick subcategory. To such a subcategory of $\mathsf{stmod}(kG)$, Rickard [1997] associates two idempotent modules[9] $E_{\mathscr{V}}$ and $F_{\mathscr{V}}$ and a triangle

$$E_{\mathscr{V}} \to k \to F_{\mathscr{V}} \to \Omega^{-1} E_{\mathscr{V}}$$

in $\mathsf{StMod}(kG)$. This triangle is characterized by the statement that $E_{\mathscr{V}}$ can be written as a filtered colimit of modules in $\mathscr{M}$, and for any M in $\mathscr{M}$, we have $\underline{\mathrm{Hom}}_{kG}(M, F_{\mathscr{V}}) = 0$. This construction is the analog in representation theory of Bousfield localization [Bousfield 1979] in algebraic topology.

As an example, if $\zeta \in H^n(G, k)$ defines a hypersurface V in V_G and $\mathscr{V}$ is the set of subvarieties of V then we write E_ζ and F_ζ instead of $E_{\mathscr{V}}$ and $F_{\mathscr{V}}$. If ζ is represented by a cocycle $\hat{\zeta} : \Omega^n(k) \to k$, then the module F_ζ can be constructed as follows. We can dimension shift $\hat{\zeta}$ to give maps

$$k \xrightarrow{\Omega^{-n}\hat{\zeta}} \Omega^{-n}(k) \xrightarrow{\Omega^{-2n}\hat{\zeta}} \Omega^{-2n}(k) \xrightarrow{\Omega^{-3n}\hat{\zeta}} \cdots$$

and the colimit is F_ζ. So for example the cohomology of F_ζ,

$$\hat{H}^*(G, F_\zeta) \cong \hat{H}^*(G, k)_\zeta \cong H^*(G, k)_\zeta$$

is the localization of either Tate or ordinary cohomology with respect to ζ.

If we take the map from the first term in the sequence to the colimit and complete to a triangle, we get the module E_ζ and the triangle

$$(9.2.1) \qquad\qquad E_\zeta \to k \to F_\zeta \to \Omega^{-1} E_\zeta.$$

If L_{ζ^i} is the kernel of $\hat{\zeta}^i$ then E_ζ can be written as the colimit of

$$\Omega^{-n} L_\zeta \to \Omega^{-2n} L_{\zeta^2} \to \Omega^{-3n} L_{\zeta^3} \to \cdots$$

More generally, if V is a closed homogeneous subvariety of V_G defined by the vanishing of elements $\zeta_1, \ldots, \zeta_t \in H^*(G, k)$ and $\mathscr{V}$ is the set of subvarieties of V, we write E_V and F_V for $E_{\mathscr{V}}$ and $F_{\mathscr{V}}$. In this case, E_V can be obtained as $E_{\zeta_1} \otimes \cdots \otimes E_{\zeta_t}$, and F_V can be obtained by completing the map $E_V \to k$ to a triangle.

---

[9]A module M is said to be idempotent if $M \otimes M$ is isomorphic to M in $\mathsf{StMod}(kG)$. The only finite dimensional idempotent module is the trivial kG-module k.

Now if V is a closed homogeneous irreducible subvariety of V_G, let $\mathscr{W}$ be the set of subvarieties of V_G which do not contain V. Then we define

$$(9.2.2) \qquad\qquad \kappa_V = E_V \otimes F_{\mathscr{W}}.$$

This is an idempotent module which corresponds to a layer of $\mathsf{stmod}(kG)$ consisting of modules with variety exactly V.

If M is a kG-module, not necessarily finitely generated, we associate to M a collection of varieties

$$\mathscr{V}_G(M) = \{V \mid M \otimes \kappa_V \text{ is not projective}\} \subseteq \mathsf{Proj}\, H^*(G, k).$$

If M happens to be finitely generated, then $\mathscr{V}_G(M)$ is just the collection of all subvarieties of $V_G(M)$. But for infinitely generated modules, the collection is not necessarily closed under specialization. The properties of $\mathscr{V}_G(M)$ include:

(9.2.3) $\mathscr{V}_G(M) = \varnothing$ if and only if M is projective,

(9.2.4) $\mathscr{V}_G(\bigoplus_\alpha M_\alpha) = \bigcup_\alpha \mathscr{V}_G(M_\alpha)$,

(9.2.5) $\mathscr{V}_G(M \otimes N) = \mathscr{V}_G(M) \cap \mathscr{V}_G(N)$,

(9.2.6) $\mathscr{V}_G(\kappa_V) = \{V\}$.

(9.2.7) It follows from (9.2.4) and (9.2.6) that every subset of $\mathsf{Proj}\, H^*(G, k)$ occurs as $\mathscr{V}_G(M)$ for some M.

10. Modules with Injective Cohomology

In this section, we continue with our assumption that G is a finite group and k is a field.

(10.1)   A better understanding of the modules κ_V comes from understanding modules whose Tate cohomology is injective as a module over the Tate cohomology of the group. In this section, we shall see that there is an essentially unique module with a given injective as its cohomology [Benson and Krause 2002]. Conjecturally, these are the translates of the modules κ_V described in the last section. This has been proved under some restrictive hypotheses in [Benson 2001], but at least it is true for elementary abelian groups, an important special case. The connection between the κ_V and modules with injective cohomology involves the study of the local cohomology of the cohomology ring, $H_{\mathfrak{p}}^{**} H^*(G, k)$. This is the subject of the next section.

(10.2)   It is well known that the indecomposable injective modules over a commutative Noetherian ring R are precisely the injective hulls $E(R/\mathfrak{p})$ of the modules $R/\mathfrak{p}$, as $\mathfrak{p}$ ranges over the prime ideals of R, and that a general injective module can be written in an essentially unique way as a direct sum of indecomposable injectives.[10] For a Noetherian graded commutative ring, the classification of injective graded modules is the same, except that we must restrict our attention to homogeneous prime ideals, and we must allow degree shift. If $\mathfrak{p}$ is a homogeneous prime ideal in $H^*(G, k)$ and d is an integer, we define $I_\mathfrak{p}$ to be $E(H^*(G, k)/\mathfrak{p})$ and $I_\mathfrak{p}[n]$ to be the result of shifting degrees by n. The notation here is that a shift of $[n]$ in a graded module means that the degree d part of the shifted module is the same as the degree $(n + d)$ part of the original module.

(10.3) Recall that the ordinary cohomology $H^*(G, k)$ is Noetherian, whereas Tate cohomology $\hat{H}^*(G, k)$ is usually not. The way to get from injective modules over ordinary cohomology to injectives over Tate cohomology is by *coinduction*. If I is an injective $H^*(G, k)$-module, we define $\hat{I}$ to be the injective $\hat{H}^*(G, k)$-module

$$\hat{I} = \operatorname{Hom}^*_{H^*(G,k)}(\hat{H}^*(G, k), I).$$

The notation here is that Hom^n denotes the graded homomorphisms which increase degree by n. If there is no superscript, it is assumed that $n = 0$.

THEOREM 10.3.1. *Every injective $\hat{H}^*(G, k)$-module is of the form $\hat{I}$ for some injective $H^*(G, k)$-module I.*

(10.4) In order to understand what happens when we coinduce $I = I_\mathfrak{p}[n]$, we consider two cases. Let $\mathfrak{m} = H^+(G, k)$ be the maximal ideal of positive degree elements in $H^*(G, k)$. If $\mathfrak{p}$ is not equal to $\mathfrak{m}$ then $\hat{I} = I$. More precisely, the restriction of $\hat{I}$ to an $H^*(G, k)$-module is naturally isomorphic to I. On the other hand, if $\mathfrak{p} = \mathfrak{m}$ then Tate duality (3.6.1) says that $I = \hat{H}^-(G, k)[n - 1]$. In this case, $\hat{I} = \hat{H}^*(G, k)[n - 1]$.

Since $\hat{H}^*(G, k)$ is usually not Noetherian, coinduction does not preserve direct sums. Actually, it does preserve direct sums as long as the injective only has copies of $I_\mathfrak{p}[n]$ with $\mathfrak{p} \neq \mathfrak{m}$, and no copies of $I_\mathfrak{m}[n]$; in other words if $\operatorname{Hom}^*_{H^*(G,k)}(k, I) = 0$. Although $\hat{H}^*(G, k)$ is not Noetherian, it is shown in [Benson and Krause 2002] that the general injective $\hat{H}^*(G, k)$-module has the form

(10.4.1) $$\bigoplus \widehat{I_\mathfrak{p}[n]} \quad \oplus \quad E\left(\bigoplus \hat{H}^*(G, k)[n - 1]\right),$$

where $\mathfrak{p} \neq \mathfrak{m}$ for each of the left hand summands, and $E(-)$ stands for injective hull over $\hat{H}^*(G, k)$.

A way to construct modules with injective cohomology is to use the Brown representability theorem [Brown 1965; Neeman 1996]. If I is an injective $H^*(G, k)$-module, the functor from $\mathsf{StMod}(kG)$ to vector spaces, which takes a kG-module

---

[10]A ring R is Noetherian if and only if an arbitrary direct sum of injective R-modules is injective.

M to the degree preserving homomorphisms

$$\mathrm{Hom}_{H^*(G,k)}(\hat{H}^*(G,M), I)$$

is *exact* (in other words it takes triangles in $\mathsf{StMod}(kG)$ to long exact sequences) and takes direct sums to direct products. Brown's representability theorem says that any such functor is representable. In other words, there exists a kG-module $T(I)$ and a functorial isomorphism

(10.4.2) $\mathrm{Hom}_{H^*(G,k)}(\hat{H}^*(G,M), I) \cong \underline{\mathrm{Hom}}_{kG}(M, T(I))$.

We remark that we could just as easily have replaced the left hand side of this isomorphism with $\mathrm{Hom}_{\hat{H}^*(G,k)}(\hat{H}^*(G,M), \hat{I})$, because this gives the same answer, and because by Theorem 10.3.1, every injective $\hat{H}^*(G,k)$-module is coinduced from $H^*(G,k)$.

(10.5)   Yoneda's lemma says that all natural transformations from a representable functor are representable. Applying this to natural transformations arising from homomorphisms between injective $H^*(G,k)$-modules, this makes T into a functor from the full subcategory $\mathsf{Inj}\, H^*(G,k)$ of injective $H^*(G,k)$-modules to the stable module category $\mathsf{StMod}(kG)$. Here are some properties of the modules $T(I)$, proved in [Benson and Krause 2002].

(10.5.1) $T(I_\mathfrak{m}[n]) \cong \Omega^{-n+1}(k)$.

(10.5.2)   If $\mathfrak{p} \neq \mathfrak{m}$ then $T(I_\mathfrak{p}[n])$ is an infinitely generated module whose variety is given by $\mathscr{V}_G(T(I_\mathfrak{p}[n])) = \{V\}$, where V is the irreducible variety corresponding to $\mathfrak{p}$.

(10.5.3) $\mathrm{Hom}^*_{H^*(G,k)}(\hat{H}^*(G,M), I) \cong \widehat{\mathrm{Ext}}^*_{kG}(M, T(I))$. This isomorphism follows by dimension shifting the defining isomorphism (10.4.2).

(10.5.4) $\hat{H}^*(G, T(I)) \cong \hat{I}$. This is the special case of (iii) where $M = k$.

(10.5.5)   If $\mathfrak{p} \neq \mathfrak{m}$ then $\widehat{\mathrm{Ext}}^*_{kG}(T(I_\mathfrak{p}[n]), T(I_\mathfrak{p}[n])) \cong \mathrm{End}^*_{H^*(G,k)}(I_\mathfrak{p})$. By a theorem of Matlis [1958], $\mathrm{End}^*_{H^*(G,k)}(I_\mathfrak{p})$ is isomorphic to the $\mathfrak{p}$-adic completion of cohomology,

$$H^*(G,k)_\mathfrak{p}^\wedge = \varprojlim_n H^*(G,k)_\mathfrak{p}/\mathfrak{p}_\mathfrak{p}^n.$$

More generally, if I has no copies of $I_\mathfrak{m}[n]$ as summands, or equivalently, if there are no homomorphisms from any degree shifted copy of k to I, then

$$\underline{\mathrm{Hom}}_{kG}(T(I), T(I')) \cong \mathrm{Hom}_{H^*(G,k)}(I, I').$$

So T is a *fully faithful* functor on the full subcategory $\mathsf{Inj}_0\, H^*(G,k)$ of injective modules I satisfying $\mathrm{Hom}^*_{H^*(G,k)}(k, I) = 0$.

(10.5.6)   The modules $T(I)$ are *pure injective*. This means that if a short exact sequence of kG-modules

$$0 \to T(I) \to X \to Y \to 0$$

has the property that every morphism from any finitely generated module to Y lifts to X, then it splits. The reason for this is that $T(I)$ is a direct summand of a direct product of finitely generated kG-modules. In fact,

(10.5.7)   The modules $T(I)$ are precisely the direct summands of direct products of modules isomorphic to $\Omega^n(k)$ for $n \in \mathbb{Z}$. The indecomposable ones are exactly $T(I_{\mathfrak{p}}[n])$. So we obtain an embedding of $\mathsf{Proj}\, H^*(G, k)$ into the Ziegler spectrum [Ziegler 1984] of pure injective kG-modules with the Zariski topology (modulo the translation Ω), and we can recover $\mathsf{Proj}\, H^*(G, k)$ from the category $\mathsf{StMod}(kG)$ if we know where the translates of the trivial module are.

(10.6)   The following conjecture from [Benson 2001] relates the modules $T(I_{\mathfrak{p}}[n])$ described in this section and the modules κ_V described in the previous section.

CONJECTURE 10.6.1. If $\mathfrak{p}$ is the homogeneous prime ideal corresponding to a closed homogeneous irreducible subvariety V of dimension d in V_G, then there is an isomorphism $T(I_{\mathfrak{p}}[n]) \cong \Omega^{-n-d}\kappa_V$ in $\mathsf{StMod}(kG)$.

This conjecture is known to hold if $H^*(G, k)_{\mathfrak{p}}$ is Cohen–Macaulay. In the next few sections, we describe how to view this conjecture in terms of local cohomology and Grothendieck's local duality. See Conjecture 13.2.2.

(10.7)   Another conjecture, related in philosophy to Conjecture 10.6.1, comes from an idea of Amnon Neeman. If M is a kG-module, not necessarily finitely generated, then for each simple module S, consider a minimal injective resolution of $\widehat{\mathrm{Ext}}^*_{kG}(S, M)$ as a module over $\hat{H}^*(G, k)$,

$$0 \to \widehat{\mathrm{Ext}}^*_{kG}(S, M) \to \hat{I}_0 \to \hat{I}_1 \to \hat{I}_2 \to \cdots$$

Each injective in such a resolution can be written in the form (10.4.1), and we can ask which nonmaximal prime ideals occur in such a decomposition. Since each $\hat{I}_j$ is coinduced from some injective $H^*(G, k)$-module I_j, we have

$$\mathrm{Hom}^*_{\hat{H}^*(G,k)}(H^*(G, k)/\mathfrak{p}, \hat{I}_j) \cong \mathrm{Hom}^*_{H^*(G,k)}(H^*(G, k)/\mathfrak{p}, I_j),$$

and this is nonzero exactly for the primes appearing in this minimal resolution. Since coinduction is exact on injectives away from the maximal ideal, it does not matter whether we resolve over $\hat{H}^*(G, k)$ or $H^*(G, k)$. The following conjecture says that the primes appearing in these minimal resolutions, as S runs over the simple kG-modules, correspond exactly to the varieties in $\mathcal{V}_G(M)$.

CONJECTURE 10.7.1. Let M be a kG-module. If $\mathfrak{p}$ is a homogeneous prime ideal in $H^*(G, k)$, we define $k(\mathfrak{p})$ to be the homogeneous field of fractions of

$H^*(G,k)_\mathfrak{p}$. Then the nonmaximal homogeneous primes $\mathfrak{p}$ for which[11]

$$\mathrm{Ext}^{**}_{H^*(G,k)_\mathfrak{p}}(k(\mathfrak{p}), \widehat{\mathrm{Ext}}^*_{kG}(S,M)_\mathfrak{p}) \neq 0$$

for some simple kG-module S are exactly the primes corresponding to the varieties in $\mathcal{V}_G(M)$.

The point of this conjecture is that it provides a method for characterizing $\mathcal{V}_G(M)$ just in terms of $\widehat{\mathrm{Ext}}^*_{kG}(S,M)$, without having to tensor M with the rather mysterious modules κ_V.

11. Duality Theorems

In this section, we describe various spectral sequences which can be interpreted as duality theorems for group cohomology. It is these theorems which demonstrate that most finitely generated graded commutative algebras are not candidates for group cohomology. The original version of the spectral sequence for finite groups appeared in [Benson and Carlson 1994b], and used multiple complexes and related finite Poincaré duality complexes of projective kG-modules. One consequence of the existence of this spectral sequence is that if $H^*(G,k)$ is Cohen–Macaulay then it is Gorenstein, with a-invariant zero (2.5). Even if $H^*(G,k)$ is not Cohen–Macaulay, the spectral sequence gives severe restrictions on the possibilities for the ring structure.

Greenlees [1995] discovered a way of using the same techniques to construct a cleaner spectral sequence of the form

$$H^{s,t}_{\mathfrak{m}} H^*(G,k) \implies H_{-s-t}(G,k)$$

giving essentially equivalent information. We present here an alternative construction [Benson 2001] of Greenlees' spectral sequence using Rickard's idempotent modules.

(11.1) Choose a homogeneous set of parameters $\zeta_1, \ldots, \zeta_r$ for $H^*(G,k)$. For each ζ_i, we truncate the triangle (9.2.1) to give a cochain complex of the form

(11.1.1) $$\cdots \to 0 \to k \to F_{\zeta_i} \to 0 \to \cdots$$

where k is in degree zero and F_{ζ_i} is in degree one, and the remaining terms are zero. The cohomology of this complex is $\Omega^{-1} E_{\zeta_i}$ concentrated in degree one. Tensoring these complexes together gives a complex Λ^* of the form

$$0 \to k \to \bigoplus_{1 \leq i \leq r} F_{\zeta_i} \to \cdots \to \bigotimes_{1 \leq i \leq r} F_{\zeta_i} \to 0,$$

---

[11]In commutative algebra, the ranks over $k(\mathfrak{p})$ of these Ext groups are called the Bass numbers.

which is exact except in degree r, where its cohomology is

$$\Omega^{-1}E_{\zeta_1} \otimes \cdots \otimes \Omega^{-1}E_{\zeta_r}.$$

Now $\mathcal{V}_G(\Omega^{-1}E_{\zeta_i})$ is the set of subvarieties of the hypersurface determined by ζ_i. So using (9.2.5), the variety of this tensor product is the intersection of these sets, which is empty. It follows using (9.2.3) that this tensor product is a projective kG-module.

Now let $\hat{P}_*$ be a Tate resolution of k as a kG-module, and consider the double complex $\hat{E}_0^{**} = \operatorname{Hom}_{kG}(\hat{P}_*, \Lambda^*)$. This double complex gives rise to two spectral sequences. If we take cohomology with respect to the differential coming from Λ^* first, the E_1 page is $\operatorname{Hom}_{kG}(\hat{P}_*, H^*(\Lambda))$. Since $H^*(\Lambda)$ is projective, the E_2 page is zero, and so the cohomology of the total complex $\operatorname{Tot}\hat{E}_0^{**}$ is zero.

On the other hand, if we first take cohomology with respect to the differential coming from $\hat{P}_*$, we obtain a spectral sequence whose E_1 page is $\hat{E}_1^{s,t} = \hat{H}^t(G, \Lambda^s)$. Now each Λ^s is a direct sum of modules of the form F_ζ, where ζ is the product of a subset of size s of $\zeta_1, \ldots, \zeta_r$. The cohomology of F_ζ is $\hat{H}^*(G, k)_\zeta \cong H^*(G, k)_\zeta$, and the maps are exactly the maps in the stable Koszul complex[12] for $\hat{H}^*(G, k)$ over $\zeta_1, \ldots \zeta_r$,

$$\hat{E}_1^{**} = C^*(\hat{H}^*(G, k); \zeta_1, \ldots, \zeta_r).$$

The stable Koszul complex computes local cohomology with respect to the maximal ideal $\mathfrak{m}$, and so we have a spectral sequence

$$(11.1.2) \qquad\qquad \hat{E}_2^{s,t} = H_{\mathfrak{m}}^{s,t}\hat{H}^*(G, k) \implies 0.$$

(11.2)   The spectral sequence (11.1.2) is almost, but not quite, the Greenlees spectral sequence, so we modify it as follows. Consider the subcomplex E_0^{**} consisting of all the terms in $\hat{E}_0^{**}$ except the ones of the form $\hat{E}_0^{0,t}$ with $t < 0$. Then we have a short exact sequence of complexes

$$0 \to \operatorname{Tot} E_0^{**} \to \operatorname{Tot}\hat{E}_0^{**} \to \operatorname{Hom}_{kG}(\hat{P}_*^-, k) \to 0.$$

Since $\operatorname{Tot}\hat{E}_0^{**}$ is exact, the long exact sequence in cohomology and Tate duality give

$$H^n\operatorname{Tot} E_0^{**} \cong H^{n+1}(\operatorname{Hom}_{kG}(\hat{P}_*^-, k)) \cong H_{-n}(G, k).$$

So the spectral sequence of the double complex E_0^{**} has

$$E_1^{**} = C^*(H^*(G, k); \zeta_1, \ldots, \zeta_r)$$

$$(11.2.1) \qquad E_2^{s,t} = H_{\mathfrak{m}}^{s,t}H^*(G, k) \implies H_{-s-t}(G, k).$$

---

[12]See for example [Bruns and Herzog 1993, §3.5]. Sometimes the stable Koszul complex is called the Čech complex, but strictly speaking the latter name should be reserved for the complex where the degree zero term is deleted and the degrees of the remaining terms are decreased by one. This is the complex used to derive the spectral sequence (11.3.1).

This is a spectral sequence of $H^*(G, k)$-modules; that is, the differentials

$$d_n \colon E_n^{s,t} \to E_n^{s+n,t-n+1}$$

are $H^*(G, k)$-linear. It converges because there are only a finite number of nonzero columns. This is because $E_2^{s,t} = 0$ unless s lies between the depth and the Krull dimension of $H^*(G, k)$. The spectral sequence (11.2.1) is isomorphic to the one constructed by Greenlees [1995], although the construction given there is slightly different. Note also that $H_*(G, k)$ is just the injective module $I_{\mathfrak{m}}$, so we can write this as

$$H_{\mathfrak{m}}^{s,t} H^*(G, k) \implies I_{\mathfrak{m}}.$$

The local cohomology can only be nonzero for s at least the depth and at most the Krull dimension, so this spectral sequence often has only a few nonvanishing columns.

(11.3)   Another variation on the construction (11.2) is as follows. Instead of eliminating just the negative part of the $s = 0$ line of the E_0 page, we eliminate the whole of the $s = 0$ line. Then after reindexing, we obtain a spectral sequence whose E_2 page is the Čech cohomology of the cohomology ring, and converging to Tate cohomology,

(11.3.1) $$\check{H}_{\mathfrak{m}}^{s,t} H^*(G, k) \implies \hat{H}^{s+t}(G, k).$$

This is the spectral sequence described in [Greenlees 1995, Theorem 4.1].

(11.4)   As an example of an application of the spectral sequence (11.2.1), consider the case where $H^*(G, k)$ is Cohen–Macaulay (2.5). In this case, the local cohomology $H_{\mathfrak{m}}^{s,*} H^*(G, k)$ is zero unless $s = r$, and the graded dual of $H_{\mathfrak{m}}^{r,*} H^*(G, k)$ is the canonical module $\Omega_{H^*(G,k)}$. So the E_2 page is only nonzero on the column $s = r$, and there is no room for differentials. It follows that the spectral sequence converges to $(\Omega_{H^*(G,k)}[r])^*$, and so $\Omega_{H^*(G,k)}[r]$ is isomorphic to the graded dual of $H_*(G, k)$, which in turn is isomorphic to $H^*(G, k)$. We deduce that $\Omega_{H^*(G,k)} \cong H^*(G, k)[-r]$. It follows that $H^*(G, k)$ is Gorenstein with a-invariant zero. This gives the following theorem, which was first proved in [Benson and Carlson 1994b], using the original version of the spectral sequence.

THEOREM 11.4.1.   *Let G be a finite group and k be a field of characteristic p. If $H^*(G, k)$ is Cohen–Macaulay, then it is Gorenstein with a-invariant zero.*

This theorem may be interpreted in terms of Poincaré series as follows. If we set $p_G(t) = \sum_{i=0}^{\infty} t^i \dim H^i(G, k)$ then the finite generation theorem says that $p_G(t)$ is a rational function of t whose poles are at roots of unity. If $H^*(G, k)$ is Cohen–Macaulay, the theorem above implies that this rational function satisfies the functional equation $p_G(1/t) = (-t)^r p_G(t)$. For this and related functional equations, see [Benson and Carlson 1994a].

(11.5) Another interpretation of Theorem 11.4.1 is as follows. If $\zeta_1, \ldots, \zeta_r$ is a homogeneous system of parameters for $H^*(G, k)$ with $|\zeta_i| = n_i$ then the quotient ring $H^*(G, k)/(\zeta_1, \ldots, \zeta_r)$ satisfies Poincaré duality with dualizing degree $a = \sum_{i=1}^r (n_i - 1)$. See the description of the quaternion group of order eight (2.4) for an explicit example of this phenomenon. Whether or not $H^*(G, k)$ is Cohen–Macaulay, it is shown in [Benson and Carlson 1994b] that there is always a nonzero element of the quotient $H^*(G, k)/(\zeta_1, \ldots, \zeta_r)$ in the dualizing degree a. In particular, we get the following corollary from that paper:

COROLLARY 11.5.1. *Let G be a finite group and k be a field of characteristic p. If $H^*(G, k)$ is a polynomial ring, then the generators are in degree one. This forces p to be 2, and $G/O_{2'}(G)$ to be an elementary abelian 2-group.*

Here, $O_{p'}(G)$ denotes the largest normal subgroup of G of order not divisible by p. If k is a field of characteristic p, then the inflation map (3.2) from $H^*(G/O_{p'}(G), k)$ to $H^*(G, k)$ is an isomorphism, so we would expect to get information only about the structure of $G/O_{p'}(G)$ from information about the cohomology of G. If $p = 2$ and $G \cong (\mathbb{Z}/2)^r$ is elementary abelian then by (2.2.2), $H^*(G, k) = k[x_1, \ldots, x_r]$ is a polynomial ring on r generators of degree one. In this case, $H_\mathfrak{m}^{s,t} H^*(G, k)$ vanishes except when $s = r$, and[13]

$$H_\mathfrak{m}^{r,*} H^*(G, k) = k[x_1^{-1}, \ldots, x_r^{-1}],$$

where the right hand side is graded in such a way that the identity element is in $H_\mathfrak{m}^{r,-r} H^*(G, k)$.

There are no differentials, and it is easy to see how the spectral sequence converges to the dual of the cohomology ring.

On the other hand, if $G \cong (\mathbb{Z}/p)^r$ with p odd, then $H^*(G, k) = \Lambda(x_1, \ldots, x_r) \otimes k[y_1, \ldots, y_r]$ is a tensor product of an exterior algebra on r generators of degree one with a polynomial algebra on r generators in degree two. Taking $y_1, \ldots, y_r$ as a homogeneous sequence of parameters, the exterior algebra $\Lambda(x_1, \ldots, x_r)$ is the finite Poincaré duality piece referred to in the discussion following Theorem 11.4.1. The local cohomology is again concentrated in degree r, and $H_\mathfrak{m}^r$ consists of 2^r copies of $k[y_1^{-1}, \ldots, y_r^{-1}]$ with generators in $H_\mathfrak{m}^{r,-r-i}$, $i = 0, \ldots, r$ dual to a basis for the exterior algebra.

(11.6) As another Cohen–Macaulay example, if $G = D_{2^n}$ is dihedral of order 2^n then $H^*(G, \mathbb{F}_2) = \mathbb{F}_2[x, y, z]/(xy)$ where x and y have degree one and z has degree two. We can take $x+y$ and z as a homogeneous system of parameters, and the quotient is $H^*(G, \mathbb{F}_2)/(x+y, z) = \mathbb{F}_2[x]/(x^2)$, which satisfies Poincaré duality

---

[13]The notation $k[x_1^{-1}, \ldots, x_r^{-1}]$ is just a shorthand notation for the graded dual of the polynomial ring in $x_1, \ldots, x_r$. Beware that the notation does not transform correctly with respect to linear transformations of $x_1, \ldots, x_r$ because it depends on the choice of system of parameters for the stable Koszul complex. The action on the inverse generators should be transposed from what the notation suggests. However, the notion is called "Macaulay's inverse system," and is standard in commutative algebra.

with dualizing degree one. The local cohomology is concentrated in degree two, and consists of two copies of $\mathbb{F}_2[(x+y)^{-1}, z^{-1}]$ with generators in $H_{\mathrm{m}}^{2,-2}$ and $H_{\mathrm{m}}^{2,-3}$ dual to 1 and x.

(11.7) If $G = SD_{2^n}$ is semidihedral of order 2^n (2.8.1) and k is a field of characteristic two then $H^*(G, k) = k[x, y, z, w]/(xy, y^3, yz, z^2 + wx^2)$ where x and y have degree one, z has degree three, and w has degree four. The depth of this ring is one, and the Krull dimension is two, so there is local cohomology in degrees one and two; H_{m}^1 consists of two copies of $\mathbb{F}_2[w^{-1}]$ with generators in $H_{\mathrm{m}}^{1,-2}$ and $H_{\mathrm{m}}^{1,-3}$ which play the role of duals for y and y^2, while H_{m}^2 consists of two copies of $\mathbb{F}_2[w^{-1}, x^{-1}]$ with generators in $H_{\mathrm{m}}^{2,-2}$ and $H_{\mathrm{m}}^{2,-5}$ dual to 1 and z. This example is described in more detail in § 2.

(11.8) The cohomology of the 2-groups of order at most 32 has been calculated by Rusin [1989]. A particularly interesting example is the group

(11.8.1) $\Gamma_7 a_2 = \langle a,\ b,\ c \mid a^4 b = ba^4,\ a^4 c = ca^4,\ bc = cb,$
$$a^8 = b^2 = c^2 = 1,\ aba^{-1} = bc,\ aca^{-1} = a^4 c \rangle,$$

of order 32, whose cohomology has Krull dimension three and depth one. This is the smallest example where the Greenlees spectral sequence has a nonzero differential. The cohomology ring $H^*(\Gamma_7 a_2, \mathbb{F}_2)$ is generated by elements z, y, x, w, v, u, t and s of degrees 1, 1, 2, 2, 3, 3, 4, 4, respectively, where the ideal of relations is generated by the elements

$$zy,\ y^2,\ yx,\ yw,\ yv,\ yu,\ yt,\ xw + zu,\ z^2 w + w^2,\ wv + zt,\ zxw + wu,$$
$$z^2 t + wt,\ x^3 + zxv + z^2 s + v^2,\ vu + xt,\ x^2 u + zxt + zws + vt,$$
$$x^3 w + z^2 xt + z^2 ws + t^2,\ zxt + ut,\ x^2 w + u^2.$$

As a module over the polynomial subring $\mathbb{F}_2[z, x, s]$, the cohomology is generated by 1, y, w, v, u and t, subject to the relations $zy = 0$, $xy = 0$ and $zu = xw$. The local cohomology is nonvanishing in degrees 1, 2 and 3; both H_{m}^1 and H_{m}^2 consist of a copy of $\mathbb{F}_2[s^{-1}]$ generated in degree -3, while H_{m}^3 consists of four copies of $\mathbb{F}_2[z^{-1}, x^{-1}, s^{-1}]$ generated in degrees -3, -4, -6 and -7. The nonzero differential d_2 takes the generator in $H_{\mathrm{m}}^{1,-3}$ to the generator in $H_{\mathrm{m}}^{3,-4}$, wiping out H_{m}^1 and having a cokernel on this part of H_{m}^3 which plays the role of the dual of the summand generated by w and u. The remaining generators in $H_{\mathrm{m}}^{3,-3}$, $H_{\mathrm{m}}^{3,-6}$ and $H_{\mathrm{m}}^{3,-7}$ are dual to 1, v and t, while the generator of $H_{\mathrm{m}}^{2,-3}$ is dual to y.

12. More Duality Theorems

(12.1) There is a version of the spectral sequence (11.2.1) for compact Lie groups [Benson and Greenlees 1997a]. This involves a dimension shift, equal to the dimension d of G as a manifold. There is also an orientation issue. Namely, the adjoint representation of G gives a homomorphism $G \to O(d)$, whose image does not necessarily lie in $SO(d)$. Composing with the determinant representation of $O(d)$, which takes values ± 1, gives a one dimensional representation $\varepsilon \colon \pi_1(BG) \cong \pi_0(G) \to k^\times$. The spectral sequence then takes the form

(12.1.1) $$H^{s,t}_{\mathfrak{m}} H^*(BG; k) \Longrightarrow H_{-d-s-t}(BG; \varepsilon).$$

Dwyer, Greenlees and Iyengar [Dwyer et al. 2002] give another proof for compact Lie groups and also a version for p-compact groups.

For example, if $H^*(BG; k)$ is a polynomial ring on generators $\zeta_1, \ldots, \zeta_r$ with $|\zeta_i| = n_i$, and $\varepsilon = k$, then we have

(12.1.2) $$d = \sum_{i=1}^r (n_i - 1).$$

If $G = U(n)$, the compact unitary group of $n \times n$ matrices, and k is any commutative coefficient ring, then $H^*(BU; k) = k[c_1, \ldots, c_n]$ is a polynomial ring on Chern classes c_i of degree $2i$ (3.3.1). In accordance with equation (12.1.2), we have $\dim U(n) = n^2 = \sum_{i=1}^n (2i - 1)$.

On the other hand, if $G = O(2n)$, the compact orthogonal group of real $2n \times 2n$ matrices preserving a positive definite inner product, and k is a field of characteristic not equal to two, then $H^*(BO(2n); k) = k[p_1, \ldots, p_n]$ is a polynomial ring on Pontrjagin classes p_i of degree $4i$ (3.3.2). Since $\dim O(2n) = 2n^2 - n$ and $\sum_{i=1}^n (4i - 1) = 2n^2 + n$, we see that the two sides of equation (12.1.2) differ by $2n$. This is because the orientation representation ε is nontrivial, and

$$H^*(BO(2n); \varepsilon) = k[p_1, \ldots, p_n] \cdot e$$

where $e \in H^{2n}(BSO(2n); k) \cong H^{2n}(BO(2n); k \oplus \varepsilon)$ is the Euler class, satisfying $e^2 = p_n$ (3.3.3). The degree of the Euler class exactly accounts for the discrepancy in equation (12.1.2).

(12.2)   Another version of the spectral sequence has been developed for virtual duality groups [Benson and Greenlees 1997b]. The latter is a class of groups which includes arithmetic groups [Borel and Serre 1973], mapping class groups of orientable surfaces [Harer 1986] and automorphism groups of free groups of finite rank [Bestvina and Feighn 2000]. A discrete group G is said to be a *duality group* of dimension d over k (see [Bieri 1976]) if there is a *dualizing module*. This is defined to be a kG-module I such that there are isomorphisms $H^i(G, M) \cong H_{d-i}(G, I \otimes_k M)$ for all kG-modules M. It turns out that such isomorphisms may be taken to be functorial in M if they exist at all, and in that case, $I \cong H^d(G, kG)$. A *Poincaré duality group* is a duality group for which the dualizing module I is isomorphic to the field k with some G-action, and it is *orientable* if the action is trivial. A *virtual duality group* of dimension d is a group G with a normal subgroup N of finite index which is a duality group of dimension d. Since the Eckmann–Shapiro lemma says that $H^*(G, kG) \cong H^*(N, kN)$, the dualizing module I does not depend on which normal subgroup is used in the definition. The spectral sequence for a virtual duality group takes the form

$$H_{\mathfrak{m}}^{s,t} H^*(G, M) \implies H_{d-s-t}(G, I \otimes M).$$

Notice that the sign of the degree shift in this case is in the opposite direction to the case of a compact Lie group. So a virtual Poincaré duality group of dimension d behaves very much like a compact Lie group of dimension $-d$.

(12.3)   In [Greenlees 2002, §8.4], there is a brief discussion of the corresponding version for continuous cohomology of p-adic Lie groups, which are a particular kind of profinite groups. These include matrix groups over the p-adic integers such as $\mathrm{SL}(n, \hat{\mathbb{Z}}_p)$. The discussion for p-adic Lie groups translates into continuous cohomology the story for virtual duality groups, with the same shift in dimension. The way this works is as follows. By [Lazard 1965, Chapter V, 2.2.7.1 and 2.5.7.1], if G is a p-adic Lie group then G has a normal open subgroup H for which $H_c^*(H, \mathbb{F}_p)$ is the exterior algebra on $H_c^1(H, \mathbb{F}_p)$, so that H is a Poincaré duality group. Furthermore, $H_c^1(H, \mathbb{F}_p)$ is a finite dimensional $\mathbb{F}_p$-vector space whose dimension is equal to the dimension d of G as a p-adic manifold. So the dualizing module ε is the $\mathbb{F}_p G$-module $H_c^d(H, \mathbb{F}_p)$, which is the same as the determinant of the adjoint representation of G on its Lie algebra. The spectral sequence then takes the form

(12.3.1) $$H_{\mathfrak{m}}^{s,t} H_c^*(G, \mathbb{F}_p) \implies H_{d-s-t}^c(G, \varepsilon).$$

(12.4)   There are also versions of the spectral sequence for other cohomology theories. For example, [Bruner and Greenlees 2003] investigates the spectral sequence

$$H_I^{*,*} ku^*(BG) \implies ku_*(BG)$$

where ku denotes connective complex K-theory, I is the kernel of the augmentation map $ku^*(BG) \to ku^*$, and G is a finite group.

. (12.5)   The papers [Dwyer et al. 2002; Greenlees 2002] also explain a more general context for some of these spectral sequences. They explain the sense in which the cochains on BG and related objects are examples of Gorenstein differential graded algebras. Their notions are expressed in the language of E_∞ ring spectra, or commutative S-algebras, see [Elmendorf et al. 1997].

13. Dual Localization

(13.1)   Greenlees and Lyubeznik [2000] introduced a way of obtaining information at nonmaximal prime ideals out of the Greenlees spectral sequence. Roughly speaking, one would like to localize the spectral sequence. Attempting to do this directly turns out to be a bad move. The reason is that every element of $H_{\mathfrak{m}}^{**} H^*(G, k)$ and every element of $I_{\mathfrak{m}}$ is killed by some power of $\mathfrak{m}$. So the idea is to dualize first, then localize, and then dualize back again. The dualization process needed for this is graded Matlis duality. See [Matlis 1958] for ordinary Matlis duality, and [Bruns and Herzog 1993, § 3.6] for the graded version. If $\mathfrak{p}$ is a homogeneous prime ideal in $H^*(G, k)$ and X is a module over $H^*(G, k)_\mathfrak{p}$, then we write $D_\mathfrak{p}(X)$ for the graded Matlis dual of X

$$D_\mathfrak{p}(X) = \operatorname{Hom}_{H^*(G,k)_\mathfrak{p}}(X, I_\mathfrak{p}),$$

where $I_\mathfrak{p}$ is the injective hull of $H^*(G, k)/\mathfrak{p}$. The latter can be viewed as a module over the completion $H^*(G, k)_\mathfrak{p}^\wedge$, and so $D_\mathfrak{p}$ takes $H^*(G, k)_\mathfrak{p}$-modules to $H^*(G, k)_\mathfrak{p}^\wedge$-modules. It takes Artinian modules to Noetherian modules, and vice-versa. Applying $D_\mathfrak{p}$ twice to an Artinian module returns the same module, and applying $D_\mathfrak{p}$ twice to a Noetherian module returns its $\mathfrak{p}$-adic completion. In this language, we can rewrite equation (10.5.3) as

$$D_\mathfrak{p}\hat{H}^*(G, M) \cong \widehat{\operatorname{Ext}}_{kG}^*(M, T(I_\mathfrak{p})).$$

Tate duality is the special case of this statement where $\mathfrak{p} = \mathfrak{m}$, because $D_\mathfrak{m}$ can be interpreted as taking the graded dual of a graded vector space, and $T(I_\mathfrak{m}) = \Omega(k)$.

Grothendieck duality [Grothendieck 1965; 1967] says that if we choose a polynomial subring $R = k[\zeta_1, \ldots, \zeta_r]$ over which $H^*(G, k)$ is finitely generated as a module, and M is a graded $H^*(G, k)$-module, then the graded Matlis dual of local cohomology is Ext over R in complementary degrees,

(13.1.1) $$D_\mathfrak{m} H_\mathfrak{m}^{s,t} M \cong \operatorname{Ext}_R^{r-s,-t}(M, R[-a])$$

where $a = \sum_{i=1}^r |\zeta_i|$ and $R[-a]$ is the canonical module for R. So the graded Matlis dual of the Greenlees spectral sequence is

$$\operatorname{Ext}_R^{r-s,-t}(H^*(G, k), R[-a]) \implies H^{-s-t}(G, k).$$

Localizing this spectral sequence with respect to a homogeneous prime ideal $\mathfrak{p} \neq \mathfrak{m}$ of dimension d gives a spectral sequence

$$\operatorname{Ext}_{R_\mathfrak{q}}^{r-s,-t}(H^*(G, k)_\mathfrak{p}, R_\mathfrak{q}[-a]) \implies H^{-s-t}(G, k)_\mathfrak{p}$$

where $\mathfrak{q} = \mathfrak{p} \cap R$. Since $R_\mathfrak{q}$ has Krull dimension $r - d$ instead of r, applying $D_\mathfrak{p}$ to this spectral sequence and using Grothendieck duality again gives a spectral sequence of the form $H_\mathfrak{p}^{s-d,t} H^*(G, k)_\mathfrak{p} \implies I_\mathfrak{p}$, or reindexing,

(13.1.2) $$H_\mathfrak{p}^{s,t} H^*(G, k)_\mathfrak{p} \implies I_\mathfrak{p}[d].$$

This is the Greenlees–Lyubeznik dual localized form of the Greenlees spectral sequence. So for example, taking $\mathfrak{p}$ to be a minimal prime in $H^*(G, k)$, this spectral sequence has only one nonvanishing column, and it follows that $H^*(G, k)_\mathfrak{p}$ is Gorenstein. This gives the following theorem.

THEOREM 13.1.3. *If G is a finite group and k is a field then $H^*(G,k)$ is generically Gorenstein.*

(13.2) There is another, more module theoretic method for getting a spectral sequence with the same E_2 page as (13.1.2), described in [Benson 2001]. Let $V \subseteq V_G$ be the closed homogeneous irreducible subvariety corresponding to $\mathfrak{p}$, and let $\mathscr{W}$ be the subset of $\operatorname{Proj} H^*(G,k)$ used in the definition (9.2.2) of κ_V. Since the maximal elements of $\mathscr{W}$ have codimension one in V_G, the cohomology of the F-idempotent is just the homogeneous localization, $\hat{H}^*(G, F_{\mathscr{W}}) = H^*(G,k)_{\mathfrak{p}}$. Let h be the height of $\mathfrak{p}$, namely the Krull dimension of $H^*(G,k)_{\mathfrak{p}}$. Then by the version of the Noether normalization theorem described in [Nagata 1962], we can choose a homogeneous set of parameters $\zeta_1, \ldots, \zeta_r$ for $H^*(G,k)$ so that $\zeta_1, \ldots, \zeta_h$ lie in $\mathfrak{p}$. So $\zeta_1, \ldots, \zeta_h$ is a system of parameters for $H^*(G,k)_{\mathfrak{p}}$. We tensor together the complexes (11.1.1) for $\zeta_1, \ldots, \zeta_h$ to obtain a complex $\Lambda^*(\zeta_1, \ldots, \zeta_h)$ of the form

$$0 \to k \to \bigoplus_{i=1}^{h} F_{\zeta_i} \to \cdots \to \bigotimes_{i=1}^{h} F_{\zeta_i} \to 0$$

and then tensor the answer with the module $F_{\mathscr{W}}$ to obtain a complex

$$\Lambda_{\mathfrak{p}}^* = \Lambda^*(\zeta_1, \ldots, \zeta_h) \otimes F_{\mathscr{W}}$$

whose cohomology is $\Omega^{-h} E_V \otimes F_{\mathscr{W}} = \Omega^{-h} \kappa_V$ concentrated in degree h. The spectral sequence of the double complex $E_0^{**}(\mathfrak{p}) = \operatorname{Hom}_{kG}(\hat{P}_*, \Lambda_{\mathfrak{p}}^*)$ gives

(13.2.1) $$E_2^{s,t}(\mathfrak{p}) = H_{\mathfrak{p}}^{s,t} H^*(G,k)_{\mathfrak{p}} \implies \hat{H}^{s+t}(G, \kappa_V).$$

CONJECTURE 13.2.2. The spectral sequences 13.1.2 and 13.2.1 are isomorphic from the E_2 page onwards.

It is proved in [Benson 2001] that Conjecture 13.2.2 implies Conjecture 10.6.1. Furthermore, Conjecture 13.2.2 clearly holds in the case where $H^*(G,k)$ is Cohen–Macaulay, because there is no room for nontrivial differentials or ungrading problems.

14. Quasiregular Sequences

In this section, we describe the theory of quasiregular sequences, first introduced in [Benson and Carlson 1994b], and describe their relationship with the local cohomology of $H^*(G,k)$. The material of this section is further developed in a companion paper [Benson 2004], written during the month following the MSRI workshop.

(14.1) Let G be a finite group of p-rank r, and let k be a field of characteristic p. A homogeneous sequence of parameters $\zeta_1, \ldots, \zeta_r$ for $H^*(G,k)$ with $|\zeta_i| = n_i$ is said to be *filter-regular* if for each $i = 0, \ldots, r-1$, the map

(14.1.1) $$(H^*(G,k)/(\zeta_1, \ldots, \zeta_i))^j \to (H^*(G,k)/(\zeta_1, \ldots, \zeta_i))^{j+n_{i+1}}$$

induced by multiplication by ζ_{i+1} is injective for j large enough. The existence of a filter-regular sequence is guaranteed by the standard method of prime avoidance.

In [Benson and Carlson 1994b, § 10], the following terminology was introduced. A sequence of parameters $\zeta_1, \ldots, \zeta_r$ is said to be *quasiregular*[14] if the map (14.1.1) is injective for $i = 0, \ldots, r - 1$ whenever $j \geq n_1 + \cdots + n_i$, and $H^*(G, k)/(\zeta_1, \ldots, \zeta_r)$ is zero in degrees at least $n_1 + \cdots + n_r$. For $i = 0$ this is the same as saying that ζ_1 is a regular element, but for $i > 0$ it allows some low degree kernel.

CONJECTURE 14.1.2. *For any finite group G and field k, there exists a quasiregular sequence in $H^*(G, k)$.*

It is proved in [Benson and Carlson 1994b] that the conjecture is true if $r \leq 2$, and Okuyama and Sasaki [2000] have a proof for $r \leq 3$. These proofs work more generally when the depth and Krull dimension differ by at most one, respectively two. In this section, I shall try to explain the ideas behind these proofs, and the relevance of quasiregular sequences for the computation of group cohomology.

(14.2) We can reinterpret the definition of quasiregular sequence in terms of cohomology of modules as follows, and in the process give some sort of explanation of where the condition $j \geq n_1 + \cdots + n_i$ comes from. We can always take our first parameter ζ_1 to be a regular element, by Duflot's Theorem 6.2.1. Consider the short exact sequence

$$0 \to L_{\zeta_1} \to \Omega^{n_1}(k) \to k \to 0.$$

The long exact sequence in cohomology gives (for $j \geq n_1$) an exact sequence

$$\cdots \to H^j(G, L_{\zeta_1}) \to H^{j-n_1}(G, k) \xrightarrow{\zeta_1} H^j(G, k) \to H^{j+1}(G, L_{\zeta_1}) \to \cdots$$

So we have $H^{j+1}(G, L_{\zeta_1}) \cong (H^*(G, k)/(\zeta_1))^j$ for $j \geq n_1$.

Working inductively, for each $i = 0, \ldots, r - 1$, if we tensor the short exact sequence

$$0 \to L_{\zeta_{i+1}} \to \Omega^{n_{i+1}}(k) \to k \to 0$$

with the module $M_i = L_{\zeta_1} \otimes \cdots \otimes L_{\zeta_i}$ and take the long exact sequence in cohomology, then we obtain the following.

PROPOSITION 14.2.1. *A homogeneous sequence of parameters $\zeta_1, \ldots, \zeta_r$ is quasiregular if and only if for each $i = 0, \ldots, r - 1$, multiplication by ζ_{i+1} is injective on $H^j(G, M_i)$ for $j \geq n_1 + \cdots + n_i + i$.*   □

---

[14]This terminology has nothing to do with the terminology of quasiregular sequences used in [Matsumura 1989]. The definition in [Benson and Carlson 1994b] omits the condition on $H^*(G, k)/(\zeta_1, \ldots, \zeta_r)$, but this condition turns out to be automatic, see Corollary 14.2.2.

COROLLARY 14.2.2. *If $\zeta_1, \ldots, \zeta_r$ is a homogeneous system of parameters and the condition for quasiregularity is satisfied for $i = 0, \ldots, r - 2$, then it is also satisfied for $i = r - 1$, and the quotient $H^*(G, k)/(\zeta_1, \ldots, \zeta_r)$ is zero in degree at least $n_1 + \cdots + n_r$.*

PROOF. $H^{j+r-1}(G, M_{r-1}) = (H^*(G, k)/(\zeta_1, \ldots, \zeta_{r-1}))^j$ for $j \geq n_1 + \cdots + n_{r-1}$. Tensoring the sequence $0 \to L_{\zeta_r} \to \Omega^{n_r} k \xrightarrow{\hat{\zeta}_r} k \to 0$ with M_{r-1}, and using the fact that $M_{r-1} \otimes L_{\zeta_r}$ is projective, we see that ζ_r induces an isomorphism $\Omega^{n_r} M_{r-1} \cong M_{r-1}$, and hence an isomorphism on $H^*(G, M_{r-1})$ in positive degrees.   □

(14.3)   To go further, we make use of the transfer map. If we choose the parameters to be the Dickson invariants (see § 7), then the restriction to each elementary abelian p-subgroup E of rank $r - 1$ of the sequence $\zeta_1, \ldots, \zeta_{r-1}$ is a homogeneous sequence of parameters in $H^*(E, k)$. It follows that $V_G(M_{r-1})$ has trivial intersection with the image of $V_E \to V_G$ for each such E. Theorem 1.5 of [Benson 1994/95] (see also Corollary 4.5 of [Carlson et al. 1998]) then shows that $M_{r-2} \otimes L_{\zeta_{r-1}} = M_{r-1}$ is projective relative to[15] the set $\mathscr{H}_r$ of centralizers $C_G(E)$ of elementary abelian p-subgroups E of rank r. So the sum of the transfers from these subgroups gives a surjective map in cohomology. Furthermore, the restrictions of $\zeta_1, \ldots, \zeta_r$ to a subgroup in $\mathscr{H}_r$ form a regular sequence, by Duflot's theorem. Now examine the diagram

$$\begin{array}{ccccccc}
\cdots \longrightarrow & H^{j+n_{r-1}}(G, M_{r-1}) & \longrightarrow & H^j(G, M_{r-2}) & \xrightarrow{\zeta_{r-1}} & H^{j+n_{r-1}}(G, M_{r-2}) \\
& \uparrow & & \uparrow & & \uparrow \\
\cdots \longrightarrow & \displaystyle\bigoplus_{H \in \mathscr{H}_r} H^{j+n_{r-1}}(H, M_{r-1}) & \longrightarrow & \displaystyle\bigoplus_{H \in \mathscr{H}_r} H^j(H, M_{r-2}) & \xrightarrow[\zeta_{r-1}]{} & \displaystyle\bigoplus_{H \in \mathscr{H}_r} H^{j+n_{r-1}}(H, M_{r-2})
\end{array}$$

where the vertical maps are given by $\sum_{H \in \mathscr{H}_r} \mathrm{Tr}_{H,G}$. The map marked ζ_{r-1} on the bottom row is injective, and a diagram chase shows that the corresponding map on the top row is therefore also injective. So ζ_{r-1} is quasiregular. Finally, the argument of the previous paragraph shows that the last parameter ζ_r is also quasiregular. This completes the argument of Okuyama and Sasaki, proving Conjecture 14.1.2 in the case where the depth and the Krull dimension differ by at most two.

It looks as though the argument above ought to admit a modification which makes it work inductively and prove the conjecture, but so far nobody has succeeded in doing this.

Carlson [1999; 2001] has developed some conjectures related to Conjecture 14.1.2, which allow a machine computation of group cohomology by computing

---

[15]A module M is said to be *projective relative to* a set of subgroups $\mathscr{H}$ of G if it is a direct summand of a direct sum of modules induced from elements of $\mathscr{H}$. This is equivalent to the statement that the sum of the transfer maps $\mathrm{Tr}_{H,G} \colon \mathrm{End}_{kH}(M) \to \mathrm{End}_{kG}(M)$ from the subgroups $H \in \mathscr{H}$ is surjective. For further details, see [Benson 1991a, § 3.6].

just a finite part at the beginning of a projective resolution. The usefulness of the conjectures depends on the fact that during the course of the calculation for a particular group, it is proved that the cohomology ring really does satisfy the conjectures, so there is no uncertainty about the answer. Condition G of [Carlson 2001] is related to the existence of a quasiregular sequence, while Condition R of that paper is a weak form of Conjecture 7.2.1. If the existence of a quasiregular sequence could be verified *a priori*, then the computational method could be guaranteed to work. This is explained in Theorem 14.5.2 below.

The cohomology of the groups of order 64 can be found in [Carlson ≥ 2004]. In the course of the computations, Conditions G and R of [Carlson 2001] were verified for these groups.

(14.4) The existence of a quasiregular sequence in group cohomology can be reformulated in terms of local cohomology as follows. If

$$H = \bigoplus_{i \geq 0} H^i = k \oplus \mathfrak{m}$$

is a graded commutative ring with $H^0 = k$ a field and $\mathfrak{m} = \bigoplus_{i>0} H^i$, and M is a graded H-module, we set

$$a_\mathfrak{m}^i(M) = \max\{n \in \mathbb{Z} \mid H_\mathfrak{m}^{i,n}(M) \neq 0\}$$

(or $a_\mathfrak{m}^i(M) = -\infty$ if $H_\mathfrak{m}^i(M) = 0$).

The following is proved in Corollary 3.7 of [Benson 2004].

THEOREM 14.4.1. *If G is a finite group and k is a field, then the following are equivalent.*

(i) *There is a quasiregular sequence in $H^*(G, k)$,*

(ii) *Every filter-regular sequence of parameters in $H^*(G, k)$ is quasiregular,*

(iii) *The Dickson invariants (see §7) in $H^*(G, k)$ are quasiregular,*

(iv) *For all $i \geq 0$ we have $a_\mathfrak{m}^i(H^*(G, k)) < 0$.*

(14.5) It is shown in [Benson 2004, §5] that we can interpret the invariants $a_\mathfrak{m}^i$ in terms of resolutions. If $R = k[\zeta_1, \ldots, \zeta_r]$ is a polynomial subring over which H is finitely generated as a module, and M is a graded H-module, let

$$0 \to F_r \to \cdots \to F_0 \to M \to 0$$

be a minimal resolution of M over R. We define $\beta_j^R(M)$ to be the largest degree of a generator of F_j as an R-module (or $\beta_j^R(M) = -\infty$ if $F_j = 0$). Then we have

(14.5.1) $$\max_{j \geq 0}\{a_\mathfrak{m}^j(M)\} = \max_{j \geq 0}\{\beta_j^R(M) - \sum_{i=1}^r |\zeta_i|\}.$$

This equation, together with Theorem 14.4.1, explains the relevance of the existence of quasiregular sequences to finding bounds for the degrees of generators

and relations for the group cohomology. Better ways of bounding the degrees of
the relations can be found in [Carlson 2001] and in [Benson 2004].

THEOREM 14.5.2. *Let G be a finite group and k be a field. If $\zeta_1, \ldots, \zeta_r$ is
a quasiregular sequence in $H^*(G, k)$, then all the generators for $H^*(G, k)$ have
degree at most $\sum_{i=1}^r |\zeta_i|$, and the relations have degree at most $2(\sum_{i=1}^r |\zeta_i|) - 2$.*

PROOF. Set $R = k[\zeta_1, \ldots, \zeta_r]$ as above. By Theorem 14.4.1, the existence of
a quasiregular sequence implies that $a_{\mathfrak{m}}^j(H^*(G, k)) < 0$ for all $j \geq 0$. So by
equation (14.5.1), we have

$$\beta_j^R(H^*(G, k)) < \sum_{i=1}^r |\zeta_i|$$

for all $j \geq 0$. The numbers β_0^R and β_1^R are the largest degrees for generators
and relations respectively of $H^*(G, k)$ as an R-module. The ring generators
have degree at most $\max(\beta_0^R, |\zeta_1|, \ldots, |\zeta_r|)$. For the ring relations, we need the
R-module relations together with relations saying how the products of pairs of
R-module generators can be written as R-linear combinations of generators. $\square$

(14.6)   The Castelnuovo–Mumford regularity of a graded H-module M (see
[Eisenbud 1995, § 20.5], for example) is defined as

$$\mathsf{Reg}\, M = \max_{j \geq 0} \{a_{\mathfrak{m}}^j(M) + j\} = \max_{j \geq 0} \left\{ \beta_j^R(M) - j - \sum_{i=1}^r (|\zeta_i| - 1) \right\}.$$

The second equality here is proved in [Benson 2004]. Usually the summation
term does not appear, because much of the literature on the subject assumes
that the graded ring H is generated over H^0 by elements of degree one; in this
context the above equality was proved in [Eisenbud and Goto 1984].

The "last survivor" described in [Benson and Carlson 1994b, Theorem 1.3]
and reinterpreted in terms of local cohomology in [Benson 2001, Theorem 4.1]
says that for a finite group G over a field k we have $H_{\mathfrak{m}}^{r,-r} H^*(G, k) \neq 0$, so
that $\mathsf{Reg}\, H^*(G, k) \geq 0$. One might strengthen Conjecture 14.1.2 to the following
statement, which has been checked for the 2-groups of order at most 64 using
Carlson's calculations [$\geq$ 2004].

CONJECTURE 14.6.1. *If G is a finite group and k is a field then $\mathsf{Reg}\, H^*(G, k) = 0$.*

This conjecture is equivalent to a strengthening of the bound given in the definition of a quasiregular sequence to $j > n_1 + \cdots + n_i - i$. Pushing the argument
given in the proof of Proposition 14.2.1 to its limits, and using some subtle information about $\hat{H}^{-1}(G, k)$, one can translate this into a strengthening of the
module theoretic bounds given in that proposition to $j > n_1 + \cdots + n_i$. For the
details, see [Benson 2004].

EXAMPLE 14.6.2. Let G be the Sylow 2-subgroup of $\mathrm{PSL}(3, \mathbb{F}_4)$, of order 64 (this
is group number 183 in the Appendix). Then $H^*(G, \mathbb{F}_2)$ has Krull dimension four

and depth two, with $a_m^2 = -3$, $a_m^3 = -5$ and $a_m^4 = -4$. So the regularity is zero. This is the only example of a 2-group of order at most 64 where $H_m^{r,-r} H^*(G, \mathbb{F}_2)$ has dimension bigger than one; in this example it has dimension two.

(14.7) Conjecture 14.6.1 can be interpreted in terms of the Greenlees spectral sequence (11.2.1). It says that the E_2 page vanishes above the line $s + t = 0$. Of course, this part of the E_2 page dies by the time the E_∞ page is reached, in order for the spectral sequence to be able to converge to a negatively graded target. In the above example, the extra dimension in $H_m^{4,-4}$ has to be hit in the spectral sequence by $H_m^{2,-3}$.

The conjecture can be generalized to compact Lie groups, virtual duality groups and p-adic Lie groups as follows.

CONJECTURE 14.7.1. If G is a compact Lie group of dimension d and k is a field then $\mathrm{Reg}\, H^*(BG; \varepsilon) = -d$. Here, ε is the orientation representation of (12.1.1).

CONJECTURE 14.7.2. If G is an orientable virtual Poincaré duality group of dimension d over a field k then $\mathrm{Reg}\, H^*(G, k) = d$.

CONJECTURE 14.7.3. If G is a p-adic Lie group of dimension d then over $\mathbb{F}_p$ we have $\mathrm{Reg}\, H^*(G, \varepsilon) = d$. Here, ε is the the orientation representation of (12.3.1).

As a nontrivial example, for the compact simply connected Lie group E_6 of dimension 78, the calculations of Kono and Mimura [1975] (see also [Benson and Greenlees 1997a]) imply that $H^*(BE_6; \mathbb{F}_2)$ has Krull dimension six and depth five, with $a_m^5 = -90$ and $a_m^6 = -84$, so that $\mathrm{Reg}\, H^*(BE_6; \mathbb{F}_2) = -78$.

Appendix: Two-Groups of Order 64 and Their mod 2 Cohomology

The table on the next page lists the Krull dimension of $H^*(G, \mathbb{F}_2)$, the depth of $H^*(G, \mathbb{F}_2)$, and the rank of the center of G (see Duflot's Theorem 6.2.1), for each of the 2-groups G of order 64. The numbering of the groups follows that of Hall and Senior [1964], who classified these groups. Underlined entries have

$$\text{Krull dimension} - \text{depth} = 2;$$

otherwise the difference is 1 or 0.

A separate table on page 45 gives the invariants $a_m^i(H^*(G, \mathbb{F}_2))$ defined in (14.4), for the entries where the difference is 2, with the rows of the table arranged in decreasing order of Krull dimension. Note that Duflot's Theorem 6.2.1 implies that a_m^0 is always zero, so the tables begin with the entry a_m^1.

All this information has been extracted from [Carlson $\geq$ 2004].

gp	K	d	r	gp	K	d	r	gp	K	d	r	gp	K	d	r	gp	K	d	r	gp	K	d	r
001	6	6	6	046	3	3	2	091	4	3	3	136	2	2	2	<u>181</u>	3	1	1	226	3	2	2
002	5	5	5	047	4	3	3	092	3	3	3	137	2	2	1	182	2	1	1	227	2	2	2
003	4	4	4	048	3	3	3	093	3	3	3	138	3	2	2	<u>183</u>	4	2	2	228	3	2	2
004	4	4	4	049	3	3	3	<u>094</u>	4	2	2	139	2	2	2	184	4	3	2	229	3	2	2
005	3	3	3	050	3	3	3	095	3	2	2	140	2	2	2	185	3	2	2	230	3	2	2
006	3	3	3	051	3	3	2	096	3	3	2	141	2	2	2	186	3	2	2	231	3	2	2
007	3	3	3	052	3	2	2	097	3	2	2	142	2	2	1	187	2	2	2	232	3	2	2
008	2	2	2	053	3	3	2	098	3	2	2	143	2	1	1	188	3	3	2	233	2	2	2
009	2	2	2	054	3	2	2	099	3	2	2	144	4	4	3	189	3	2	2	234	3	2	2
010	2	2	2	055	3	3	2	100	3	2	2	145	3	3	3	190	2	2	2	235	2	2	2
011	1	1	1	056	3	2	2	101	2	2	2	146	4	3	3	191	2	2	2	236	2	2	2
012	5	5	4	057	2	2	2	102	3	2	2	147	4	3	3	192	2	2	2	237	3	3	2
013	4	4	4	058	2	2	1	103	4	4	2	148	4	3	3	193	3	2	2	238	3	2	2
014	4	4	3	059	3	3	3	104	3	3	2	149	3	3	3	194	2	2	2	239	3	2	2
015	5	4	4	060	3	2	2	105	3	3	1	150	4	3	3	195	3	3	2	240	2	2	2
016	4	4	4	061	3	2	2	106	4	3	2	151	4	3	3	196	3	2	2	241	3	3	1
017	4	3	3	062	3	2	2	107	3	3	2	152	3	3	3	197	3	2	2	242	3	2	1
018	4	4	3	063	2	2	2	108	3	2	2	153	3	3	3	198	3	2	2	243	2	2	1
019	3	3	3	064	2	2	2	109	3	2	1	154	4	4	2	199	2	2	2	244	3	2	1
020	4	3	3	065	2	2	2	110	4	3	2	155	3	3	2	200	3	2	2	245	2	1	1
021	3	3	2	066	2	2	1	111	3	2	2	156	2	2	2	201	4	3	2	246	2	1	1
022	4	4	4	067	2	1	1	112	3	2	1	157	4	3	2	<u>202</u>	4	2	2	<u>247</u>	3	1	1
023	3	3	3	068	5	4	3	<u>113</u>	4	2	2	158	3	3	2	203	3	3	2	248	2	2	1
024	4	3	3	069	4	4	3	114	3	3	2	159	3	3	2	204	3	2	2	249	2	1	1
025	3	3	3	070	3	3	3	115	3	2	2	160	3	3	2	205	3	3	2	<u>250</u>	4	2	1
026	3	2	2	071	4	4	3	116	3	2	2	161	3	2	2	206	3	2	2	<u>251</u>	3	1	1
027	3	3	2	072	4	3	3	117	3	3	2	162	2	2	2	207	3	2	2	252	3	2	1
028	4	3	3	073	4	3	3	118	3	2	2	<u>163</u>	4	2	2	208	3	2	2	<u>253</u>	3	1	1
029	3	3	3	074	4	3	3	119	2	2	2	164	3	3	2	209	3	2	2	<u>254</u>	3	1	1
030	3	3	3	075	3	3	3	120	3	2	1	165	3	2	2	210	2	2	2	255	2	1	1
031	3	2	2	076	4	3	3	<u>121</u>	3	1	1	166	3	2	2	211	2	2	2	256	3	2	1
032	3	2	2	077	4	3	2	122	2	2	1	167	3	2	2	212	2	2	2	<u>257</u>	3	1	1
033	3	2	2	078	3	3	2	123	4	3	2	168	3	2	2	213	3	2	2	258	2	1	1
034	3	3	2	079	3	2	2	<u>124</u>	4	2	2	169	4	3	2	214	3	2	2	259	4	3	1
035	2	2	2	080	3	2	2	125	3	2	2	170	4	3	2	215	3	2	2	260	3	2	1
036	2	2	1	<u>081</u>	5	3	3	126	3	2	1	171	3	3	2	216	3	2	2	261	3	3	1
037	3	3	3	082	3	3	3	127	3	2	1	172	3	2	2	217	3	3	2	<u>262</u>	3	1	1
038	2	2	2	083	4	3	3	128	4	3	2	<u>173</u>	4	2	2	218	3	2	2	263	3	2	1
039	2	2	2	084	4	4	3	129	3	2	2	174	3	2	2	219	3	2	2	264	2	2	1
040	3	2	2	085	4	3	3	<u>130</u>	3	1	1	<u>175</u>	4	2	2	220	3	2	2	265	2	2	1
041	2	2	2	086	4	3	3	<u>131</u>	4	2	2	176	3	3	2	221	3	2	2	266	2	1	1
042	2	1	1	087	3	3	3	132	3	2	2	177	3	2	2	222	2	2	2	267	1	1	1
043	4	4	3	088	3	3	3	<u>133</u>	3	1	1	178	3	2	2	223	3	2	2				
044	4	3	3	089	4	3	3	134	3	3	2	179	3	2	2	224	3	2	2				
045	3	3	3	090	3	3	3	135	3	2	2	<u>180</u>	3	1	1	225	3	2	2				

Table 1. For each 2-group G of order 64, identified by its number in the notation of [Hall and Senior 1964], we give the Krull dimension K of $H^*(G, \mathbb{F}_2)$, the depth d of $H^*(G, \mathbb{F}_2)$, and the rank r of the center of G. Underlined entries have $K - d = 2$. Data taken from [Carlson $\geq$ 2004].

Group	a_m^1	a_m^2	a_m^3	a_m^4	a_m^5
081	$-\infty$	$-\infty$	-5	-5	-5
094	$-\infty$	-4	-4	-4	
113	$-\infty$	-4	-4	-4	
124	$-\infty$	-4	-4	-4	
131	$-\infty$	-5	-4	-4	
163	$-\infty$	-5	-4	-4	
173	$-\infty$	-5	-4	-4	
175	$-\infty$	-5	-4	-4	
183	$-\infty$	-3	-5	-4	
202	$-\infty$	-4	-4	-4	
250	$-\infty$	-5	-4	-4	

Group	a_m^1	a_m^2	a_m^3
121	-5	-3	-3
130	-5	-3	-3
133	-5	-3	-3
180	-5	-3	-3
181	-5	-3	-3
247	-3	-3	-3
251	-4	-3	-3
253	-5	-3	-3
254	-4	-3	-3
257	-5	-3	-3
262	-5	-3	-3

Table 2. Invariants $a_m^i(H^*(G, \mathbb{F}_2))$ defined in (14.4), for the underlined entries of Table 1.

Acknowledgements

I thank Luchezar Avramov and Srikanth Iyengar for their many valuable comments on these notes, which have improved the exposition in a number of places. I also thank the Mathematical Sciences Research Institute for its hospitality during the Commutative Algebra Program, 2002–03.

References

[Adem and Milgram 1994] A. Adem and R. J. Milgram, *Cohomology of finite groups*, Grundlehren der Mathematischen Wissenschaften **309**, Springer, Berlin, 1994. Second edition, 2004.

[Adem and Milgram 1995] A. Adem and R. J. Milgram, "The mod 2 cohomology rings of rank 3 simple groups are Cohen–Macaulay", pp. 3–12 in *Prospects in topology* (Princeton, NJ, 1994), edited by F. Quinn, Ann. of Math. Stud. **138**, Princeton Univ. Press, Princeton, NJ, 1995.

[Becker and Gottlieb 1975] J. C. Becker and D. H. Gottlieb, "The transfer map and fiber bundles", *Topology* **14** (1975), 1–12.

[Benson 1991a] D. J. Benson, *Representations and cohomology, I: Basic representation theory of finite groups and associative algebras*, Cambridge Studies in Advanced Mathematics **30**, Cambridge University Press, Cambridge, 1991. Paperback reprint, 1998.

[Benson 1991b] D. J. Benson, *Representations and cohomology, II: Cohomology of groups and modules*, Cambridge Studies in Advanced Mathematics **31**, Cambridge University Press, Cambridge, 1991. Paperback reprint, 1998.

[Benson 1993] D. J. Benson, "The image of the transfer map", *Arch. Math. (Basel)* **61**:1 (1993), 7–11.

[Benson 1994/95] D. J. Benson, "Cohomology of modules in the principal block of a finite group", *New York J. Math.* **1** (1994/95), 196–205.

[Benson 2001] D. Benson, "Modules with injective cohomology, and local duality for a finite group", *New York J. Math.* **7** (2001), 201–215.

[Benson 2004] D. Benson, "Dickson invariants, regularity and computation in group cohomology", *Illinois J. Math.* **48**:1 (2004), 171–197.

[Benson and Carlson 1992] D. J. Benson and J. F. Carlson, "Products in negative cohomology", *J. Pure Appl. Algebra* **82**:2 (1992), 107–129.

[Benson and Carlson 1994a] D. J. Benson and J. F. Carlson, "Functional equations for Poincaré series in group cohomology", *Bull. London Math. Soc.* **26**:5 (1994), 438–448.

[Benson and Carlson 1994b] D. J. Benson and J. F. Carlson, "Projective resolutions and Poincaré duality complexes", *Trans. Amer. Math. Soc.* **342**:2 (1994), 447–488.

[Benson and Greenlees 1997a] D. J. Benson and J. P. C. Greenlees, "Commutative algebra for cohomology rings of classifying spaces of compact Lie groups", *J. Pure Appl. Algebra* **122**:1-2 (1997), 41–53.

[Benson and Greenlees 1997b] D. J. Benson and J. P. C. Greenlees, "Commutative algebra for cohomology rings of virtual duality groups", *J. Algebra* **192**:2 (1997), 678–700.

[Benson and Krause 2002] D. Benson and H. Krause, "Pure injectives and the spectrum of the cohomology ring of a finite group", *J. Reine Angew. Math.* **542** (2002), 23–51.

[Benson and Wilkerson 1995] D. J. Benson and C. W. Wilkerson, "Finite simple groups and Dickson invariants", pp. 39–50 in *Homotopy theory and its applications* (Cocoyoc, Mexico, 1993), edited by R. J. M. Alejandro Adem and D. C. Ravenel, Contemp. Math. **188**, Amer. Math. Soc., Providence, RI, 1995.

[Benson et al. 1995] D. J. Benson, J. F. Carlson, and J. Rickard, "Complexity and varieties for infinitely generated modules", *Math. Proc. Cambridge Philos. Soc.* **118**:2 (1995), 223–243.

[Benson et al. 1996] D. J. Benson, J. F. Carlson, and J. Rickard, "Complexity and varieties for infinitely generated modules, II", *Math. Proc. Cambridge Philos. Soc.* **120**:4 (1996), 597–615.

[Bestvina and Feighn 2000] M. Bestvina and M. Feighn, "The topology at infinity of $\mathrm{Out}(F_n)$", *Invent. Math.* **140**:3 (2000), 651–692.

[Bieri 1976] R. Bieri, *Homological dimension of discrete groups*, Queen Mary College Mathematics Notes, Mathematics Department, Queen Mary College, London, 1976.

[Bøgvad and Halperin 1986] R. Bøgvad and S. Halperin, "On a conjecture of Roos", pp. 120–127 in *Algebra, algebraic topology and their interactions* (Stockholm, 1983), edited by J.-E. Roos, Lecture Notes in Math. **1183**, Springer, Berlin, 1986.

[Borel and Serre 1973] A. Borel and J.-P. Serre, "Corners and arithmetic groups", *Comment. Math. Helv.* **48** (1973), 436–491.

[Bourguiba and Zarati 1997] D. Bourguiba and S. Zarati, "Depth and the Steenrod algebra", *Invent. Math.* **128**:3 (1997), 589–602.

[Bousfield 1979] A. K. Bousfield, "The localization of spectra with respect to homology", *Topology* **18**:4 (1979), 257–281.

[Bousfield and Kan 1972] A. K. Bousfield and D. M. Kan, *Homotopy limits, completions and localizations*, Lecture Notes in Mathematics **304**, Springer, Berlin, 1972.

[Broto and Henn 1993] C. Broto and H.-W. Henn, "Some remarks on central elementary abelian p-subgroups and cohomology of classifying spaces", *Quart. J. Math. Oxford Ser.* (2) **44**:174 (1993), 155–163.

[Brown 1965] E. H. Brown, Jr., "Abstract homotopy theory", *Trans. Amer. Math. Soc.* **119** (1965), 79–85.

[Brown 1982] K. S. Brown, *Cohomology of groups*, Graduate Texts in Mathematics **87**, Springer, New York, 1982. Corrected reprint, 1994.

[Bruner and Greenlees 2003] R. R. Bruner and J. P. C. Greenlees, *The connective K-theory of finite groups*, Mem. Amer. Math. Soc. **785**, Amer. Math. Soc., Providence, RI, 2003.

[Bruns and Herzog 1993] W. Bruns and J. Herzog, *Cohen–Macaulay rings*, Cambridge Studies in Advanced Mathematics **39**, Cambridge University Press, Cambridge, 1993.

[Carlson 1981a] J. F. Carlson, "Complexity and Krull dimension", pp. 62–67 in *Representations of algebras* (Puebla, 1980), edited by M. Auslander and E. Lluis, Lecture Notes in Math. **903**, Springer, Berlin, 1981.

[Carlson 1981b] J. F. Carlson, "The complexity and varieties of modules", pp. 415–422 in *Integral representations and applications* (Oberwolfach, 1980), edited by K. W. Roggenkamp, Lecture Notes in Math. **882**, Springer, Berlin, 1981.

[Carlson 1983] J. F. Carlson, "The varieties and the cohomology ring of a module", *J. Algebra* **85**:1 (1983), 104–143.

[Carlson 1987] J. F. Carlson, "Varieties and transfers", *J. Pure Appl. Algebra* **44**:1-3 (1987), 99–105.

[Carlson 1995] J. F. Carlson, "Depth and transfer maps in the cohomology of groups", *Math. Z.* **218**:3 (1995), 461–468.

[Carlson 1999] J. F. Carlson, "Problems in the calculation of group cohomology", pp. 107–120 in *Computational methods for representations of groups and algebras* (Essen, 1997), edited by P. Dräxler et al., Progr. Math. **173**, Birkhäuser, Basel, 1999.

[Carlson 2001] J. F. Carlson, "Calculating group cohomology: tests for completion", *J. Symbolic Comput.* **31**:1-2 (2001), 229–242.

[Carlson ≥ 2004] J. F. Carlson, "Cohomology of 2-groups of order ≤ 64". Available at http://www.math.uga.edu/~jfc/groups/cohomology.html. These tables will be published in *Cohomology rings of finite groups*, by J. F. Carlson and L. Townsley.

[Carlson et al. 1998] J. F. Carlson, C. Peng, and W. W. Wheeler, "Transfer maps and virtual projectivity", *J. Algebra* **204**:1 (1998), 286–311.

[Cartan and Eilenberg 1956] H. Cartan and S. Eilenberg, *Homological algebra*, Princeton University Press, Princeton, NJ, 1956.

[Dickson 1911] L. E. Dickson, "A fundamental system of invariants of the general modular linear group with a solution of the form problem", *Trans. Amer. Math. Soc.* **12** (1911), 75–98.

[Duflot 1981] J. Duflot, "Depth and equivariant cohomology", *Comment. Math. Helv.*
56:4 (1981), 627–637.

[Dwyer and Wilkerson 1994] W. G. Dwyer and C. W. Wilkerson, "Homotopy fixed-
point methods for Lie groups and finite loop spaces", *Ann. of Math.* (2) **139**:2 (1994),
395–442.

[Dwyer et al. 2002] W. G. Dwyer, J. P. C. Greenlees, and S. Iyengar, "Duality in
algebra and topology", Preprint, University of Notre Dame, 2002. Available at
http://www.nd.edu/~wgd.

[Eisenbud 1995] D. Eisenbud, *Commutative algebra, with a view toward algebraic
geometry*, Graduate Texts in Mathematics **150**, Springer, New York, 1995.

[Eisenbud and Goto 1984] D. Eisenbud and S. Goto, "Linear free resolutions and
minimal multiplicity", *J. Algebra* **88**:1 (1984), 89–133.

[Elmendorf et al. 1997] A. D. Elmendorf, I. Kříž, M. A. Mandell, and J. P. May,
Rings, modules, and algebras in stable homotopy theory, Mathematical Surveys and
Monographs, American Mathematical Society, Providence, RI, 1997.

[Evens 1961] L. Evens, "The cohomology ring of a finite group", *Trans. Amer. Math.
Soc.* **101** (1961), 224–239.

[Evens 1991] L. Evens, *The cohomology of groups*, Oxford Mathematical Monographs,
The Clarendon Press Oxford University Press, New York, 1991. Oxford Science
Publications.

[Fiedorowicz and Priddy 1978] Z. Fiedorowicz and S. Priddy, *Homology of classical
groups over finite fields and their associated infinite loop spaces*, Lecture Notes in
Mathematics **674**, Springer, Berlin, 1978.

[Friedlander and Suslin 1997] E. M. Friedlander and A. Suslin, "Cohomology of finite
group schemes over a field", *Invent. Math.* **127**:2 (1997), 209–270.

[Golod 1959] E. Golod, "The cohomology ring of a finite p-group", *Dokl. Akad. Nauk
SSSR* **125** (1959), 703–706. In Russian.

[Green 2003] D. J. Green, "On Carlson's depth conjecture in group cohomology", *Math.
Z.* **244**:4 (2003), 711–723.

[Greenlees 1995] J. P. C. Greenlees, "Commutative algebra in group cohomology", *J.
Pure Appl. Algebra* **98**:2 (1995), 151–162.

[Greenlees 2002] J. P. C. Greenlees, "Local cohomology in equivariant topology", pp.
1–38 in *Local cohomology and its applications* (Guanajuato, 1999), edited by G.
Lyubeznik, Lecture Notes in Pure and Appl. Math. **226**, Dekker, New York, 2002.

[Greenlees and Lyubeznik 2000] J. P. C. Greenlees and G. Lyubeznik, "Rings with a
local cohomology theorem and applications to cohomology rings of groups", *J. Pure
Appl. Algebra* **149**:3 (2000), 267–285.

[Grothendieck 1965] A. Grothendieck, *Cohomologie locale des faisceaux cohérents et
théorèmes de Lefschetz locaux et globaux*, Séminaire de géométrie algébrique **2**,
Institut des Hautes Études Scientifiques, Paris, 1965. Reprinted North-Holland,
Amsterdam, 1968, Advanced Studies in Pure Math **2**.

[Grothendieck 1967] A. Grothendieck, *Local cohomology*, Lecture Notes in Math. **41**,
Springer, Berlin, 1967.

[Hall and Senior 1964] M. Hall, Jr. and J. K. Senior, *The groups of order 2^n ($n \leq 6$)*, The Macmillan Co., New York, 1964.

[Harer 1986] J. L. Harer, "The virtual cohomological dimension of the mapping class group of an orientable surface", *Invent. Math.* **84**:1 (1986), 157–176.

[Iyengar 2004] S. Iyengar, "Modules and cohomology over group algebras: One commutative algebraist's perspective", pp. 51–86 in *Trends in Commutative Algebra*, edited by L. Avramov et al., Math. Sci. Res. Inst. Publ. **51**, Cambridge University Press, New York, 2004.

[Kleinerman 1982] S. N. Kleinerman, *The cohomology of Chevalley groups of exceptional Lie type*, Mem. Amer. Math. Soc. **268**, Amer. Math. Soc., Providence, RI, 1982.

[Kono and Mimura 1975] A. Kono and M. Mimura, "Cohomology mod 2 of the classifying space of the compact connected Lie group of type E_6", *J. Pure Appl. Algebra* **6** (1975), 61–81.

[Lazard 1965] M. Lazard, "Groupes analytiques p-adiques", *Inst. Hautes Études Sci. Publ. Math.* **26** (1965), 389–603.

[Matlis 1958] E. Matlis, "Injective modules over Noetherian rings", *Pacific J. Math.* **8** (1958), 511–528.

[Matsumura 1989] H. Matsumura, *Commutative ring theory*, Second ed., Cambridge Studies in Advanced Mathematics **8**, Cambridge University Press, Cambridge, 1989. Translated from the Japanese by M. Reid.

[Milnor 1956] J. Milnor, "Construction of universal bundles, II", *Ann. of Math.* (2) **63** (1956), 430–436.

[Minh and Symonds 2004] P. A. Minh and P. Symonds, "Cohomology and finite subgroups of profinite groups", *Proc. Amer. Math. Soc.* **132** (2004), 1581–1588.

[Nagata 1962] M. Nagata, *Local rings*, Interscience Tracts in Pure and Applied Mathematics, No. 13, Interscience Publishers a division of John Wiley & Sons New York-London, 1962. Corrected reprint, Robert E. Krieger Publishing Co., Huntington, N.Y., 1975.

[Neeman 1996] A. Neeman, "The Grothendieck duality theorem via Bousfield's techniques and Brown representability", *J. Amer. Math. Soc.* **9**:1 (1996), 205–236.

[Neusel 2000] M. D. Neusel, *Inverse invariant theory and Steenrod operations*, Mem. Amer. Math. Soc. **692**, Amer. Math. Soc., Providence, RI, 2000.

[Okuyama and Sasaki 2000] T. Okuyama and H. Sasaki, 2000. Private communication, Chiba, Japan.

[Quillen 1971a] D. Quillen, "The mod 2 cohomology rings of extra-special 2-groups and the spinor groups", *Math. Ann.* **194** (1971), 197–212.

[Quillen 1971b] D. Quillen, "The spectrum of an equivariant cohomology ring, I", *Ann. of Math.* (2) **94** (1971), 549–572.

[Quillen 1971c] D. Quillen, "The spectrum of an equivariant cohomology ring, II", *Ann. of Math.* (2) **94** (1971), 573–602.

[Quillen 1972] D. Quillen, "On the cohomology and K-theory of the general linear groups over a finite field", *Ann. of Math.* (2) **96** (1972), 552–586.

[Rickard 1997] J. Rickard, "Idempotent modules in the stable category", *J. London Math. Soc.* (2) **56**:1 (1997), 149–170.

[Rusin 1989] D. J. Rusin, "The cohomology of the groups of order 32", *Math. Comp.* **53**:187 (1989), 359–385.

[Schwartz 1994] L. Schwartz, *Unstable modules over the Steenrod algebra and Sullivan's fixed point set conjecture*, Chicago Lectures in Mathematics, University of Chicago Press, Chicago, IL, 1994.

[Serre 1965a] J.-P. Serre, *Cohomologie galoisienne*, 3rd ed., Lecture Notes in Mathematics **5**, Springer, Berlin, 1965.

[Serre 1965b] J.-P. Serre, "Sur la dimension cohomologique des groupes profinis", *Topology* **3** (1965), 413–420.

[Sri] Srikanth Iyengar's article in this volume; details under [Iyengar 2004].

[Steenrod 1962] N. E. Steenrod, *Cohomology operations*, Annals of Mathematics Studies **50**, Princeton University Press, Princeton, 1962. Notes written and revised by D. B. A. Epstein.

[Venkov 1959] B. B. Venkov, "Cohomology algebras for some classifying spaces", *Dokl. Akad. Nauk SSSR* **127** (1959), 943–944. in Russian.

[Wilkerson 1982] C. Wilkerson, Spring 1982. Letter to Frank Adams.

[Wilkerson 1983] C. Wilkerson, "A primer on the Dickson invariants", pp. 421–434 in *Proceedings of the Northwestern Homotopy Theory Conference* (Evanston, IL, 1982), edited by H. R. Miller and S. B. Priddy, Contemp. Math. **19**, Amer. Math. Soc., Providence, RI, 1983.

[Ziegler 1984] M. Ziegler, "Model theory of modules", *Ann. Pure Appl. Logic* **26**:2 (1984), 149–213.

DAVE BENSON
DEPARTMENT OF MATHEMATICS
UNIVERSITY OF GEORGIA
ATHENS GA 30602
UNITED STATES
bensondj@math.uga.edu

Trends in Commutative Algebra
MSRI Publications
Volume **51**, 2004

Modules and Cohomology over Group Algebras: One Commutative Algebraist's Perspective

SRIKANTH IYENGAR

ABSTRACT. This article explains basic constructions and results on group algebras and their cohomology, starting from the point of view of commutative algebra. It provides the background necessary for a novice in this subject to begin reading Dave Benson's article in this volume.

CONTENTS

Introduction

The available accounts of group algebras and group cohomology [Benson 1991a; 1991b; Brown 1982; Evens 1991] are all written for the mathematician on the street. This one is written for commutative algebraists by one of their own. There is a point to such an exercise: though group algebras are typically noncommutative, module theory over them shares many properties with that over commutative rings. Thus, an exposition that draws on these parallels could benefit an algebraist familiar with the commutative world. However, such an endeavour is not without its pitfalls, for often there are subtle differences between the two situations. I have tried to draw attention to similarities and to

Mathematics Subject Classification: Primary 13C15, 13C25. Secondary 18G15, 13D45.

Part of this article was written while the author was funded by a grant from the NSF.

discrepancies between the two subjects in a series of commentaries on the text that appear under the rubric Ramble[1].

The approach I have adopted toward group cohomology is entirely algebraic. However, one cannot go too far into it without some familiarity with algebraic topology. To gain an appreciation of the connections between these two subjects, and for a history of group cohomology, one might read [Benson and Kropholler 1995; Mac Lane 1978].

In preparing this article, I had the good fortune of having innumerable 'chalk-and-board' conversations with Lucho Avramov and Dave Benson. My thanks to them for all these, and to the Mathematical Sciences Research Institute for giving me an opportunity to share a roof with them, and many others, during the Spring of 2003. It is also a pleasure to thank Kasper Andersen, Graham Leuschke, and Claudia Miller for their remarks and suggestions.

1. The Group Algebra

Let G be a group, with identity element 1, and let k be a field. Much of what is said in this section is valid, with suitable modifications, more generally when k is a commutative ring. Let $k[G]$ denote the k-vector space with basis the elements of G; thus $k[G] = \bigoplus_{g \in G} kg$. The product on G extends to an associative multiplication on $k[G]$: for basis elements g and h, one has $g \cdot h = gh$, where the product on the right is taken in G, while the product of arbitrary elements is specified by the distributive law and the rule $a \cdot g = g \cdot a$ for $a \in k$. The identity element 1 is the identity in $k[G]$. The k-linear ring homomorphism $\eta \colon k \to k[G]$ with $\eta(1) = 1$ makes $k[G]$ a k-algebra. This is the *group algebra* of G with coefficients in k.

Note that $k[G]$ is commutative if and only if the group G is abelian. Moreover, it is finite-dimensional as a k-vector space precisely when G is finite.

An important part of the structure on $k[G]$ is the augmentation of k-algebras $\varepsilon \colon k[G] \to k$ defined by $\varepsilon(g) = 1$ for each $g \in G$. Through ε one can view k as a $k[G]$-bimodule. The kernel of ε, denoted $\mathrm{I}(G)$, is the k-subspace of $k[G]$ with basis $\{g-1 \mid g \in G\}$; it is a two-sided ideal, called the *augmentation ideal* of G. For every pair of elements g, h in G, the following relations hold in the group algebra:

$$g^{-1} - 1 = g^{-1}(1 - g),$$
$$gh - 1 = g(h - 1) + (g - 1) = (g - 1)h + (h - 1).$$

Thus, if a subset $\{g_\lambda\}_{\lambda \in \Lambda}$ of G, with Λ an index set, generates the group, the subset $\{g_\lambda - 1\}_{\lambda \in \Lambda}$ of $k[G]$ generates $\mathrm{I}(G)$ both as a left ideal and as a right ideal.

---

[1] This word has at least two meanings: "a leisurely walk", or "to talk or write in a discursive, aimless way"; you can decide which applies. By the by, its etymology, at least according to www.dictionary.com, might amuse you.

(1.1) **Functoriality.** The construction of the group algebra is functorial: given a group homomorphism $\varphi \colon G_1 \to G_2$, the k-linear map

$$k[\varphi] \colon k[G_1] \to k[G_2], \quad \text{where } g \mapsto \varphi(g),$$

is a homomorphism of k-algebras, compatible with augmentations. Its kernel is generated both as a left ideal and as a right ideal by the set $\{g - 1 \mid g \in \operatorname{Ker} \varphi\}$.

For example, when N is a normal subgroup of a group G, the canonical surjection $G \to G/N$ induces the surjection of k-algebras $k[G] \to k[G/N]$. Since its kernel is generated by the set $\{n - 1 \mid n \in N\}$, there is a natural isomorphism of k-algebras

$$k[G/N] \cong k \otimes_{k[N]} k[G] = \frac{k[G]}{I(N)k[G]}.$$

Let me illustrate these ideas on a few simple examples.

(1.2) **Cyclic groups.** The group algebra of the infinite cyclic group is $k[x^{\pm 1}]$, the algebra of Laurent polynomials in the variable x. Here x is a generator of the group; its inverse is x^{-1}. The augmentation maps x to 1, and the augmentation ideal is generated, as an ideal, by $x - 1$.

In view of (1.1), the group algebra of the cyclic group of order d is $k[x]/(x^d - 1)$, and the augmentation ideal is again generated by $x - 1$.

(1.3) **Products of groups.** Let G_1 and G_2 be groups. By (1.1), for $n = 1, 2$ the canonical inclusions $\iota_n \colon G_n \to G_1 \times G_2$ induce homomorphisms of k-algebras $k[\iota_n] \colon k[G_n] \to k[G_1 \times G_2]$. Since the elements in the image of $k[\iota_1]$ commute with those in the image of $k[\iota_2]$, one obtains a homomorphism of augmented k-algebras

$$k[G_1] \otimes_k k[G_2] \to k[G_1 \times G_2],$$
$$g_1 \otimes_k g_2 \quad \mapsto (g_1, g_2).$$

This is an isomorphism since it maps the basis $\{g_1 \otimes_k g_2 \mid g_i \in G_i\}$ of the k-vector space $k[G_1] \otimes_k k[G_2]$ bijectively to the basis $\{(g_1, g_2) \mid g_i \in G_i\}$ of $k[G_1 \times G_2]$. For this reason, the group algebra of $G_1 \times G_2$ is usually identified with $k[G_1] \otimes_k k[G_2]$.

(1.4) **Abelian groups.** Let G be a finitely generated abelian group. The structure theorem for such groups tells us that there are nonnegative numbers n and $d_1, \ldots, d_m$, with $d_j \geq 2$ and $d_{i+1} \mid d_i$, such that

$$G = \mathbb{Z}^n \oplus \frac{\mathbb{Z}}{(d_1 \mathbb{Z})} \oplus \cdots \oplus \frac{\mathbb{Z}}{(d_m \mathbb{Z})}.$$

The description of the group algebra of cyclic groups given in (1.2), in conjunction with the discussion in (1.3), yields

$$k[G] = \frac{k[x_1^{\pm 1}, \ldots, x_n^{\pm 1}, y_1, \ldots, y_m]}{(y_1^{d_1} - 1, \ldots, y_m^{d_m} - 1)}$$

The augmentation is given by $x_i \mapsto 1$ and $y_j \mapsto 1$, the augmentation ideal is generated by $\{x_1-1, \ldots, x_n-1, y_1-1, \ldots, y_m-1\}$.

RAMBLE. Observe: the group algebra in (1.4) above is a complete intersection.

(1.5) **Finite p-groups.** Let R be a ring; it need not be commutative. Recall that the intersection of all its left maximal ideals is equal to the intersection of all its right maximal ideals, and called the Jacobson radical of R. Thus, R has a unique left maximal ideal exactly when it has a unique right maximal ideal, and then these ideals coincide. In this case, one says that R is *local*; note that the corresponding residue ring is a division ring; for details see [Lang 2002, XVII § 6], for example.

Suppose that the characteristic of k is p, with $p \geq 2$. Let G be a finite p-*group*, so that the order of G is a power of p. I claim:

The group algebra $k[G]$ is local with maximal ideal $\mathrm{I}(G)$.

Indeed, it suffices to prove (and the claim is equivalent to): the augmentation ideal $\mathrm{I}(G)$ is nilpotent. Now, since G is a p-group, its centre Z is nontrivial, so (1.1) yields an isomorphism of k-algebras

$$\frac{k[G]}{\mathrm{I}(Z)k[G]} \cong k[G/Z].$$

Since the order of G/Z is strictly less than that of G, one can assume that $\mathrm{I}(G/Z)$ is nilpotent. By the isomorphism above, this entails $\mathrm{I}(G)^n \subseteq \mathrm{I}(Z)k[G]$, for some positive integer n. Now Z is an abelian p-group, so $\mathrm{I}(Z)$ is nilpotent, by (1.4). Since $\mathrm{I}(Z)$ is in the centre of $k[G]$, one obtains that $\mathrm{I}(G)$ is nilpotent, as claimed.

The converse also holds:

(1.6) EXERCISE. Let G be a finite group and p the characteristic of k. Prove that if the ring $k[G]$ is local, then G is a p-group. (Hint: $k[G]$ has finite rank over k, so its nilradical is equal to its Jacobson radical.)

(1.7) **The diagonal map.** Let G be a group and let $G \to G \times G$ be the diagonal homomorphism, given by $g \mapsto (g, g)$. Following (1.3), one identifies the group ring of $G \times G$ with $k[G] \otimes_k k[G]$, and then the diagonal homomorphism induces a homomorphism of augmented k-algebras

$$\Delta\colon k[G] \to k[G] \otimes_k k[G], \quad \text{where } g \mapsto g \otimes_k g.$$

This is called the *diagonal* homomorphism, or *coproduct*, of the group algebra $k[G]$.

There is another piece of structure on the group algebra: the map $G \to G$ given by $g \mapsto g^{-1}$ is an anti-isomorphism of groups, and hence induces an anti-isomorphism of group algebras

$$\sigma\colon k[G] \to k[G],$$

that is to say, σ is an isomorphism of additive groups with $\sigma(rs) = \sigma(s)\sigma(r)$. The map σ is referred to as the *antipode* of the group algebra. It commutes with the diagonal map, in the sense that

$$\sigma^{(G \times G)} \circ \Delta^G = \Delta^G \circ \sigma^G.$$

Here are the salient properties of the diagonal and the antipode:

(a) Δ is a homomorphism of augmented k-algebras;
(b) Δ is co-associative, in that the following diagram commutes:

$$
\begin{array}{ccc}
k[G] & \xrightarrow{\;\;\Delta\;\;} & k[G] \otimes_k k[G] \\
\Big\downarrow{\scriptstyle \Delta} & & \Big\downarrow{\scriptstyle \Delta \otimes_k 1} \\
k[G] \otimes_k k[G] & \xrightarrow{\;1 \otimes_k \Delta\;} & k[G] \otimes_k k[G] \otimes_k k[G]
\end{array}
$$

(c) The following diagram commutes:

$$
\begin{array}{ccc}
 & k[G] & \\
{\scriptstyle\cong}\nearrow & \Big\downarrow{\scriptstyle \Delta} & \searrow{\scriptstyle\cong} \\
k \otimes_k k[G] \xleftarrow{\;\varepsilon \otimes_k 1\;} & k[G] \otimes_k k[G] & \xrightarrow{\;1 \otimes_k \varepsilon\;} k[G] \otimes_k k
\end{array}
$$

This property is paraphrased as: ε is a co-unit for Δ.

(d) For each element $r \in k[G]$, if $\Delta(r) = \sum_{i=1}^n (r'_i \otimes_k r''_i)$, then

$$\sum_{i=1}^n \sigma(r'_i) r''_i = \eta(\varepsilon(r)) = \sum_{i=1}^n r'_i \sigma(r''_i)$$

Taking these properties as the starting point, one arrives at the following notion.

(1.8) **Hopf algebras.** An augmented k-algebra H, with unit $\eta \colon k \to H$ and augmentation $\varepsilon \colon H \to k$ with k-linear homomorphisms $\Delta \colon H \to H \otimes_k H$ and $\sigma \colon H \to H$ satisfying conditions (a)–(d) listed above, is said to be a *Hopf algebra*. Among these, (b) and (c) are the defining properties of a *coalgebra* with diagonal Δ; see [Montgomery 1993] or [Sweedler 1969]. Property (a) says that the algebra and coalgebra structures are compatible. At first — and perhaps second and third — glance, property (d) appears mysterious. Here is one explanation that appeals to me: The diagonal homomorphism endows the k-vector space $\mathrm{Hom}_k(H, H)$ with the structure of a k-algebra, with the product of elements f and g given by

$$(f \star g)(r) = \sum_{i=1}^n f(r'_i) g(r''_i), \quad \text{where } \Delta(r) = \sum_{i=1}^n (r'_i \otimes_k r''_i).$$

This is called the *convolution product* on $\mathrm{Hom}_k(H, H)$; its unit is the element $\eta \circ \varepsilon$. Condition (d) asserts that σ is the inverse of the identity on H.

The group algebra is the prototypical example of a Hopf algebra, and many constructions and results pertaining to them are best viewed in that generality; see [Benson 1991a, Chapter 3]. There is another good source of Hopf algebras, close to home: the coordinate rings of algebraic groups. You might, as I did, find it entertaining and illuminating to write down the Hopf structure on the coordinate ring of the circle $x^2 + y^2 = 1$.

If this all too brief foray into Hopf algebras has piqued your curiosity and you wish to know more, you could start by reading Bergman's charming introduction [Bergman 1985]; if you prefer to jump right into the thick of things, then [Montgomery 1993] is the one for you.

2. Modules over Group Algebras

This section is an introduction to modules over group algebras. When G is a *finite* group, the k-algebra $k[G]$ is finite-dimensional, that is to say, of finite rank over k. Much of the basic theory for modules over finite group algebras is just a specialization of the theory for finite-dimensional algebras. For example, I hinted in Exercise (1.6) that for finite group algebras, the nilradical coincides with the Jacobson radical; this holds, more generally, for any finite-dimensional k-algebra. Here I will focus on two crucial concepts: the Jordan–Hölder theorem and the Krull–Schmidt property.

(2.1) **The Jordan–Hölder theorem.** Let R be a ring and M an R-module. It is clear that when M is both artinian and noetherian it has a *composition series*: a series of submodules $0 = M_l \subset M_{l-1} \subset \cdots \subset M_1 \subset M_0 = M$ with the property that the subfactors M_i/M_{i+1} are *simple*, that is to say, they have no proper submodules. It turns out that if $0 = M'_{l'} \subset M'_{l'-1} \subset \cdots \subset M'_1 \subset M'_0 = M$ is another composition series, then $l = l'$ and, for $1 \leq i, j \leq l$, the factors M_i/M_{i-1} are a permutation of the factors M'_j/M'_{j-1}. This is a consequence of the Jordan–Hölder theorem, which says that for each R-module, any two series (not necessarily composition series) of submodules can be refined to series of the same length and with the same subfactors.

Suppose that R is artinian; for example, R may be a finite-dimensional k-algebra, or, more specifically, a finite group algebra. In this case every finite, by which I mean 'finitely generated', module over it is both artinian and noetherian and so has a composition series. Here is one consequence: since every simple module is a quotient of R, all the simple modules appear in a composition series for R, and so there can only be finitely many of them.

(2.2) **Indecomposable modules.** Recall that a module is said to be *indecomposable* if it has no nontrivial direct summands. It is clear that a simple module is indecomposable, but an indecomposable module may be far from simple — in either sense of the word. For example, over a commutative ring, the only simple modules are the residue fields, whereas it is usually not possible to classify all the indecomposable modules; I will pick up on this point a few paragraphs down the road. For now, here are a couple of remarks that are useful to keep in mind when dealing with indecomposability; see the discussion in (2.10).

In this sequel, when I say $(R, \mathfrak{m}, k)$ is a local ring, I mean that R is local, with maximal ideal $\mathfrak{m}$ and residue ring k.

(2.3) EXERCISE. Let $(R, \mathfrak{m}, k)$ be a commutative local ring. Prove that if M is indecomposable, then $\operatorname{socle}(M) \subseteq \mathfrak{m}M$.

(2.4) EXERCISE. Let R be a commutative local Gorenstein ring and M an indecomposable R-module. Prove that if $\operatorname{socle}(R) \cdot M \neq 0$, then $M \cong R$.

(2.5) **The Krull–Schmidt property.** Let R be a ring. It is not hard to see that each finite R-module can be broken down into a finite direct sum of indecomposables. The ring R has the *Krull–Schmidt property* if for each finite R-module such a decomposition is unique up to a permutation of the indecomposable factors: if

$$\bigoplus_{i=1}^{m} M_i \cong \bigoplus_{j=1}^{n} N_j,$$

with each M_i and N_j indecomposable, then $m = n$, and, with a possible rearrangement of the N_j, one has $M_i \cong N_i$ for each i.

For example, complete commutative noetherian local rings have this property; see [Swan 1968, (2.22)]. In the present context, the relevant result is that artinian rings have the Krull–Schmidt property [Benson 1991a, (1.4.6)]. When G is a finite group, $k[G]$ is artinian; in particular, it has the Krull–Schmidt property.

The Krull–Schmidt property is of great help in studying modules over group algebras, for it allows one to focus on the indecomposables. The natural question arises: when does the group algebra have only finitely many isomorphism classes of indecomposable modules? In other words, when is the group algebra of *finite representation type*? This is the case, for example, when every indecomposable module is simple, for there are only finitely many of them; see (2.1). There is an important context when this happens: when the characteristic of k is coprime to the order of the group. This is a consequence of Maschke's Theorem:

(2.6) THEOREM (MASCHKE). *Let G be a finite group such that $|G|$ is coprime to the the characteristic of k. Each short exact sequence of $k[G]$-modules splits.*

PROOF. Let $0 \to L \to M \xrightarrow{\pi} N \to 0$ be an exact sequence of $k[G]$-modules. Since k is a field, π admits a k-linear section; let $\sigma \colon N \to M$ be one such. It is not hard to verify that the map

$$\tilde{\sigma} \colon N \to M, \qquad \text{where } \tilde{\sigma}(n) = \frac{1}{|G|} \sum_{g \in G} g\sigma(g^{-1}n) \quad \text{for all } n \in N,$$

is $k[G]$-linear, and that $\pi \circ \tilde{\sigma} = \operatorname{id}^N$. Thus, the exact sequence splits, as desired. $\square$

This theorem has a perfect converse: if each short exact sequence of $k[G]$-modules splits, the characteristic of k is coprime to $|G|$. In fact, it suffices that the exact sequence $0 \to I(G) \to k[G] \xrightarrow{\varepsilon} k \to 0$ splits. The proof is elementary, and is recommended as an exercise; I will offer a solution in the proof of Theorem (3.1).

A group algebra can have finite representation type even if not every indecomposable module is simple:

(2.7) **Finite cyclic groups.** In describing this example, it is convenient to let p denote 1 when the characteristic of k is 0, and the characteristic of k otherwise.

Let G be a finite cyclic group. Write $|G|$ as $p^n q$, where n is a nonnegative integer and p and q are coprime. Let $R = k[x]/(x^{p^n q} - 1)$, the group algebra. The binomial theorem in characteristic p yields $x^{p^n q} - 1 = (x^q - 1)^{p^n}$, so the Jacobson radical of R is $(x^q - 1)$. In $k[x]$, the polynomial $x^q - 1$ breaks up into a product of distinct irreducible polynomials:

$$x^q - 1 = \prod_{i=1}^{d} f_i(x), \quad \text{with} \quad \sum_{i=1}^{d} \deg(f_i(x)) = q.$$

Since the ideals $(f_i(x)^{p^n})$, where $1 \le i \le d$, in $k[x]$ are pairwise comaximal, the Chinese Remainder Theorem yields

$$R \cong \prod_{i=1}^{d} R_i, \quad \text{where } R_i = \frac{k[x]}{(f_i(x)^{p^n})}.$$

This implies that each R-module M decomposes uniquely as $M = \bigoplus_{i=1}^{d} M_i$, where M_i is an R_i-module. Furthermore, it is easy to see that $R_i/(f_i(x)^s)$, for $1 \le s \le p^n$, is a complete list of indecomposable modules over R_i, and that each M_i has a unique decomposition into a direct sum of such modules. This is exactly as predicted by the Krull–Schmidt theory. The upshot is that we know 'everything' about the modules over the group algebras of finite cyclic groups.

All this is subsumed in the structure theory of modules over principal ideal rings. By the by, the finite cyclic groups are the source of group algebras of finite representation type, in the following sense; see [Benson 1991a, (4.4)] for the appropriate references.

(2.8) THEOREM. *If k is an infinite field of characteristic p and G a finite group, then $k[G]$ has finite representation type exactly when G has cyclic Sylow p-subgroups.*   □

In some cases of infinite representation type, it is still possible to classify all the indecomposable modules. The Klein group is one such. Let me give you a flavour of the modules that arise over its group algebra. For the calculations, it is helpful to recall a result on syzygies of indecomposable modules.

(2.9)   Let $(R, \mathfrak{m}, k)$ be a commutative artinian local ring and E the injective hull of the R-module k. Let M be a finite R-module. Write $\Omega^1 M$ for the first syzygy of M, and $\Omega^{-1}M$ for the first co-syzygy of M. These are defined by exact sequences

(†) $0 \to \Omega^1 M \to R^b \to M \to 0$ and $0 \to M \to E^c \to \Omega^{-1}M \to 0$,

with $b = \mathrm{rank}_k(M/\mathfrak{m}M)$ and $c = \mathrm{rank}_k \mathrm{socle}(M)$.

The conclusion of the following exercise is valid for the syzygy module even when R is a Gorenstein ring of higher (Krull) dimension, as long as M is also maximal Cohen–Macaulay; this was first proved by J. Herzog [1978].

EXERCISE. Assume that R is Gorenstein. Prove that when M is indecomposable, so are $\Omega^1 M$ and $\Omega^{-1}M$.

I cannot resist giving a sketch of the argument: Suppose $\Omega^1 M = U \oplus V$, with U and V nonzero. Since R is self-injective, neither U nor V can be free: if U is free, then it is injective and hence splits from R^b in the exact sequence (†) above, and that cannot happen. Now, $\mathrm{Hom}_R(-, R)$ applied to (†) yields an exact sequence

$$0 \to M^* \to R^b \to U^* \oplus V^* \to 0.$$

This presents M^* as the first syzygy of $U^* \oplus V^*$ (why?); that is,

$$M^* = \Omega^1(U^* \oplus V^*) = \Omega^1(U^*) \oplus \Omega^1(V^*).$$

Note that the modules $\Omega^1(U^*)$ and $\Omega^1(V^*)$ are nonzero: if $\Omega^1(U^*) = 0$, then $\mathrm{pdim}_R(U^*)$ is finite, so U^* is free, and hence U is free, a contradiction. It follows that the same is true even after we dualize them. Applying $\mathrm{Hom}_R(-, R)$ to the equality above gives us

$$M^{**} = \Omega^1(U^*)^* \oplus \Omega^1(V^*)^*$$

Since $M \cong (M^*)^*$, one obtains that M is indecomposable.

Now we turn to indecomposable modules over the Klein group.

(2.10) **The Klein group.** Let k be a field of characteristic 2 and let G be $\mathbb{Z}_2 \times \mathbb{Z}_2$, the Klein group. Let R denote its group algebra over k, so $R = k[y_1, y_2]/(y_1^2-1, y_2^2-1)$.

This k-algebra looks more familiar once we change variables: setting $x_i = y_i - 1$ one sees that $R = k[x_1, x_2]/(x_1^2, x_2^2)$; a local zero dimensional complete intersection with maximal ideal $\mathfrak{m} = (x_1, x_2)$. Note that R is Gorenstein, so $R \cong \mathrm{Hom}_k(R, k)$ and, for any R-module M, one has $M^* \cong \mathrm{Hom}_k(M, k)$, where $(-)^* = \mathrm{Hom}_R(M, R)$. I will use these remarks without ado.

For each positive integer n, let M_n denote $\Omega^n(k)$, the n-th syzygy of k. I claim that in the infinite family $\{\ldots, M_2, M_1, k, (M_1)^*, (M_2)^*, \ldots\}$ no two modules are isomorphic and that each is indecomposable.

Indeed, a repeated application of Exercise (2.9) yields that each M_n is indecomposable, and hence also that $(M_n)^*$ is indecomposable, since $(M_n)^{**} \cong M_n$.

As to the remaining assertion: for $i = 1, 2$, let $R_i = k[x_i]/(x_i^2)$. The minimal R_i-free resolution of k is

$$F_i = \quad \cdots \xrightarrow{x_i} R_i \xrightarrow{x_i} R_i \xrightarrow{x_i} R_i \to 0$$

Since $R = R_1 \otimes_k R_2$, the complex of R-modules $F_1 \otimes_k F_2$ is the minimal free resolution of the R-module k. It follows that the n-th Betti number of k is $n+1$. Thus, for any positive integer n, the n-th syzygy M_n of k is defined by an exact sequence

$$(\dagger) \qquad 0 \to M_n \to R^n \xrightarrow{\partial_{n-1}} R^{n-1} \to \cdots \to R^2 \xrightarrow{\partial_1} R \to k \to 0,$$

with $\partial_i(R^{i+1}) \subseteq \mathfrak{m}R^i$ for each i. It follows that $\mathrm{rank}_k M_n = 2n + 1$, and hence also that $\mathrm{rank}_k (M_n)^* = 2n + 1$. Therefore, to settle the claim that the modules in question are all distinct, it remains to verify that the R-modules M_n and $(M_n)^*$ are not isomorphic. These modules appear in exact sequences

$$0 \to M_n \to R^n \xrightarrow{\partial_{n-1}} R^{n-1} \qquad \text{and} \qquad 0 \to (M_n)^* \to R^{n+1} \xrightarrow{\partial_{n+1}^*} R^{n+2}.$$

The one on the right is obtained from

$$R^{n+2} \xrightarrow{\partial_{n+1}} R^{n+1} \to M_n \to 0,$$

keeping in mind that $R^* \cong R$. Since $\partial_{n-1}(R^n) \subseteq \mathfrak{m}R^{n-1}$ and $\partial_{n+1}^*(R^{n+1}) \subseteq \mathfrak{m}R^{n+2}$, the desired conclusion is a consequence of:

EXERCISE. Let $(R, \mathfrak{m}, k)$ be a local ring. If $0 \to K \to R^b \xrightarrow{f} R^c$ is an exact sequence of R-modules with $f(R^b) \subseteq \mathfrak{m}R^c$, then

$$\mathrm{socle}(K) = \mathrm{socle}(R^b) = \mathrm{socle}(R)^b.$$

This completes the justification that the given family consists of nonisomorphic indecomposables. In this process we found that $\mathrm{rank}_k M_n = 2n + 1 = \mathrm{rank}_k (M_n)^*$. It turns out that the M_n, their k-duals, and k are the only indecomposables of odd rank; here is a sketch of the proof. Exercise: fill in the details.

Let M be an indecomposable R-module with $\mathrm{rank}_k M = 2n + 1$ for some integer n. In particular, $M \not\cong R$, and so Exercise (2.4) tells us that $(xy)M = 0$, so $\mathfrak{m}^2 M = 0$ and hence $\mathfrak{m}M \subseteq \mathrm{socle}(M)$; the opposite inclusion also holds, by Exercise (2.3), hence $\mathfrak{m}M = \mathrm{socle}(M)$. Thus, one has an exact sequence of R-modules

$$0 \to \mathrm{socle}(M) \to M \to M/\mathfrak{m}M \to 0$$

Now we use Exercise (2.9); in the notation there, from the exact sequence above one deduces that either $b \leq n$ or $c \leq n$. In the former case $\mathrm{rank}_k(\Omega^1 M) \leq 2n-1$ and in the latter $\mathrm{rank}_k(\Omega^{-1}M) \leq 2n - 1$. In any case, the ranks of $\Omega^1 M$ and

$\Omega^{-1}M$ are odd. Now an induction on rank yields that M belongs to the family of indecomposable R-modules that we have already constructed.

At this point, we know all the indecomposable R-modules of odd rank. The ones of even rank are harder to deal with. To get an idea of what goes on here, solve:

EXERCISE. Prove that every rank 2 indecomposable R-module is isomorphic to a member of the family of cyclic R-modules

$$V_{(\alpha_1, \alpha_2)} = \frac{R}{(\alpha_1 x_1 + \alpha_2 x_2, xy)}, \quad \text{where } (\alpha_1, \alpha_2) \neq (0, 0).$$

Moreover, $V_{(\alpha_1, \alpha_2)} \cong V_{(\beta_1, \beta_2)}$ if and only if (α_1, α_2) and (β_1, β_2) are proportional.

Thus, the nonisomorphic indecomposable R-modules of rank 2 are parametrized by the projective line over k; it turns out that this is the case in any even rank, at least when k is algebraically closed. This classification of the indecomposable modules over the Klein group goes back to Kronecker; see [Alperin 1986] or [Benson 1991a, (4.3)] for a modern treatment.

This discussion shows that while the group algebra of $\mathbb{Z}_2 \times \mathbb{Z}_2$ in characteristic 2 is not of finite type, in any given rank all but finitely many of its indecomposable modules are contained in a one-parameter family. More generally, by allowing for finitely many one-parameter families in each rank, one obtains the notion of a *tame* algebra. Tame group algebras $k[G]$ are completely classified: the characteristic of k is 2, and the Sylow 2-subgroups of G are isomorphic to one of the following groups: Klein, dihedral, semidihedral, or generalized quaternion. See [Benson 1991a, (4.4.4)]. The significance of this result lies in that every finite-dimensional k-algebra that is neither of finite type nor tame is *wild*, which implies that the set of isomorphism classes of its finite-rank indecomposable modules contains representatives of the indecomposable modules over a tensor algebra in two variables.

RAMBLE. There is a significant parallel between module theory over finite group algebras and over artinian commutative Gorenstein rings; see the discussion around Theorem (3.6). In fact, this parallel extends to the category of maximal Cohen–Macaulay modules over commutative complete local Gorenstein rings. For example, analogous to Theorem (2.8), among this class of rings those of *finite Cohen–Macaulay type* (which means that there are only finitely many isomorphism classes of indecomposable maximal Cohen–Macaulay modules) have been completely classified, at least when the ring contains a field. A systematic exposition of this result can be found in [Yoshino 1990]. The next order of complexity beyond finite Cohen–Macaulay type is bounded Cohen–Macaulay type, which is a topic of current research: see [Leuschke and Wiegand $\geq$ 2004].

The rest of this section describes a few basic constructions, like tensor products and homomorphisms, involving modules over group algebras.

(2.11) **Conjugation.** Over a noncommutative ring, the category of left modules can be drastically different from that of right modules. For example, there exist rings over which every left module has a finite projective resolution, but not every right module does. Thus, in general, one has to be very careful vis-à-vis left and right module structures.

However, in the case of group algebras, each left module can be endowed with a natural structure of a right module, and vice versa. More precisely, if M is a *left* $k[G]$-module, then the k-vector space underlying M may be viewed as a *right* $k[G]$-module by setting

$$m \cdot g = g^{-1}m \quad \text{for each } g \in G \text{ and } m \in M.$$

For this reason, when dealing with modules over group algebras, one can afford to be lax about whether they are left modules or right modules. This also means, for instance, that a left module is projective (or injective) if and only if the corresponding right module has the same property.

This is similar to the situation over commutative rings: each left module N over a commutative ring R is a right module with multiplication

$$n \cdot r = rn \quad \text{for each } r \in R \text{ and } n \in N.$$

There is an important distinction between the two situations: over R, the module N becomes an R-bimodule with right module structure as above. However, over $k[G]$, the module M with prescribed right module structure is not a bimodule.

(2.12) **Tensor products.** Over an arbitrary ring, one cannot define the tensor product of two left modules. However, if M and N are two left modules over a group algebra $k[G]$, one can view M as a right module via conjugation (2.11) and make sense of $M \otimes_{k[G]} N$. But then this tensor product is *not* a $k[G]$-module, because M and N are not bimodules. In this respect, the group ring behaves like any old ring.

There is another tensor product construction, a lot more important when dealing with group algebras than the one above, that gives us back a $k[G]$-module. To describe it, we return briefly to the world of arbitrary k-algebras.

Let R and S be k-algebras and let M and N be (left) modules over R and S, respectively. There is a natural left $(R \otimes_k S)$-module structure on $M \otimes_k N$ with

$$(r \otimes_k s) \cdot (m \otimes_k n) = rm \otimes_k sn.$$

Now let M and N be left $k[G]$-modules. The preceding recipe provides an action of $k[G] \otimes_k k[G]$ on $M \otimes_k N$. This restricts, via the diagonal map (1.7), to a left $k[G]$-module structure on $M \otimes_k N$. Going through the definitions one finds that

$$g \cdot (m \otimes_k n) = gm \otimes_k gn,$$

for all $g \in G$, $m \in M$ and $n \in N$. It is worth remarking that the 'twisting' map

$$M \otimes_k N \xrightarrow{\cong} N \otimes_k M,$$
$$(m \otimes_k n) \mapsto (n \otimes_k m),$$

which is bijective, is $k[G]$-linear.

RAMBLE. To a commutative algebraist, the tensor product $M \otimes_k N$ has an unsettling feature: it is taken over k, rather than over $k[G]$. However, bear in mind that the $k[G]$-module structure on $M \otimes_k N$ uses the diagonal homomorphism. The other possibilities, namely acquiring the structure from M or from N, don't give us anything nearly as useful. For instance, $M \otimes_k N$ viewed as a $k[G]$-module via its left-hand factor is just a direct sum of copies of M.

(2.13) **Homomorphisms.** Let M and N be left $k[G]$-modules. One can then consider $\mathrm{Hom}_{k[G]}(M, N)$, the k-vector space of $k[G]$-linear maps from M to N. Like the tensor product over $k[G]$, this is not, in general, a $k[G]$-module. Note that since the $k[G]$-module k is cyclic with annihilator $\mathrm{I}(G)$, and $\mathrm{I}(G)$ is generated as an ideal by elements $g - 1$, one has

$$\mathrm{Hom}_{k[G]}(k, M) = \{m \in M \mid gm = m\}.$$

The k-subspace on the right is of course M^G, the set of G-invariant elements in M.

As with $M \otimes_k N$, one can endow the k-vector space $\mathrm{Hom}_k(M, N)$ with a canonical left $k[G]$-structure. This is given by the following prescription: for each $g \in G$, $\alpha \in \mathrm{Hom}_k(M, N)$, and $m \in M$, one has

$$(g \cdot \alpha)(m) = g\alpha(g^{-1}m).$$

In particular, $g \cdot \alpha = \alpha$ if and only if $\alpha(gm) = g\alpha(m)$; that is to say,

$$\mathrm{Hom}_{k[G]}(M, N) = \mathrm{Hom}_k(M, N)^G.$$

Thus the homomorphisms functor $\mathrm{Hom}_{k[G]}(M, N)$ is recovered as the k-subspace of G-invariant elements in $\mathrm{Hom}_k(M, N)$. This identification leads to the following Hom-Tensor adjunction formula:

$$\mathrm{Hom}_{k[G]}(L \otimes_k M, N) \cong \mathrm{Hom}_{k[G]}(L, \mathrm{Hom}_k(M, N)).$$

This avatar of Hom-Tensor adjunction is very useful in the study of modules over group algebras; see, for example, the proof of (3.2).

RAMBLE. Let G be a finite group such that the characteristic of k is coprime to $|G|$, and let $0 \to L \to M \to N \to 0$ be an exact sequence $k[G]$-modules. Applying $\mathrm{Hom}_{k[G]}(k, -)$ to it yields, in view of Maschke's theorem (2.6), an exact sequence

$$0 \to L^G \to M^G \to N^G \to 0.$$

This is why invariant theory in characteristics coprime to $|G|$ is so drastically different from that in the case where the characteristic of k divides $|G|$.

(2.14) **A technical point.** Let M be a left $k[G]$-module and set $M^*=\mathrm{Hom}_k(M,k)$. One has two choices for a left $k[G]$-module structure on M^*: one given by specializing the discussion in (2.13) to the case where $N = k$, and the other by conjugation — see (2.11) — from the natural *right* module structure on M^*. A direct calculation reveals that they coincide. What is more, these modules have the property that the canonical maps of k-vector spaces

$$M \to M^{**} \qquad\qquad N \otimes_k M^* \to \mathrm{Hom}_k(M,N)$$
$$m \mapsto \big(f \mapsto f(m)\big) \qquad\qquad n \otimes_k f \mapsto \big(m \mapsto f(m)n\big)$$

are $k[G]$-linear. These maps are bijective when $\mathrm{rank}_k M$ is finite.

RAMBLE. Most of what I said from (2.11) onward applies, with appropriate modifications, to arbitrary Hopf algebras. For example, given modules M and N over a Hopf algebra H, the tensor product $M \otimes_k N$ is also an H-module with

$$h \cdot (m \otimes_k n) = \sum_{i=1}^{n} h'_i m \otimes_k h''_i n, \quad \text{where } \Delta(h) = \sum_{i=1}^{n} h'_i \otimes_k h''_i.$$

There are exceptions; for example, over a group algebra $M \otimes_k N \cong N \otimes_k M$; see (2.12). This holds over H only when $\sum_{i=1}^{n} h'_i \otimes_k h''_i = \sum_{i=1}^{n} h''_i \otimes_k h'_i$, that is to say, when the diagram

$$
\begin{array}{ccc}
 & H & \\
\Delta \swarrow & & \searrow \Delta \\
H \otimes_k H & \xrightarrow{\ \ \tau\ \ } & H \otimes_k H
\end{array}
$$

commutes, where $\tau(h' \otimes_k h'') = (h'' \otimes_k h')$. Such an H is said to be *cocommutative*.

3. Projective Modules

The section focuses on projective modules over group algebras. First, I address the question: When is every module over the group algebra projective? In other words, when is the group algebra *semisimple*? Here is a complete answer, at least in the case of a finite group.

(3.1) THEOREM. *Let G be a finite group. The following conditions are equivalent:*

(i) *The group ring $k[G]$ is semisimple.*
(ii) *k, viewed as a $k[G]$-module via the augmentation, is projective.*
(iii) *The characteristic of k is coprime to $|G|$.*

PROOF. (i) $\implies$ (ii) is a tautology.

(ii) $\implies$ (iii): As k is projective, the augmentation homomorphism $\varepsilon \colon k[G] \to k$, being a surjection, has a $k[G]$-linear section $\sigma \colon k \to k[G]$. Write $\sigma(1) = \sum_{g \in G} a_g g$, with a_g in k. Fix an element $h \in G$. Note that $\sigma(1) = \sigma(h \cdot 1) = h \cdot \sigma(1)$, where the first equality holds because $k[G]$ acts on k via ε, the second by the $k[G]$-linearity of σ. This explains the first equality below:

$$\sum_{g \in G} a_g g = \sum_{g \in G} a_g (hg) = \sum_{g \in G} a_{h^{-1}g} g.$$

The second is just a reindexing. The elements of G are a basis for the group algebra, so the equality above entails $a_{h^{-1}} = a_1$. This holds for each $h \in G$, so

$$1 = \varepsilon(\sigma(1)) = a_1 \sum_{g \in G} \varepsilon(g) = a_1 \sum_{g \in G} 1 = a_1 |G|.$$

In particular, the characteristic of k is coprime to $|G|$.

(iii) $\implies$ (i): Let M be a $k[G]$-module, and pick a surjection $P \twoheadrightarrow M$ with P projective. Maschke's theorem (2.6) provides that every short exact sequence of $k[G]$-modules splits; equivalently, that every surjective homomorphism is split. In particular, $P \twoheadrightarrow M$ splits, so M is a direct summand of P, and hence projective. $\qquad\square$

EXERCISE. A commutative ring is semisimple if and only if it is a product of fields.

The last result dealt with modules en masse; now the focus is shifted to individual modules.

Stability properties of projective modules. The gist of the following paragraphs is that many of the standard functors of interest preserve projectivity. A crucial, and remarkable, result in this direction is

(3.2) THEOREM. *Let G be a group and P a projective $k[G]$-module. For any $k[G]$-module X, the $k[G]$-modules $P \otimes_k X$ and $X \otimes_k P$ are projective.*

Take note that the tensor product is over k, as it must be, for such a conclusion is utterly wrong were it over $k[G]$. This theorem underscores the point raised in (2.12) about the importance of this tensor product in the module theory of group algebras; the other results in this section are all formal consequences of this one.

RAMBLE. There is another way to think about Theorem (3.2): one may view the entire category of $k[G]$-modules as a 'ring' with direct sum and tensor product over k playing the role of addition and multiplication respectively; the unit is k, and the commutativity of the tensor product means that this is even a 'commutative' ring. (With suitable compatibility conditions, such data define a *symmetric monoidal category*.) In this language, the theorem above is equivalent to the statement that the subcategory of projective modules is an ideal.

PROOF OF THEOREM (3.2). I will prove that $P \otimes_k X$ is projective. A similar argument works for $X \otimes_k P$; alternatively, note that it is isomorphic to $P \otimes_k X$, by (2.12).

One way to deduce that $P \otimes_k X$ is projective is to invoke the following isomorphism from (2.13), which is natural on the category of left $k[G]$-modules:

$$\mathrm{Hom}_{k[G]}(P \otimes_k X, -) \cong \mathrm{Hom}_{k[G]}(P, \mathrm{Hom}_k(X, -)).$$

Perhaps the following proof is more illuminating: by standard arguments one reduces to the case where $P = k[G]$. Write $X^\natural$ for the k-vector space underlying X. Now, by general principles, the inclusion of k-vector spaces $X^\natural \subset k[G] \otimes_k X$, defined by $x \mapsto 1 \otimes_k x$, induces a $k[G]$-linear map

$$k[G] \otimes_k X^\natural \to k[G] \otimes_k X, \quad \text{where } g \otimes_k x \mapsto g(1 \otimes_k x) = g \otimes_k gx.$$

The action of $k[G]$ on $k[G] \otimes_k X^\natural$ is *via the left-hand factor*. An elementary calculation verifies that the map below, which is $k[G]$-linear, is its inverse:

$$k[G] \otimes_k X \to k[G] \otimes_k X^\natural, \quad \text{where } g \otimes_k x \mapsto g \otimes_k (g^{-1}x).$$

Therefore, the $k[G]$-modules $k[G] \otimes_k X$ and $k[G] \otimes_k X^\natural$ are isomorphic. It remains to note that the latter module is a direct sum of copies of $k[G]$.   □

One corollary of Theorem (3.2) is the following recognition principle for semisimplicity of the group algebra; it extends to arbitrary groups the equivalence of conditions (i) and (ii) in Theorem (3.1).

(3.3) LEMMA. *Let G be a group. The following conditions are equivalent.*

(i) $k[G]$ *is semisimple*;
(ii) *the $k[G]$-module k is projective.*

PROOF. The nontrivial implication is that (ii) $\Longrightarrow$ (i). As to that, it follows from Theorem (3.2) that $k \otimes_k M$ is projective for each $k[G]$-module M, so it remains to check that the canonical isomorphism $k \otimes_k M \to M$ is $k[G]$-linear. Note that this is something that needs checking for the $k[G]$-action on $k \otimes_k M$ is via the diagonal; see (2.12).   □

RAMBLE. Lemma (3.3), although not its proof, is reminiscent of a phenomenon encountered in the theory of commutative local rings: Over such a ring, the residue field is often a 'test' module. The Auslander–Buchsbaum–Serre characterization of regularity is no doubt the most celebrated example. It says that a noetherian commutative local ring R, with residue field k, is regular if and only if the R-module k has finite projective dimension.

There are analogous results that characterize complete intersections (Avramov and Gulliksen) and Gorenstein rings (Auslander and Bridger).

There is however an important distinction between a group algebra over k and a local ring with residue field k: over the latter, k is the only simple module, whilst the former can have many others. From this perspective, Lemma (3.3) is rather surprising. The point is that an arbitrary finite-dimensional algebra is semisimple if and only if *every* simple module is projective; the nontrivial implication holds because each finite module has a composition series.

(3.4) THEOREM. *Let G be a finite group. For each finite $k[G]$-module M, the following $k[G]$-modules are projective simultaneously:* M, $M \otimes_k M$, $M^* \otimes_k M$, $M \otimes_k M^*$, $\mathrm{Hom}_k(M, M)$, *and* M^*.

PROOF. It suffices to verify: M, $M \otimes_k M$, and $M^* \otimes_k M$ are simultaneously projective.

Indeed, applied to M^* that would imply, in particular, that M^* and $(M^*)^* \otimes_k M^*$ are simultaneously projective. Now, $(M^*)^* \cong M$, since $\mathrm{rank}_k M$ is finite, and $M \otimes_k M^* \cong M^* \otimes_k M$, by the discussion in (2.12). Thus, one obtains the simultaneous projectivity of all the modules in question, except for $\mathrm{Hom}_k(M, M)$. However, the finiteness of $\mathrm{rank}_k M$ implies this last module is isomorphic to $M \otimes_k M^*$.

As to the desired simultaneous projectivity, it is justified by the diagram

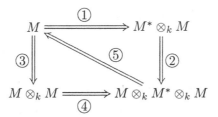

where $X \implies Y$ should be read as 'if X is projective, then so is Y'. Implications (1)–(4) hold by Theorem (3.2). As to (5), the natural maps of k-vector spaces

$$M \to \mathrm{Hom}_k(M, M) \otimes_k M \to M$$
$$m \mapsto 1 \otimes_k m \text{ and } \alpha \otimes_k m \mapsto \alpha(m)$$

are $k[G]$-linear, and exhibit M as a direct summand of $\mathrm{Hom}_k(M, M) \otimes_k M$. However, as remarked before, the $k[G]$-modules $\mathrm{Hom}_k(M, M)$ and $M \otimes_k M^*$ are isomorphic, so M is a direct summand of $M \otimes_k M^* \otimes_k M$.   □

Projective versus Injectives. So far, I have focused on projective modules, without saying anything at all about injective, or flat, modules. Now, a commutative algebraist well knows that projective modules and injective modules are very different beasts. There is, however, one exception.

(3.5) EXERCISE. Let R be a commutative noetherian local ring. Prove that when R is zero-dimensional and Gorenstein, an R-module is projective if and only if it is injective. Conversely, if there is a nonzero R-module that is both projective and injective, then R is zero-dimensional and Gorenstein.

The preceding exercise should be compared with the next two results.

(3.6) THEOREM. *Let G be a finite group and M a finite $k[G]$-module. The following conditions are equivalent*:

(i) M *is projective*;
(ii) *the flat dimension of M is finite*;
(iii) M *is injective*;
(iv) *the injective dimension of M is finite*.

These equivalences hold for any $k[G]$-module, finite or not; see [Benson 1999].
  The preceding theorem has an important corollary.

(3.7) COROLLARY. *The group algebra of a finite group is self-injective.*   □

There are many other proofs, long and short, of this corollary; see [Benson 1991a, (3.1.2)]. Moreover, it is an easy exercise (do it) to deduce Theorem (3.6) from it.

RAMBLE. Let G be a finite group. Thus, the group algebra $k[G]$ is finite-dimensional and, by the preceding corollary, injective as a module over itself. These properties may tempt us commutative algebraists to proclaim: $k[G]$ is a zero-dimensional Gorenstein ring. And, for many purposes, this is a useful point of view, since module theory over a group algebra resembles that over a Gorenstein ring; Theorem (3.6) is one manifestation of this phenomenon. By the by, there are diverse extensions of the Gorenstein property for commutative rings to the noncommutative setting: Frobenius rings, quasi-Frobenius rings, symmetric rings, self-injective rings, etc.

The proof of Theorem (3.6) is based on Theorem (3.4) and an elementary observation about modules over finite-dimensional algebras.

(3.8) LEMMA. *Let R be a k-algebra with $\mathrm{rank}_k R$ finite. For each finite left R-module M, one has $\mathrm{pdim}_R M = \mathrm{fdim}_R M = \mathrm{injdim}_{R^{\mathrm{op}}} M^*$.*

PROOF. Since $\mathrm{rank}_k M$ is finite, $(M^*)^* \cong M$, so it suffices to prove the equivalence of the conditions

(i) M is projective;
(ii) M is flat;
(iii) the right R-module M^* is injective.

The implication (i) $\implies$ (ii) is immediate and hold for all rings. The equivalence (ii) $\iff$ (iii) is a consequence of the standard adjunction isomorphism

$$\mathrm{Hom}_k(- \otimes_R M, k) \cong \mathrm{Hom}_R(-, M^*)$$

and is valid for arbitrary k-algebras.
  (iii) $\implies$ (i): Since M is finite over R, one can construct a surjective map $\pi\colon R^n \twoheadrightarrow M$. Dualizing this yields an inclusion $\pi^*\colon M^* \hookrightarrow (R^n)^*$ of right R-modules. This map is split because M^* is injective, and hence π^{**} is split. Since $\mathrm{rank}_k R$ and $\mathrm{rank}_k M$ are both finite, $\pi^{**} = \pi$, so that π is split as well. Thus, M is projective, as claimed.   □

PROOF OF THEOREM (3.6). Theorem (3.4) yields that M is projective if and only if M^* is projective, while the lemma above implies that M^* is projective if and only if $(M^*)^*$ is injective, i.e., M is injective. This settles (i) $\iff$ (iii).

That (i) $\implies$ (ii) needs no comment. The lemma above contains (ii) $\iff$ (iv); moreover, it implies that to verify (ii) $\implies$ (i), one may assume $\mathrm{pdim}_R M$ finite, that is to say, there is an exact sequence

$$0 \to P_n \xrightarrow{\partial_n} P_{n-1} \xrightarrow{\partial_{n-1}} \cdots \to P_0 \to M \to 0,$$

where each P_i is finite and projective; see (6.6). If $n \geq 1$, then, since P_n is injective by the already verified implication (i) $\implies$ (iii), the homomorphism ∂_n splits, and one obtains an exact sequence

$$0 \to \partial_{n-1}(P_{n-1}) \to P_{n-2} \to \cdots \to P_0 \to M \to 0.$$

In this sequence $\partial_{n-1}(P_{n-1})$, being a direct summand of P_{n-1}, is projective, and hence injective. An iteration of the preceding argument yields that M is a direct summand of P_0, and hence projective.   □

RAMBLE. The small finitistic left global dimension of a ring R is defined as

$$\sup \{\mathrm{pdim}_R M \mid M \text{ a finite left } R\text{-module with } \mathrm{pdim}_R M < \infty.\}$$

One way of rephrasing Theorem (3.6) is to say that this number is zero when R is a finite group algebra. Exercise: Prove that a similar result holds also for modules over commutative artinian rings. However, over arbitrary finite-dimensional algebras, the small finitistic global dimension can be any nonnegative integer. A conjecture of Bass [1960] and Jans [1961], which remains open, asserts that this number is finite; look up [Happel 1990] for more information on this topic.

Hopf algebras. Theorem (3.2) holds also for modules over any finite-dimensional Hopf algebra; the proof via the adjunction isomorphism does not work, but the other one does. However, I found it a nontrivial task to pin down the details, and I can recommend it to you as a good way to gain familiarity with Hopf algebras. Given this, it is not hard to see that for *cocommutative* Hopf algebras, the analogues of theorems (3.4) and (3.6), and Corollary (3.7), all hold; the cocommutativity comes in because in the proof of (3.4) I used the fact that tensor products are symmetric; confer with the discussion in (2.14).

4. Structure of Projectives

So far, I have not addressed the natural question: what are the projective modules over the group algebra? In this section, I tabulate some crucial facts concerning these. Most are valid for arbitrary finite-dimensional algebras and are easier to state in that generality; [Alperin 1986] is an excellent reference for this circle of ideas.

(4.1) **Projective covers.** Let R be a ring and M a finite R-module. A *projective cover* of M is a surjective homomorphism $\pi\colon P \to M$ with P a projective R-module and such that each homomorphism $\sigma\colon P \to P$ that fits in a commutative diagram

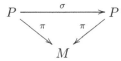

is bijective, and hence an automorphism. It is clear that projective covers, when they exist, are unique up to isomorphism. Thus, one speaks of *the* projective cover of M. Often P, rather than π, is thought as being the projective cover of M, although this is an abuse of terminology.

Among surjective homomorphisms $\kappa\colon Q \to M$ with Q a projective R-module, projective covers can be characterized by either of the properties:

(i) $Q/JQ \cong M/JM$, where J is the Jacobson radical of R;
(ii) Q is minimal with respect to direct sum decompositions.

When R is a noetherian ring over which every finite R-module has a projective cover, it is easy to see that a projective resolution

$$\mathbf{P}\colon \cdots \to P_n \xrightarrow{\partial_n} P_{n-1} \xrightarrow{\partial_{n-1}} \cdots \xrightarrow{\partial_1} P_0 \to 0$$

of M so constructed that P_n is a projective cover of $\mathrm{Ker}(\partial_{n-1})$ is unique up to isomorphism of complexes of R-modules. Such a $\mathbf{P}$ is called the *minimal projective resolution* of M. Following conditions (i) and (ii) above, the minimality can also be characterized by either the property that $\partial(\mathbf{P}) \subseteq J\mathbf{P}$, or that $\mathbf{P}$ splits off from any projective resolution of M.

Projective covers exist for each finite M in two cases of interest: when R is a finite-dimensional k-algebra, and when R is a (commutative) local ring. This is why these two classes of rings have a parallel theory of minimal resolutions.

(4.2) **Simple modules.** Let R be a finite-dimensional k-algebra with Jacobson radical J, and let $\mathscr{P}$ and $\mathscr{S}$ be the isomorphism classes of indecomposable projective R-modules and of simple R-modules, respectively.

(a) The Krull–Schimdt property holds for R, so every P in $\mathscr{P}$ occurs as a direct summand of R, and there is a unique decomposition

$$R \cong \bigoplus_{P \in \mathscr{P}} P^{e_R(P)}, \quad \text{with } e_R(P) \geq 1.$$

In particular, R has only finitely many indecomposable projective modules.

(b) The simple R-modules are precisely the indecomposable modules of the semisimple ring $\widetilde{R} = R/J$ (verify this) so property (a) specialized to $\widetilde{R}$ reads

$$\widetilde{R} \cong \bigoplus_{S \in \mathscr{S}} S^{e_{\widetilde{R}}(S)}, \quad \text{with } e_{\widetilde{R}}(S) \geq 1.$$

(c) The ring $\widetilde{R}$ in (b), being semisimple, is a direct sum of matrix rings over finite-dimensional division algebras over k; see [Lang 2002, XVII]. Moreover, when k is algebraically closed, these division algebras coincide with k (why?), and we obtain that $e_{\widetilde{R}}(S) = \operatorname{rank}_k S$ for each $S \in \mathscr{S}$.

(d) From (a)–(c) one obtains that the assignment $P \mapsto P/JP$ is a bijection between $\mathscr{P}$ and $\mathscr{S}$; in other words, there are as many indecomposable projective R-modules as there are simple R-modules. Moreover, $e_R(P) = e_{\widetilde{R}}(P/JP)$. When k is algebraically closed, combining the last equality with that in (c) and the decomposition in (a) yields

$$\operatorname{rank}_k R = \sum_{P \in \mathscr{P}} \operatorname{rank}_k(P/JP)\operatorname{rank}_k P.$$

I will illustrate the preceding remarks by describing the indecomposable projective modules over certain finite group algebras.

(4.3) **Cyclic groups.** This example builds on the description in (2.7) of modules over the group algebra of a finite cyclic group G. We saw there that

$$k[G] \cong \prod_{i=1}^{d} \frac{k[x]}{(f_i(x)^{p^n})}.$$

This is the decomposition that for general finite-dimensional algebras is a consequence of the Krull–Schmidt property; see (4.2.a). For each $1 \le i \le d$, set $P_i = k[x]/(f_i(x)^{p^n})$. These $k[G]$-modules are all projective, being summands of $k[G]$, indecomposable (why?), and no two of them are isomorphic (count ranks, or look at their annihilators). Moreover, as a consequence of the decomposition above, any projective $k[G]$-module is a direct sum of the P_i. Thus, there are exactly d distinct isomorphism classes of indecomposable projective R-modules.

Over any commutative ring, the only simple modules are the residue fields. Thus, the simple modules over $k[G]$ are $k[x]/(f_i(x))$ where $1 \le i \le d$; in particular, there are as many as there are indecomposable projectives, exactly as (4.2.d) predicts.

Now I will describe the situation over finite abelian groups. Most of what I have to say can be deduced from:

(4.4) LEMMA. *Let R and S be finite-dimensional k-algebras, and set $T = R \otimes_k S$. Let M and N be R-modules. If S is local with residue ring is k, and the induced map $k \to S \to k$ is the identity, then*

(a) $M \cong N$ *as R-modules if and only if $M \otimes_k S \cong N \otimes_k S$ as T-modules;*
(b) *the R-module M is indecomposable if and only if the T-module $M \otimes_k S$ is;*
(c) M *is projective if and only if the T-module $M \otimes_k S$ is projective.*

In particular, the map $P \mapsto P \otimes_k S$ induces a bijection between the isomorphism classes of indecomposable projective modules over R and over T.

PROOF. To begin with, note that $M \otimes_k S$ and $N \otimes_k S$ are both left R-modules and also right S-modules, with the obvious actions. Moreover, because of our hypothesis that the residue ring of S is k, one has isomorphisms of R-modules

$$M \cong (M \otimes_k S) \otimes_S k \qquad \text{and} \qquad N \cong (N \otimes_k S) \otimes_S k.$$

Now, the nontrivial implication in (a) and in (c) — the one concerning descent — is settled by applying $- \otimes_S k$. As to (b), the moot point is the ascent, so assume the R-module M is indecomposable and that $M \otimes_k S \cong U \oplus V$ as T-modules. Applying $- \otimes_S k$, one obtains isomorphisms of R-modules

$$M \cong (M \otimes_k S) \otimes_S k \cong (U \otimes_S k) \oplus (V \otimes_S k)$$

Since M is indecomposable, one of $U \otimes_S k$ or $V \otimes_S k$ is zero; say, $U \otimes_S k$ is 0, that is to say, $U = U\mathfrak{n}$, where $\mathfrak{n}$ is the maximal ideal of S. This implies $U = 0$, because, S being local and finite-dimensional over k, the ideal $\mathfrak{n}$ is nilpotent. $\square$

(4.5) **Finite abelian groups.** Again, we adopt that convention that p is the characteristic of k when the latter is positive, and 1 otherwise.

Let G a finite abelian group, and write $|G|$ as $p^n q$, where n is a nonnegative integer and p and q are coprime. Via the fundamental theorem on finitely generated abelian groups this decomposition of $|G|$ translates into one of groups: $G = A \oplus B$, where A and B are abelian, $|A| = p^n$, and $|B| = q$. Hence, $k[G] \cong k[A] \otimes_k k[B]$.

Now, $A \cong \mathbb{Z}/(p^{e_1}\mathbb{Z}) \oplus \cdots \oplus \mathbb{Z}/(p^{e_m}\mathbb{Z})$, for nonnegative integers $e_1, \ldots, e_m$, so

$$k[A] \cong \frac{k[y_1, \ldots, y_m]}{(y_1^{p^{e_1}} - 1, \ldots, y_m^{p^{e_m}} - 1)}$$

The binomial theorem in characteristic p yields $y_i^{p^{e_i}} - 1 = (y_i - 1)^{p^{e_i}}$ for each i. Thus, it is clear that $k[A]$ is an artinian local ring with residue field k.

In the light of this and Lemma (4.4), to find the indecomposable projectives over $k[G]$, it suffices to find those over $k[B]$.

When B is cyclic, this information is contained in (4.3). The general case is more delicate. First, since $|B|$ is coprime to p, every $k[G]$-module is projective, so the indecomposables among them are precisely the simple $k[B]$-modules; see Theorem (3.1). Now, as noted before, over any commutative ring the only simple modules are the residue fields. Thus, the problem is to find the maximal ideals of $k[B]$. Writing B as $\mathbb{Z}/(q_1\mathbb{Z}) \oplus \cdots \oplus \mathbb{Z}/(q_n\mathbb{Z})$, one has

$$k[B] \cong \frac{k[x_1, \ldots, x_n]}{(x_1^{q_1} - 1, \ldots, x_m^{q_n} - 1)}.$$

If k is algebraically closed, there are $q_1 \cdots q_n$ distinct maximal ideals, and hence as many distinct indecomposable projectives. The general situation is trickier.

By the by, if you use the method outlined above for constructing projective modules over a cyclic group, the outcome will appear to differ from that given by (4.3). Exercise: Reconcile them.

(4.6) **p-groups.** As always, free $k[G]$-modules are projective. When the characteristic of k is p and G is a p-group, these are the only projectives over $k[G]$. This is thus akin to the situation over commutative local rings, and the proof over this latter class of rings given in [Matsumura 1989] carries over; the key ingredient is that, as noted in (1.5), the group algebra of a p-group is an artinian local ring.

In general, the structure of projective modules over the group algebra is a lot more complicated. However, the triviality of the projectives in the case of p-groups also has implications for the possible ranks of indecomposable projectives over the group algebra of an arbitrary group G.

(4.7) **Sylow subgroups.** Let p^d be the order of a p-Sylow subgroup of G. If a finite $k[G]$-module P is projective, then p^d divides $\mathrm{rank}_k P$.

Indeed, for each p-Sylow subgroup $H \subseteq G$, the restriction of P to the subring $k[H]$ of $k[G]$ is a projective module, and hence a free module. Thus, by the preceding remark, $\mathrm{rank}_k P$ is divisible by $\mathrm{rank}_k k[H]$, that is to say, by $|H|$.

The numerological restrictions in (4.2) and (4.7) can be very handy when hunting for projective modules over finite group algebras. Here is a demonstration.

(4.8) **Symmetric group on three letters.** The symmetric group on three letters, Σ_3, is generated by elements a and b, subject to the relations

$$a^2 = 1, \quad b^3 = 1, \quad \text{and} \quad ba = ab^2.$$

Thus, $\Sigma_3 = \{1, b, b^2, a, ab, ab^2\}$. It has three 2-Sylow subgroups: $\{1, a\}$, $\{1, ab\}$, and $\{1, ba\}$, and one 3-Sylow subgroup: $\{1, b, b^2\}$.

Let p be the characteristic of the field k; we allow the possibility that $p = 0$.

CASE (α). If $p \neq 2, 3$, every $k[\Sigma_3]$-module is projective, by Theorem (3.1).

CASE (β). Suppose $p = 3$. By (4.7), the rank of each finite projective $k[G]$-module is divisible by 3, since the latter is the order of the 3-Sylow subgroup. Moreover, (4.2.d) implies that the number of indecomposable projectives equals the number of simple modules, and the latter is at least 2, for example, by Exercise (1.6). These lead us to the conclusion that there are exactly two indecomposable projectives, each having rank 3.

One way to construct them is as follows: Let $H = \{1, a\}$, a 2-Sylow subgroup of Σ_3. There are two nonisomorphic $k[H]$-module structures on k: the trivial one, given by the augmentation map, and the one defined by character $\sigma \colon H \to k$ with $\sigma(a) = -1$; denote the latter ${}^\sigma k$. Plainly, both these $k[H]$-modules are simple and hence, by Theorem (3.1), projective. Consequently, base change along the canonical inclusion $k[H] \to k[\Sigma_3]$ gives us two projective $k[\Sigma_3]$-modules,

$$k[\Sigma_3] \otimes_{k[H]} k \quad \text{and} \quad k[\Sigma_3] \otimes_{k[H]} {}^\sigma k.$$

They both have rank 3. I leave it to you to verify that they are not isomorphic. Hint: calculate the Σ_3-invariants.

CASE (γ). The situation gets even more interesting when $p = 2$. I claim that there are two indecomposable projective $k[G]$-modules, of ranks 2 and 4, when $x^2 + x + 1$ is irreducible in k, and three of them, each of rank 2, otherwise.

Indeed, let $H = \{1, b, b^2\}$; this is a cyclic group of order 3. Hence, by (4.3), when $x^2 + x + 1$ is irreducible in $k[x]$, the group algebra $k[H]$ has 2 (nonisomorphic) simple modules, of ranks 1 and 2, and when $x^2 + x + 1$ factors in $k[x]$, there are 3 simple modules, each of rank 1. As the characteristic of k does not divide $|H|$, all these simple modules are projective, so base change along the inclusion $k[H] \subset k[\Sigma_3]$ gives rise to the desired number of projective modules, and of the right ranks, over $k[G]$. Note that, by (4.7), projective modules of rank 2 are indecomposable. Thus, to be sure that these are the projectives one seeks, one has to verify that in the former case the rank 4 module is indecomposable, and in the latter that the three rank 2 modules are nonisomorphic. Once again, I will let you check this.

5. Cohomology of Supplemented Algebras

This section collects basic facts concerning the cohomology of supplemented algebras. To begin with, recall that in the language of Cartan and Eilenberg [1956] a *supplemented k-algebra* is a k-algebra R with unit $\eta \colon k \to R$ and an augmentation $\varepsilon \colon R \to k$ such that $\varepsilon \circ \eta$ is the identity on k.

Group algebras are supplemented, but there are many more examples. Take, for instance, any positively (or negatively) graded k-algebra with degree 0 component equal to k. Or, for that matter, take the power series ring $k[\![x_1, \cdots, x_n]\!]$, with η the canonical inclusion, and ε the evaluation at 0. More generally, thanks to Cohen's Structure Theorem, if a complete commutative local ring R, with residue field k, contains a field, then R is a supplemented k-algebra.

Let R be a supplemented k-algebra, and view k as an R-module via the augmentation. Let M be a (left) R-module. The *cohomology of R with coefficients in M* is the graded k-vector space $\operatorname{Ext}_R^*(k, M)$. The cohomology of R with coefficients in k, that is to say, $\operatorname{Ext}_R^*(k, k)$, is usually called the *cohomology of R*.

The k-vector space structure on $\operatorname{Ext}_R^*(k, k)$ can be enriched to that of a supplemented k-algebra, and then $\operatorname{Ext}_R^*(k, M)$ can be made into a *right* module over it. There are two ways to introduce these structures: via Yoneda splicing and via compositions. They yield the same result, up to a sign; see (5.2). I have opted for composition products because it is this description that I use to calculate group cohomology in the sequel.

(5.1) **Composition products.** Let P be a projective resolution of k. Composition endows the complex of k-vector spaces $\operatorname{Hom}_R(P, P)$ with a product structure, and this product is compatible with the differential, in the sense that, for every pair of homogenous elements f, g in $\operatorname{Hom}_R(P, P)$, one has

$$\partial(fg) = \partial(f)g + (-1)^{|f|} f \partial(g).$$

In other words, $\mathrm{Hom}_R(P, P)$ is a differential graded algebra (DGA). One often refers to this as the *endomorphism DGA* of P. It is not hard to verify that the multiplication of $\mathrm{Hom}_R(P, P)$ descends to homology, that is to say, to $\mathrm{Ext}_R^*(k, k)$. This is the *composition product* on cohomology, and it makes it a graded k-algebra. It is even supplemented, since $\mathrm{Ext}_R^0(k, k) = k$.

Let F be a projective resolution of M. The endomorphism DGA $\mathrm{Hom}_R(P, P)$ acts on the complex $\mathrm{Hom}_R(P, F)$ via composition on the right, and, once again, this action is compatible with the differentials. Thus, $\mathrm{Hom}_R(P, F)$ becomes a DG *right* module over $\mathrm{Hom}_R(P, P)$. These structures are inherited by the corresponding homology vector spaces; thus does $\mathrm{Ext}_R^*(k, M)$ become a right $\mathrm{Ext}_R^*(k, k)$-module.

One has to check that the composition products defined do not depend on the choice of resolutions; [Bourbaki 1980, (7.2)] justifies this, and much more.

(5.2) REMARK. As mentioned before, one can introduce products on $\mathrm{Ext}_R^*(k, k)$ also via Yoneda multiplication, and, *up to a sign*, this agrees with the composition product; [Bourbaki 1980, (7.4)] has a careful treatment of these issues. The upshot is that one can set up an isomorphism of k-algebras between the Yoneda Ext-algebra and Ext-algebra with composition products. Thus, one has the freedom to use either structure, as long as it is done consistently.

(5.3) **Graded-commutativity.** Let E be a graded algebra. Elements x and y in E are said to commute, in the graded sense of the word, if

$$xy = (-1)^{|x||y|}\, yx.$$

If every pair of its elements commute, E is said to be graded-commutative. When E is concentrated in degree 0 or in even degrees, it is graded-commutative precisely when it is commutative in the usual sense.

An exterior algebra on a finite-dimensional vector space sitting in odd degrees is another important example of a graded-commutative algebra. More generally, given a graded vector space V, with $V_i = 0$ for $i < 0$, the tensor product of the symmetric algebra on V_{even} and exterior algebra on V_{odd}, that is to say, the k-algebra

$$\mathrm{Sym}(V_{\mathrm{even}}) \otimes_k \bigwedge V_{\mathrm{odd}},$$

is graded-commutative. If the characteristic of k happens to be 2, then $\mathrm{Sym}(V)$ is also graded-commutative even when $V_{\mathrm{odd}} \neq 0$. This fails in odd characteristics, the point being that, in a graded-commutative algebra, for an element x of odd degree, $x^2 = -x^2$, so that $x^2 = 0$ when 2 is invertible in E.

A graded-commutative algebra with the property that $x^2 = 0$ whenever the degree of x is odd is said to be *strictly graded-commutative*. An exterior algebra (with generators in odd degrees) is one example. Here is one more, closer to home: for a homomorphism of commutative rings $R \to S$, the graded S-module $\mathrm{Tor}_*^R(S, S)$ is strictly graded-commutative, with the pitchfork product (homology product) defined by Cartan and Eilenberg; see [Mac Lane 1995, VIII § 2].

(5.4) **Functoriality.** The product in cohomology is functorial, in that, given a homomorphism of supplemented k-algebras $\varphi\colon R \to R'$, the induced map of graded k-vector spaces

$$\mathrm{Ext}^*_\varphi(k,k)\colon \mathrm{Ext}^*_{R'}(k,k) \to \mathrm{Ext}^*_R(k,k)$$

is a homomorphism of supplemented k-algebras.

Now let R and S be supplemented k-algebras. The tensor product $R \otimes_k S$ is also a supplemented k-algebra, and the canonical maps

$$R \xleftarrow{\ 1 \otimes \varepsilon^S\ } R \otimes_k S \xrightarrow{\ \varepsilon^R \otimes 1\ } S$$

respect this structure. By functoriality of products, the diagram above induces homomorphisms of supplemented k-algebras

$$\mathrm{Ext}^*_R(k,k) \xrightarrow{\ \mathrm{Ext}^*_{1 \otimes \varepsilon^S}(k,k)\ } \mathrm{Ext}^*_{R \otimes_k S}(k,k) \xleftarrow{\ \mathrm{Ext}^*_{\varepsilon^R \otimes 1}(k,k)\ } \mathrm{Ext}^*_S(k,k).$$

It is not hard to check that the images of these maps commute, in the graded sense, so one has a diagram of supplemented k-algebras:

$$(*) \qquad \mathrm{Ext}^*_R(k,k) \xrightarrow{\ \mathrm{Ext}^*_{\mathrm{id} \otimes \varepsilon^S}(k,k)\ } \mathrm{Ext}^*_{R \otimes_k S}(k,k) \xleftarrow{\ \mathrm{Ext}^*_{\varepsilon^R \otimes \mathrm{id}}(k,k)\ } \mathrm{Ext}^*_S(k,k)$$

$$\mathrm{id} \otimes 1 \searrow \qquad \uparrow \qquad \swarrow 1 \otimes \mathrm{id}$$

$$\mathrm{Ext}^*_R(k,k) \otimes_k \mathrm{Ext}^*_S(k,k)$$

I should point out that the tensor product on the lower row is the *graded* tensor product and the multiplication on it is defined accordingly, that is,

$$(r \otimes_k s) \cdot (r' \otimes_k s') = (-1)^{|s||r'|}(rr' \otimes_k ss').$$

Under suitable finiteness hypotheses — for example, if R and S are noetherian — the vertical map in $(*)$ is bijective. However, this is not of importance to us.

The cohomology of Hopf algebras. The remainder of this section deals with the cohomology of Hopf algebras. So let H be a Hopf algebra, with diagonal Δ and augmentation ε; see (1.8). The main example to keep in mind is the case when H is the group algebra of a group, with the diagonal defined in (1.7).

One crucial property of the cohomology algebra of H, which distinguishes it from the cohomology of an arbitrary supplemented algebra, is the following.

(5.5) PROPOSITION. *The cohomology algebra* $\mathrm{Ext}^*_H(k,k)$ *is graded-commutative.*

Note that H is not assumed to be cocommutative. This is a striking result, and its proof is based on the diagram of k-algebra homomorphisms

$$(5\text{-}1) \qquad \mathrm{Ext}^*_H(k,k) \otimes_k \mathrm{Ext}^*_H(k,k) \to \mathrm{Ext}^*_{H \otimes_k H}(k,k) \xrightarrow{\ \mathrm{Ext}^*_\Delta(k,k)\ } \mathrm{Ext}^*_H(k,k),$$

where the one on the left is the vertical map in (5.4.1), with R and S equal to H, and the one on the right is induced by the diagonal homomorphism.

(5.6) PROPOSITION. *The composition of homomorphisms in* (5.5.5–1) *is the product map; that is to say,* $(x \otimes_k y) \mapsto xy$ *for* x *and* y *in* $\mathrm{Ext}^*_H(k, k)$.

In particular, the product map of $\mathrm{Ext}^*_H(k, k)$ *is a homomorphism of* k-*algebras.*

PROOF. The diagram in question expands to the following commutative diagram of homomorphisms of k-algebras, where the lower half is obtained from (5.4.1), the upper half is induced by property (c) of Hopf algebras — see (1.8) — to the effect that ε is a co-unit for the diagonal.

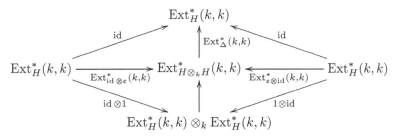

Let x and y be elements in $\mathrm{Ext}^*_H(k, k)$. The element x goes to $x \otimes_k 1$ under the map heading southeast, and to x under the map heading northeast. The commutativity of the diagram thus implies that $x \otimes_k 1 \mapsto x$ under the composed vertical map. A similar diagram chase reveals that $1 \otimes_k y \mapsto y$. Since the vertical maps are homomorphisms of k-algebras, one has

$$x \otimes_k y = (x \otimes_k 1) \cdot (1 \otimes_k y) \mapsto xy.$$

This is the conclusion we seek. □

The proof of Proposition (5.5) uses also the following elementary exercise, of which there are versions for groups, for coalgebras, etc.

(5.7) EXERCISE. A graded k-algebra R is graded-commutative precisely when the product map $R \otimes_k R \to R$ with $r \otimes s \mapsto rs$ is a homomorphism of rings.

Now one can prove that the cohomology algebra is graded-commutative.

PROOF OF PROPOSITION (5.5). By the preceding proposition, the product map $\mathrm{Ext}^*_H(k, k) \otimes_k \mathrm{Ext}^*_H(k, k) \to \mathrm{Ext}^*_H(k, k)$ given by $x \otimes_k y \mapsto xy$ is a homomorphism of rings (for a general algebra it is only k-linear). To complete the proof one has to do Exercise (5.7). □

6. Group Cohomology

In this section we return to group algebras.

(6.1) **Cohomology.** Let G be a group and let M be a $k[G]$-module. Recall that $k[G]$ is a supplemented algebra. The *cohomology of* G *with coefficients in* M is the graded k-vector space

$$\mathrm{H}^*(G, M) = \mathrm{Ext}^*_{k[G]}(k, M).$$

There is no ambiguity concerning the field k since $\mathrm{Ext}^*_{k[G]}(k, M)$ is isomorphic to $\mathrm{Ext}^*_{\mathbb{Z}[G]}(\mathbb{Z}, M)$; see [Evens 1961, (1.1)]. The *cohomology of* G is $\mathrm{H}^*(G, k)$.

Standard properties of Ext-modules carry over to the situation on hand. For instance, each short exact sequence of $k[G]$-modules $0 \to L \to M \to N \to 0$ engenders a long exact sequence of k-vector spaces

$$0 \to \mathrm{H}^0(G, L) \to \mathrm{H}^0(G, M) \to \mathrm{H}^0(G, N) \to \mathrm{H}^1(G, L) \to \mathrm{H}^1(G, M) \to \cdots .$$

Note that $\mathrm{H}^n(G, -) = 0$ for $n \geq 1$ if and only if k is projective. Therefore, one has the following cohomological avatar of Maschke's theorem (3.1):

(6.2) THEOREM. *Let G be a finite group. Then $\mathrm{H}^n(G, -) = 0$ for each integer $n \geq 1$ if and only if the characteristic of k is coprime to $|G|$.* $\qquad\square$

As is typical in homological algebra, low degree cohomology modules have nice interpretations. For a start, $\mathrm{Ext}^0_{k[G]}(k, M) = \mathrm{Hom}_{k[G]}(k, M)$, so (2.13) yields

$$\mathrm{H}^0(G, M) = M^G.$$

Thus, one can view the functors $\mathrm{H}^n(G, -)$ as the derived functors of invariants.

The degree 1 component of $\mathrm{H}^*(G, M)$ is also pretty down to earth. Recall that a map $\theta\colon G \to M$ is said to be a *derivation*, or a *crossed homomorphism*, if it satisfies the Leibniz formula: $\theta(gh) = \theta(g) + g\theta(h)$, for every g, h in G. The asymmetry in the Leibniz rule is explained when one views M, which is *a priori* only a left $k[G]$-module, as a $k[G]$-bimodule with trivial right action: $m \cdot g = m$. Using the k-vector space structure on M one can add derivations, and multiply them with elements in k, so they form a k-vector space; this is denoted $\mathrm{Der}(G; M)$. This vector space interests us because of the following

(6.3) LEMMA. *The k-vector spaces $\mathrm{Hom}_{k[G]}(\mathrm{I}(G), M)$ and $\mathrm{Der}(G; M)$ are isomorphic via the maps*

$$\mathrm{Hom}_{k[G]}(\mathrm{I}(G), M) \to \mathrm{Der}(G; M) \qquad\qquad \mathrm{Der}(G; M) \to \mathrm{Hom}_{k[G]}(\mathrm{I}(G), M)$$
$$\alpha \mapsto \big(g \mapsto \alpha(g - 1)\big), \qquad\qquad\qquad \theta \mapsto \big(g - 1 \mapsto \theta(g)\big).$$

The proof is an elegant computation and is best rediscovered on one's own. As to its bearing on $\mathrm{H}^1(G, M)$: applying $\mathrm{Hom}_{k[G]}(-, M)$ to the exact sequence

$$0 \to \mathrm{I}(G) \to k[G] \to k \to 0$$

of $k[G]$-modules leads to the exact sequence of k-vector spaces

$$0 \to M^G \to M \to \mathrm{Der}(G; M) \to \mathrm{H}^1(G, M) \to 0.$$

In this sequence, each $m \in M$ maps to a derivation: $g \mapsto (g-1)m$; these are the *inner derivations* from G to M, and their set is denoted by $\mathrm{IDer}(G; M)$. Thus,

$$\mathrm{H}^1(G, M) = \mathrm{Der}(G; M)/\mathrm{IDer}(G; M).$$

Let us specialize to the case when $M = k$. The Leibniz rule for a derivation $\theta\colon G \to k$ then reads: $\theta(gh) = \theta(g) + \theta(h)$, so $\mathrm{Der}(G; k)$ coincides with group homomorphisms from G to k. Moreover, every inner derivation from G to k is trivial. The long and short of this discussion is that $H^1(G, k)$ is precisely the set of additive characters from G to k.

There are other descriptions, some of a more group theoretic flavour, for $H^1(G, M)$; for those the reader may look in [Benson 1991a].

The discussion in Section 5 on products on cohomology applies in the special case of the cohomology of group algebras. In particular, since $k[G]$ is a Hopf algebra, Proposition (5.5) specializes thus:

(6.4) THEOREM. *The cohomology algebra $H^*(G, k)$ is graded-commutative.*   □

(6.5) **Künneth formula.** Let G_1 and G_2 be groups. Specializing (5.4.1) to the case where $R = k[G_1]$ and $S = k[G_2]$, one obtains a homomorphism of k-algebras

$$H^*(G_1, k) \otimes_k H^*(G_2, k) \to H^*(G_1 \times G_2, k).$$

This map is bijective whenever the group algebras are noetherian. This is the case when, for example, G_i is finite, or finitely generated and abelian.

(6.6) **Resolutions.** If one wants to compute cohomology from first principles, one has to first obtain a projective resolution of k over $k[G]$. In this regard, it is of interest to get as economical a resolution as possible. Fortunately, any finitely generated module over $k[G]$ has a minimal projective resolution; we discussed this point already in (4.1); unfortunately, writing down this minimal resolution is a challenge. In this the situation over group algebras is similar to that over commutative local rings. What is more difficult is calculating products from these minimal resolutions.

There is a canonical resolution for k over $k[G]$ called the *Bar resolution*; while it is never minimal, it has the merit that there is a simple formula for calculating the product of cohomology classes. The are many readable sources for this, such as [Benson 1991a, (3.4)], [Evens 1991, (2.3)], and [Mac Lane 1995, IV § 5], so I will not reproduce the details here.

7. Finite Generation of the Cohomology Algebra

In the preceding section, we noted that the cohomology algebra of a finite group is graded-commutative. From this, the natural progression is to the following theorem, contained in [Evens 1991], [Golod 1959], and [Venkov 1959].

(7.1) THEOREM. *Let G be a finite group. The k-algebra $H^*(G, k)$ is finitely generated, and hence noetherian.*   □

This result, and its analogues for other types of groups, is the starting point of Benson's article [2004]; see the discussion in Section 4 of it. There are many ways of proving Theorem (7.1), some more topological than others; one that is entirely algebraic is given in [Evens 1961, (7.4)].

In this section I prove the theorem in some special cases. But first:

RAMBLE. Theorem (7.1) has an analogue in commutative algebra: Gulliksen [1974] proves that when a commutative local ring R, with residue field k, is a complete intersection, the cohomology algebra $\mathrm{Ext}_R^*(k, k)$ is noetherian. There is a perfect converse: Bøgvad and Halperin [1986] have proved that if the k-algebra $\mathrm{Ext}_R^*(k, k)$ is noetherian, then R must be complete intersection.

There are deep connections between the cohomology of modules over complete intersections and over group algebras. This is best illustrated by the theory of support varieties. In group cohomology it was initiated by Quillen [1971a; 1971b], and developed in depth by Benson and Carlson, among others; see [Benson 1991b] for a systematic introduction. In commutative algebra, support varieties were introduced by Avramov [1989]; see also [Avramov and Buchweitz 2000].

As always, there are important distinctions between the two contexts. For example, the cohomology algebra of a complete intersection ring is generated by its elements of degree 1 and 2, which need not be the case with group algebras. More importantly, once the defining relations of the complete intersection are given, one can write down the cohomology algebra; the prescription for doing so is given in [Sjödin 1976]. Computing group cohomology is an entirely different cup of tea. Look up [Carlson 2001] for more information on the computational aspects of this topic.

Now I describe the cohomology algebra of finitely generated abelian groups. In this case, the group algebra is a complete intersection — see (1.4) — so one may view the results below as being about commutative rings or about finite groups.

(7.2) PROPOSITION. *For each positive integer n, the cohomology of $\mathbb{Z}^n$ is the exterior algebra on an n-dimensional vector space concentrated in degree 1.*

PROOF. As noted in (1.2), the group algebra of $\mathbb{Z}$ is $k[x^{\pm 1}]$, with augmentation defined by $\varepsilon(x) = 1$. The augmentation ideal is generated by $x - 1$, and since this element is regular, the Koszul complex

$$0 \to k[x^{\pm 1}] \xrightarrow{x-1} k[x^{\pm 1}] \to 0,$$

is a free resolution of k. Applying $\mathrm{Hom}_{k[x^{\pm 1}]}(-, k)$ yields the complex with trivial differentials: $0 \to k \to k \to 0$, and situated in cohomological degrees 0 and 1. Thus, $\mathrm{H}^0(\mathbb{Z}, k) = k = \mathrm{H}^1(\mathbb{Z}, k)$. Moreover, $\mathrm{H}^1(\mathbb{Z}, k) \cdot \mathrm{H}^1(\mathbb{Z}, k) = 0$, by degree considerations, so that the cohomology algebra is the exterior algebra $\wedge_k k$, where the generator for k sits in degree 1.

For $\mathbb{Z}^n$, one uses the Künneth formula (6.5) to calculate group cohomology:

$$\mathrm{H}^*(\mathbb{Z}^n, k) = \mathrm{H}^*(\mathbb{Z}, k)^{\otimes n} = \wedge_k k^n,$$

where the generators of k^n are all in (cohomological) degree 1. □

The next proposition computes the cohomology of cyclic p-groups. It turns out that one gets the same answer for all but one of them; the odd man out is the group of order two.

(7.3) PROPOSITION. *Let k be a field of characteristic p, and let $G = \mathbb{Z}/p^e\mathbb{Z}$, for some integer $e \geq 1$.*

(i) *When $p = 2$ and $e = 1$, $\mathrm{H}^*(G, k) = \mathrm{Sym}(ke_1^*)$, with $|e_1^*| = 1$.*
(ii) *Otherwise $\mathrm{H}^*(G, k) = \bigwedge(ke_1^*) \otimes_k \mathrm{Sym}(ke_2^*)$, with $|e_1^*| = 1$ and $|e_2^*| = 2$.*

PROOF. The group algebra of G is $k[x]/(x^{p^e} - 1)$, and its augmentation ideal is $(x - 1)$. Note that $x^{p^e} - 1 = (x - 1)^{p^e}$, so the substitution $y = x - 1$ presents the group algebra in the more psychologically comforting, to this commutative algebraist, form $k[y]/(y^{p^e})$. Write R for this algebra; it is a 0-dimensional hypersurface ring — the simplest example of a complete intersection — with socle generated by the element y^{p^e-1}. The R-module k has minimal free resolution

$$P: \cdots \to Re_3 \xrightarrow{y} Re_2 \xrightarrow{y^{p^e-1}} Re_1 \xrightarrow{y} Re_0 \to 0.$$

This is an elementary instance of the periodic minimal free resolution, of period 2, of the residue field of hypersurfaces constructed by Tate [1957]; see also [Eisenbud 1980]. Applying $\mathrm{Hom}_R(-, k)$ to the resolution above results in the complex

$$\mathrm{Hom}_R(P, k): 0 \to ke_0^* \xrightarrow{0} ke_1^* \xrightarrow{0} ke_2^* \xrightarrow{0} ke_3^* \xrightarrow{0} \cdots$$

Thus, one obtains $\mathrm{H}^n(G, k) = k$ for each integer $n \geq 0$.

Multiplicative structure. Next we calculate the products in group cohomology, and for this I propose to use compositions in $\mathrm{Hom}_R(P, P)$; see (5.1). More precisely: since P is a complex of free modules, the canonical map

$$\mathrm{Hom}_R(P, \varepsilon): \mathrm{Hom}_R(P, P) \to \mathrm{Hom}_R(P, k)$$

is an isomorphism in homology. Given two cycles in $\mathrm{Hom}_R(P, k)$, I will lift them to cycles in $\mathrm{Hom}_R(P, P)$, compose them there, and then push down the resultant cycle to $\mathrm{Hom}_R(P, k)$; this is their product.

For example, the cycle e_1^* of degree -1 lifts to the cycle α in $\mathrm{Hom}_R(P, P)$ given by

It is a lifting of e_1^* since $\varepsilon(\alpha(e_1)) = 1$, and a cycle since $\partial\alpha = -\alpha\partial$. Similarly, the cycle e_2^* lifts to the cycle β given by

This is all one needs in order to compute the entire cohomology rings of G. As indicated before, there are two cases to consider.

When $p = 2$ and $e = 1$, one has $y^{p^e-2} = 1$, so that $\varepsilon(\alpha^n(e_n)) = 1$ for each positive integer n. Therefore, $(e_1^*)^n = e_n^*$, and since the e_n^* form a basis for the graded k-vector space $\mathrm{H}^*(G, k)$, one obtains $\mathrm{H}^*(G, k) = k[e_1^*]$, as desired.

Suppose that either $p \geq 3$ or $e \geq 2$. In this case

$$\varepsilon(\alpha^{n+1}(e_{n+1})) = 0, \quad \varepsilon(\beta^n(e_{2n})) = 1, \quad \text{and} \quad \varepsilon(\alpha\beta^{n-1}(e_{2n-1})) = 1,$$

for each positive integer n. Passing to $\mathrm{Hom}_R(P, k)$, these relations translate to

$$(e_1^*)^{n+1} = 0, \quad (e_2^*)^n = e_{2n}^*, \quad e_1^*(e_2^*)^{n-1} = e_{2n-1}^*.$$

In particular, the homomorphism of k-algebras $k[e_1^*, e_2^*] \to \mathrm{H}^*(G, k)$ is surjective; here, $k[e_1^*, e_2^*]$ is the graded-polynomial algebra on e_1^* and e_2^*, that is to say, it is the tensor product of the exterior algebra on e_1^* and the usual polynomial algebra on e_2^*. This map is also injective: just compare Hilbert series.

This completes our calculation of the cohomology of cyclic p-groups.   □

(7.4) **Finitely generated abelian groups.** Let the characteristic of k be p, and let the group G be finitely generated and abelian. By the fundamental theorem of finitely generated abelian groups, there are integers n and $e_1, \ldots, e_m$, such that

$$G \cong \mathbb{Z}^n \oplus \frac{\mathbb{Z}}{(p^{e_1}\mathbb{Z})} \oplus \cdots \oplus \frac{\mathbb{Z}}{(p^{e_m}\mathbb{Z})} \oplus G'.$$

where G' is a finite abelian group whose order is coprime to p. By the Künneth formula (6.5), the group cohomology of G is the k-algebra

$$\mathrm{H}^*(\mathbb{Z}^n, k) \otimes_k \mathrm{H}^*(\mathbb{Z}/p^{e_1}\mathbb{Z}, k) \otimes_k \cdots \otimes_k \mathrm{H}^*(\mathbb{Z}/p^{e_m}\mathbb{Z}, k) \otimes_k \mathrm{H}^*(G', k).$$

Note that $\mathrm{H}^*(G', k) = k$, by Theorem (6.2); the remaining terms of the tensor product above are computed by propositions (7.2) and (7.3).

To give a flavour of the issues that may arise in the nonabelian case, I will calculate the cohomology of Σ_3. This gives me also an excuse to introduce an important tool in this subject:

(7.5) **The Lyndon–Hochschild–Serre spectral sequence.** Let G be a finite group and M a $k[G]$-module. Let N be a normal subgroup in G.

Via the canonical inclusion of k-algebras $k[N] \subseteq k[G]$, one can view M also as an $k[N]$-module. Since N is a normal subgroup, the k-subspace M^N of N-invariant elements of M is stable under multiplication by elements in G (check!) and hence it is a $k[G]$-submodule of M. Furthermore, $\mathrm{I}(N) \cdot M^N = 0$, so that M^N has the structure of a module over $k[G]/\mathrm{I}(N)k[G]$, that is to say, of a $k[G/N]$-module; see (1.1). It is clear from the definitions that $(M^N)^{G/N} = M^G$. In other words, one has an isomorphism of functors

$$\mathrm{Hom}_{k[G/N]}(k, \mathrm{Hom}_{k[N]}(k, -)) \cong \mathrm{Hom}_{k[G]}(k, -).$$

The functor on the left is the composition of two functors: $\mathrm{Hom}_{k[N]}(k,-)$ and $\mathrm{Hom}_{k[G/N]}(k,-)$. Thus standard homological algebra provides us with a spectral sequence that converges to its composition, that is to say, to $\mathrm{H}^*(G,M)$. In our case, the spectral sequence sits in the first quadrant and has second page

$$\mathrm{E}_2^{p,q} = \mathrm{H}^p(G/N, \mathrm{H}^q(N,M))$$

and differential

$$\partial_r^{p,q}: \mathrm{E}_r^{p,q} \to \mathrm{E}_r^{p+r,q-r+1} .$$

This is the *Lyndon–Hochschild–Serre spectral sequence* associated to N.

Here are two scenarios where the spectral sequence collapses.

(7.6) Suppose the characteristic of k does not divide $[G:N]$, the index of N in G. In this case, $\mathrm{H}^p(G/N, -) = 0$ for $p \geq 1$, by Maschke's theorem (6.2), so that the spectral sequence in (7.5) collapses to yield an isomorphism

$$\mathrm{H}^*(G,M) \cong \mathrm{H}^0(G/N, \mathrm{H}^*(N,M)) = \mathrm{H}^*(N,M)^{G/N}.$$

In particular, with $M = k$, one obtains that $\mathrm{H}^*(G,k) \cong \mathrm{H}^*(N,k)^{G/N}$; this isomorphism is compatible with the multiplicative structures. Note that the object on the right is the ring of invariants of the action of G/N on the group cohomology of N. Thus does invariant theory resurface in group cohomology.

(7.7) Suppose the characteristic of k does not divide $|N|$. Then $\mathrm{H}^q(N,M) = 0$ for $q \geq 1$, and once again the spectral sequence collapses to yield an isomorphism

$$\mathrm{H}^*(G,M) \cong \mathrm{H}^*(G/N, M^N).$$

The special case $M = k$ reads $\mathrm{H}^*(G,k) = \mathrm{H}^*(G/N, k)$.

As an application we calculate the cohomology of Σ_3:

(7.8) **The symmetric group on three elements.** In the notation in (4.8), set $N = \{1, b, b^2\}$; this is a normal subgroup of Σ_3, and the quotient group Σ_3/N is (isomorphic to) $\mathbb{Z}/2\mathbb{Z}$. We use the Hochschild–Serre spectral sequence generated by N in order to calculate the cohomology of Σ_3. There are three cases.

CASE (α). When $p \neq 2, 3$, Maschke's theorem (6.2) yields

$$\mathrm{H}^n(\Sigma_3, k) \cong \begin{cases} k & \text{if } n = 0, \\ 0 & \text{otherwise.} \end{cases}$$

CASE (γ). If $p = 2$, then

$$\mathrm{H}^*(\Sigma_3, k) = k[e_1^*], \quad \text{where } |e_1^*| = 1;$$

the polynomial ring on the variable e_1 of degree 1. Indeed, the order of N is 3, so (7.7) yields that $\mathrm{H}^*(\Sigma_3, k) = \mathrm{H}^*(\mathbb{Z}/2\mathbb{Z}, k)$. Proposition (7.3) does the rest.

CASE (β). Suppose that $p = 3$. One obtains from (7.6) that

$$H^*(\Sigma_3, k) = H^*(N, k)^{\mathbb{Z}/2\mathbb{Z}}.$$

The group N is cyclic of order 3, so its cohomology is $k[e_1^*, e_2^*]$, with $|e_1^*| = 1$ and $|e_2^*| = 2$; see Proposition (7.3). The next step is to compute the ring of invariants. The action of y, the generator of $\mathbb{Z}/2\mathbb{Z}$, on $H^*(N, k)$ is compatible with products, so it is determined entirely by its actions on e_1^* and on e_2^*. I claim that

$$y(e_1^*) = -e_1^* \quad \text{and} \quad y(e_2^*) = -e_2^*.$$

Using the description of $H^1(N, k)$ given in (6.3), it is easy to verify the assertion on the left; the one of the right is a little harder. Perhaps the best way to get this is to observe that the action of y on $H^*(N, k)$ is compatible with the *Bockstein* operator on cohomology and that this takes e_1^* to e_2^*; see [Evens 1961, (3.3)]. At any rate, given this, it is not hard to see that

$$H^*(\Sigma_3, k) = \bigwedge(ke_1^*e_2^*) \otimes_k \mathrm{Sym}(k(e_2^*)^2),$$

the tensor product of an exterior algebra on an element of degree 3 and a symmetric algebra on an element of degree 4.

Hopf algebras. In this article I have indicated at various points that much of the module theory over group algebras extends to Hopf algebras. I wrap up by mentioning a perfect generalization of Theorem (7.1), due to E. Friedlander and Suslin [1997]: If a finite-dimensional Hopf algebra H is cocommutative, its cohomology algebra $\mathrm{Ext}_H^*(k, k)$ is finitely generated.

References

[Alperin 1986] J. L. Alperin, *Local representation theory: Modular representations as an introduction to the local representation theory of finite groups*, Cambridge Studies in Advanced Mathematics **11**, Cambridge University Press, Cambridge, 1986.

[Avramov 1989] L. L. Avramov, "Modules of finite virtual projective dimension", *Invent. Math.* **96**:1 (1989), 71–101.

[Avramov and Buchweitz 2000] L. L. Avramov and R.-O. Buchweitz, "Support varieties and cohomology over complete intersections", *Invent. Math.* **142**:2 (2000), 285–318.

[Bass 1960] H. Bass, "Finitistic dimension and a homological generalization of semi-primary rings", *Trans. Amer. Math. Soc.* **95** (1960), 466–488.

[Benson 1991a] D. J. Benson, *Representations and cohomology, I: Basic representation theory of finite groups and associative algebras*, Cambridge Studies in Advanced Mathematics **30**, Cambridge University Press, Cambridge, 1991. Paperback reprint, 1998.

[Benson 1991b] D. J. Benson, *Representations and cohomology, II: Cohomology of groups and modules*, Cambridge Studies in Advanced Mathematics **31**, Cambridge University Press, Cambridge, 1991. Paperback reprint, 1998.

[Benson 1999] D. J. Benson, "Flat modules over group rings of finite groups", *Algebr. Represent. Theory* **2**:3 (1999), 287–294.

[Benson 2004] D. Benson, "Commutative algebra in the cohomology of groups", pp. 1–50 in *Trends in Commutative Algebra*, edited by L. Avramov et al., Math. Sci. Res. Inst. Publ. **51**, Cambridge University Press, New York, 2004.

[Benson and Kropholler 1995] D. J. Benson and P. H. Kropholler, "Cohomology of groups", pp. 917–950 in *Handbook of algebraic topology*, edited by I. M. James, North-Holland, Amsterdam, 1995.

[Bergman 1985] G. M. Bergman, "Everybody knows what a Hopf algebra is", pp. 25–48 in *Group actions on rings* (Brunswick, Maine, 1984), edited by S. Montgomery, Contemp. Math. **43**, Amer. Math. Soc., Providence, RI, 1985.

[Bøgvad and Halperin 1986] R. Bøgvad and S. Halperin, "On a conjecture of Roos", pp. 120–127 in *Algebra, algebraic topology and their interactions* (Stockholm, 1983), edited by J.-E. Roos, Lecture Notes in Math. **1183**, Springer, Berlin, 1986.

[Bourbaki 1980] N. Bourbaki, *Algèbre, X: Algèbre homologique*, Masson, Paris, 1980.

[Brown 1982] K. S. Brown, *Cohomology of groups*, Graduate Texts in Mathematics **87**, Springer, New York, 1982. Corrected reprint, 1994.

[Carlson 2001] J. F. Carlson, "Calculating group cohomology: tests for completion", *J. Symbolic Comput.* **31**:1-2 (2001), 229–242.

[Cartan and Eilenberg 1956] H. Cartan and S. Eilenberg, *Homological algebra*, Princeton University Press, Princeton, NJ, 1956.

[Eisenbud 1980] D. Eisenbud, "Homological algebra on a complete intersection, with an application to group representations", *Trans. Amer. Math. Soc.* **260**:1 (1980), 35–64.

[Evens 1961] L. Evens, "The cohomology ring of a finite group", *Trans. Amer. Math. Soc.* **101** (1961), 224–239.

[Evens 1991] L. Evens, *The cohomology of groups*, Oxford Mathematical Monographs, The Clarendon Press Oxford University Press, New York, 1991. Oxford Science Publications.

[Friedlander and Suslin 1997] E. M. Friedlander and A. Suslin, "Cohomology of finite group schemes over a field", *Invent. Math.* **127**:2 (1997), 209–270.

[Golod 1959] E. Golod, "The cohomology ring of a finite p-group", *Dokl. Akad. Nauk SSSR* **125** (1959), 703–706. In Russian.

[Gulliksen 1974] T. H. Gulliksen, "A change of ring theorem with applications to Poincaré series and intersection multiplicity", *Math. Scand.* **34** (1974), 167–183.

[Happel 1990] D. Happel, "Homological conjectures in representation theory of finite-dimensional algebras", preprint, Sherbrooke University, 1990. Available at http:// www.math.ntnu.no/~oyvinso/Nordfjordeid/Program/sheerbrooke.dvi.

[Herzog 1978] J. Herzog, "Ringe mit nur endlich vielen Isomorphieklassen von maximalen, unzerlegbaren Cohen–Macaulay–Moduln", *Math. Ann.* **233**:1 (1978), 21–34.

[Jans 1961] J. P. Jans, "Some generalizations of finite projective dimension", *Illinois J. Math.* **5** (1961), 334–344.

[Lang 2002] S. Lang, *Algebra*, 3rd ed., Graduate Texts in Math. **211**, Springer, New York, 2002.

[Leuschke and Wiegand ≥ 2004] G. Leuschke and R. Wiegand, "Local rings of bounded Cohen–Macaulay type", *Algebras and Representation Theory*. Available at http://www.leuschke.org/research/papers.html. To appear.

[Mac Lane 1978] S. Mac Lane, "Origins of the cohomology of groups", *Enseign. Math.* (2) **24**:1-2 (1978), 1–29.

[Mac Lane 1995] S. Mac Lane, *Homology*, Classics in Mathematics, Springer, Berlin, 1995. Reprint of the 1975 edition.

[Matsumura 1989] H. Matsumura, *Commutative ring theory*, Second ed., Cambridge Studies in Advanced Mathematics **8**, Cambridge University Press, Cambridge, 1989. Translated from the Japanese by M. Reid.

[Montgomery 1993] S. Montgomery, *Hopf algebras and their actions on rings*, CBMS Regional Conference Series in Mathematics **82**, Published for the Conference Board of the Mathematical Sciences, Washington, DC, 1993.

[Quillen 1971a] D. Quillen, "The spectrum of an equivariant cohomology ring, I", *Ann. of Math.* (2) **94** (1971), 549–572.

[Quillen 1971b] D. Quillen, "The spectrum of an equivariant cohomology ring, II", *Ann. of Math.* (2) **94** (1971), 573–602.

[Sjödin 1976] G. Sjödin, "A set of generators for $\mathrm{Ext}_R(k,k)$", *Math. Scand.* **38**:2 (1976), 199–210.

[Swan 1968] R. Swan, *Algebraic K-theory*, Lecture Notes Math **76**, Springer, Berlin, 1968.

[Sweedler 1969] M. E. Sweedler, *Hopf algebras*, Mathematics Lecture Note Series, Benjamin, New York, 1969.

[Tate 1957] J. Tate, "Homology of Noetherian rings and local rings", *Illinois J. Math.* **1** (1957), 14–27.

[Venkov 1959] B. B. Venkov, "Cohomology algebras for some classifying spaces", *Dokl. Akad. Nauk SSSR* **127** (1959), 943–944. in Russian.

[Yoshino 1990] Y. Yoshino, *Cohen–Macaulay modules over Cohen–Macaulay rings*, London Mathematical Society Lecture Note Series **146**, Cambridge University Press, Cambridge, 1990.

SRIKANTH IYENGAR
305 AVERY HALL
DEPARTMENT OF MATHEMATICS
UNIVERSITY OF NEBRASKA
LINCOLN, NE 68588
UNITED STATES
iyengar@math.unl.edu

Trends in Commutative Algebra
MSRI Publications
Volume 51, 2004

An Informal Introduction to Multiplier Ideals

MANUEL BLICKLE AND ROBERT LAZARSFELD

ABSTRACT. Multiplier ideals are associated with a complex variety and an ideal or ideal sheaf thereon, and satisfy certain vanishing theorems that have proved rich in applications, for example in local algebra. This article offers an introduction to the study of multiplier ideals, mainly adopting the geometric viewpoint.

CONTENTS

1. Introduction

Given a smooth complex variety X and an ideal (or ideal sheaf) $\mathfrak{a}$ on X, one can attach to $\mathfrak{a}$ a collection of *multiplier ideals* $\mathcal{J}(\mathfrak{a}^c)$ depending on a rational weighting parameter $c > 0$. These ideals, and the vanishing theorems they satisfy, have found many applications in recent years. In the global setting they have been used to study pluricanonical and other linear series on a projective variety [Demailly 1993; Angehrn and Siu 1995; Siu 1998; Ein and Lazarsfeld 1997; 1999; Demailly 1999]. More recently they have led to the discovery of some surprising uniform results in local algebra [Ein et al. 2001; 2003; 2004]. The purpose of these lectures is to give an easy-going and gentle introduction to the algebraically-oriented local side of the theory.

Multiplier ideals can be approached (and historically emerged) from three different viewpoints. In commutative algebra they were introduced and studied

Lazarsfeld's research was partially supported by NSF Grant DMS 0139713.

by Lipman [1993] (under the name "adjoint ideals", which now means something else), in connection with the Briançon–Skoda theorem. On the analytic side of the field, Nadel [1990] attached a multiplier ideal to any plurisubharmonic function, and proved a Kodaira-type vanishing theorem for them. (In fact, the "multiplier" in the name refers to their analytic construction; see Section 2.4.) This machine was developed and applied with great success by Demailly, Siu and others. Algebro-geometrically, the foundations were laid in passing by Esnault and Viehweg in connection with their work involving the Kawamata–Viehweg vanishing theorem. More systematic developments of the geometric theory were subsequently undertaken by Ein, Kawamata and Lazarsfeld. We will take the geometric approach here.

The present notes follow closely a short course on multiplier ideals given by Lazarsfeld at the Introductory Workshop for the Commutative Algebra Program at the MSRI in September 2002. The three main lectures were supplemented with a presentation by Blickle on multiplier ideals associated to monomial ideals (which appears here in Section 3). We have tried to preserve in this write-up the informal tone of these talks: thus we emphasize simplicity over generality in statements of results, and we present very few proofs. Our primary hope is to give the reader a feeling for what multiplier ideals are and how they are used. For a detailed development of the theory from an algebro-geometric perspective we refer to Part Three of the forthcoming book [Lazarsfeld 2004]. The analytic picture is covered in Demailly's lectures [2001].

We conclude this introduction by fixing the set-up in which we work and giving a brief preview of what is to come. Throughout these notes, X denotes a smooth affine variety over an algebraically closed field k of characteristic zero and $R = k[X]$ is the coordinate ring of X, so that $X = \operatorname{Spec} R$. We consider a nonzero ideal $\mathfrak{a} \subseteq k[X]$ (or equivalently a sheaf of ideals $\mathfrak{a} \subseteq \mathscr{O}_X$). Given a rational number $c \geq 0$ our plan is to define and study the multiplier ideal

$$\mathcal{J}(c \cdot \mathfrak{a}) = \mathcal{J}(\mathfrak{a}^c) \subseteq k[X].$$

As we proceed, there are two ideas to keep in mind. The first is that $\mathcal{J}(\mathfrak{a}^c)$ measures in a somewhat subtle manner the singularities of the divisor of a typical function f in $\mathfrak{a}$: for fixed c, "nastier" singularities are reflected by "deeper" multiplier ideals. Secondly, $\mathcal{J}(\mathfrak{a}^c)$ enjoys remarkable formal properties arising from the Kawamata–Viehweg–Nadel vanishing theorem. One can view the power of multiplier ideals as arising from the confluence of these facts.

The theory of multiplier ideals described here has striking parallels with the theory of tight closure developed by Hochster and Huneke in positive characteristic. Many of the uniform local results that can be established geometrically via multiplier ideals can also be proven (in more general algebraic settings) via tight closure. For some time the actual connections between the two theories were not well understood. However very recent work of Hara and Yoshida [2003] and Takagi [2004] has generalized tight closure theory to define a so called test ideal

$\tau(\mathfrak{a})$, which corresponds to the multiplier ideal $\mathcal{J}(\mathfrak{a})$ under reduction to positive characteristic. This provides a first big step towards identifying concretely the links between these theories.

Concerning the organization of these notes, we start in Section 2 by giving the basic definition and examples. Section 3 discusses in detail multiplier ideals of monomial ideals. Invariants arising from multiplier ideals, with some applications to uniform Artin–Rees numbers, are taken up in Section 4. Section 5 is devoted to a discussion of some basic results about multiplier ideals, notably Skoda's theorem and the restriction and subadditivity theorems. We consider asymptotic constructions in Section 6, with applications to uniform bounds for symbolic powers following [Ein et al. 2001].

We are grateful to Karen Smith for suggestions concerning these notes.

2. Definition and Examples

As just stated, X is a smooth affine variety of dimension n over an algebraically closed field of characteristic zero, and we fix an ideal $\mathfrak{a} \subseteq k[X]$ in the coordinate ring of X. Very little is lost by focusing on the case $X = \mathbb{C}^n$ of affine n-space over the complex numbers $\mathbb{C}$, so that $\mathfrak{a} \subseteq \mathbb{C}[x_1, \ldots, x_n]$ is an ideal in the polynomial ring in n variables.

2.1. Log resolution of an ideal. The starting point is to realize the ideal $\mathfrak{a}$ geometrically.

DEFINITION 2.1. A *log resolution* of an ideal sheaf $\mathfrak{a} \subseteq \mathcal{O}_X$ is a proper, birational map $\mu : Y \to X$ whose exceptional locus is a divisor E, satisfying the following conditions:

(i) Y is nonsingular.
(ii) $\mathfrak{a} \cdot \mathcal{O}_Y = \mu^{-1}\mathfrak{a} = \mathcal{O}_Y(-F)$, with $F = \sum r_i E_i$ an effective divisor.
(iii) $F + E$ has simple normal crossing support.

Recall that a (Weil) divisor $D = \sum \alpha_i D_i$ has simple normal crossing support if each of its irreducible components D_i is smooth, and if locally analytically one has coordinates $x_1, \ldots, x_n$ of Y such that $\mathrm{Supp}\, D = \sum D_i$ is defined by $x_1 \cdots \cdots x_a$ for some a between 1 and n. In other words, all the irreducible components of D are smooth and intersect transversally. The existence of a log resolution for any sheaf of ideals in any variety over a field of characteristic zero is essentially Hironaka's celebrated result [1964] on resolution of singularities. Nowadays there are more elementary constructions of such resolutions, for instance [Bierstone and Milman 1997; Encinas and Villamayor 2000; Paranjape 1999].

EXAMPLE 2.2. Let $X = \mathbb{A}^2 = \mathrm{Spec}\, k[x, y]$ and $\mathfrak{a} = (x^2, y^2)$. Blowing up the origin in $\mathbb{A}^2$ yields

$$Y = Bl_0(\mathbb{A}^2) \xrightarrow{\mu} \mathbb{A}^2 = X.$$

Clearly, Y is nonsingular. Computing on the chart for which the blowup μ is a map $\mathbb{A}^2 \to \mathbb{A}^2$ given by $(u, v) \mapsto (u, uv)$ shows that $\mathfrak{a} \cdot \mathscr{O}_Y = \mathscr{O}_Y(-2E)$. On the same chart we have $\mathfrak{a} \cdot \mathscr{O}_Y = (u^2, u^2 v^2) = (u^2)$ and $(u = 0)$ is the equation of the exceptional divisor. This resolution is illustrated in Figure 1, where we have drawn schematically the curves in $\mathbb{A}^2$ defined by typical k-linear combinations of generators of $\mathfrak{a}$, and the proper transforms of these curves on Y. Note that these proper transforms do not meet: this reflects the fact that $\mathfrak{a}$ has become principal on Y.

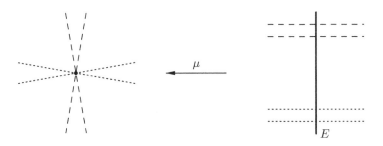

Figure 1. Log resolution of (x^2, y^2).

EXAMPLE 2.3. Now let $\mathfrak{a} = (x^3, y^2)$. Here a log resolution is constructed by the familiar sequence of three blowups used to resolve a cuspidal curve (Figure 2). We have $\mathfrak{a} \cdot \mathscr{O}_Y = \mathscr{O}_Y(-2E_1 - 3E_2 - 6E_3)$, where E_i is the exceptional divisor of the i-th blowup.

These examples illustrate the principle that a log resolution of an ideal $\mathfrak{a}$ is very close to being the same as a resolution of singularities of a divisor of a general function in $\mathfrak{a}$.

2.2. Definition of multiplier ideals. Besides a log resolution of $\mu : Y \to X$ of the ideal $\mathfrak{a}$, the other ingredient for defining the multiplier ideal is the relative canonical divisor

$$K_{Y/X} = K_Y - \mu^* K_X = \operatorname{div}(\det(\operatorname{Jac} \mu)).$$

It is unique as a divisor (and not just as a divisor class) if one requires its support to be contained in the exceptional locus of μ. Alternatively, $K_{Y/X}$ is the effective divisor defined by the vanishing of the determinant of the Jacobian of μ. The canonical divisor K_X is the class corresponding to the canonical line bundle ω_X. If X is smooth, ω_X is just the sheaf of top differential forms Ω_X^n on X.

The next proposition is extremely useful for basic computations of multiplier ideals; see [Hartshorne 1977, Exercise II.8.5].

PROPOSITION 2.4. *Let $Y = \operatorname{Bl}_Z X$, where Z is a smooth subvariety of the smooth variety X of codimension c. Then the relative canonical divisor $K_{Y/X}$ is $(c-1)E$, E being the exceptional divisor of the blowup.*

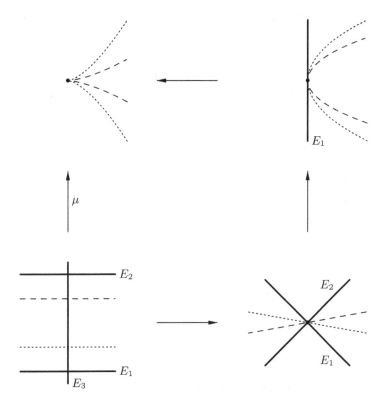

Figure 2. Log resolution of (x^3, y^2).

Now we can give a provisional definition of the multiplier ideal of an ideal $\mathfrak{a}$: it coincides in our setting with Lipman's construction [1993].

DEFINITION 2.5. Let $\mathfrak{a} \subseteq k[X]$ be an ideal. Fix a log resolution $\mu : Y \to X$ of $\mathfrak{a}$ such that $\mathfrak{a} \cdot \mathscr{O}_Y = \mathscr{O}_Y(-F)$, where $F = \sum r_i E_i$, and $K_{Y/X} = \sum b_i E_i$. The *multiplier ideal* of $\mathfrak{a}$ is

$$\mathscr{J}(\mathfrak{a}) = \mu_* \mathscr{O}_Y(K_{Y/X} - F)$$
$$= \{h \in k[X] \mid \operatorname{div}(\mu^* h) + K_{Y/X} - F \geq 0\}$$
$$= \{h \in k[X] \mid \operatorname{ord}_{E_i}(\mu^* h) \geq r_i - b_i \text{ for all } i\}.$$

(We will observe later that this is independent of the choice of resolution.)

The definition may seem at first blush a little mysterious. One way to motivate it is to note that $\mathscr{J}(\mathfrak{a})$ is the push-forward of a bundle that is very natural from the viewpoint of vanishing theorems. In fact, the bundle $\mathscr{O}_Y(-F)$ appearing above is (close to being) ample for the map μ. Therefore $K_{Y/X} - F$ has the shape to which Kodaira-type vanishing results will apply. In any event, the definition will justify itself before long through the properties of the ideals so defined.

EXERCISE 2.6. Use the fact that $\mu_*\omega_Y = \omega_X$ to show that $\mathcal{J}(\mathfrak{a})$ is indeed an ideal in $k[X]$.

EXERCISE 2.7. Show that the integral closure $\bar{\mathfrak{a}}$ of $\mathfrak{a}$ is equal to $\mu_*\mathcal{O}_Y(-F)$. Use this to conclude that $\mathfrak{a} \subseteq \bar{\mathfrak{a}} \subseteq \mathcal{J}(\mathfrak{a}) = \overline{\mathcal{J}(\mathfrak{a})}$. (Recall that the integral closure of an ideal $\mathfrak{a}$ consists of all elements f such that $v(f) \geq v(a)$ for all valuations v of $\mathcal{O}_X$.)

EXERCISE 2.8. Verify that for ideals $\mathfrak{a} \subseteq \mathfrak{b}$ one has $\mathcal{J}(\mathfrak{a}) \subseteq \mathcal{J}(\mathfrak{b})$. Use this and the previous exercise to show that $\mathcal{J}(\mathfrak{a}) = \mathcal{J}(\bar{\mathfrak{a}})$.

The above definition of the multiplier ideal is not general enough for the most interesting applications. As it turns out, allowing an additional rational (or real) parameter c considerably increases the power of the theory.

Note that a log resolution of an ideal $\mathfrak{a}$ is at the same time a log resolution of any integer power $\mathfrak{a}^n$ of that ideal. Thus we extend the last definition, using the same log resolution for every $c \geq 0$:

DEFINITION 2.9. For every rational number $c \geq 0$, the *multiplier ideal* of the ideal $\mathfrak{a}$ with exponent (or coefficient) c is

$$\mathcal{J}(\mathfrak{a}^c) = \mathcal{J}(c \cdot \mathfrak{a}) = \mu_*\mathcal{O}_Y(K_{Y/X} - \lfloor c \cdot F \rfloor)$$
$$= \{h \in k[X] \mid \mathrm{ord}_{E_i}(\mu^*h) \geq \lfloor cr_i \rfloor - b_i \text{ for all } i\},$$

where $\mu : Y \to X$ is a log resolution of $\mathfrak{a}$ such that $\mathfrak{a} \cdot \mathcal{O}_Y = \mathcal{O}_Y(-F)$.

Note that we do not assign any meaning to $\mathfrak{a}^c$ itself, only to $\mathcal{J}(\mathfrak{a}^c)$.[1] The round-down operation $\lfloor \cdot \rfloor$ applied to a $\mathbb{Q}$-divisor $D = \sum a_i D_i$ for distinct prime divisors D_i is just rounding down the coefficients. That is, $\lfloor D \rfloor = \sum \lfloor a_i \rfloor D_i$. The round up $\lceil D \rceil = -\lfloor -D \rfloor$ is defined analogously.

EXERCISE 2.10 (CAUTION WITH ROUNDING). Show that rounding does not in general commute with restriction or pullback.

EXERCISE 2.11. Let $\mathfrak{m}$ be the maximal ideal of a point $x \in X$. Show that

$$\mathcal{J}(\mathfrak{m}^c) = \begin{cases} \mathfrak{m}^{\lfloor c \rfloor + 1 - n} & \text{for } c \geq n = \dim X. \\ \mathcal{O}_X & \text{otherwise.} \end{cases}$$

EXAMPLE 2.12. Let $\mathfrak{a} = (x^2, y^2) \subseteq k[x,y]$. For the log resolution of $\mathfrak{a}$ as calculated above we have $K_{Y/X} = E$. Therefore,

$$\mathcal{J}(\mathfrak{a}^c) = \mu_*\big(\mathcal{O}_Y(E - \lfloor 2c \rfloor E)\big) = (x,y)^{\lfloor 2c \rfloor - 1}.$$

(In view of Exercise 2.8, this is a special case of Exercise 2.11.)

---

[1]There is a way to define the integral closure of an ideal $\mathfrak{a}^c$, for $c \geq 0$ rational, such that it is consistent with the definition of the multiplier ideal. For $c = p/q$ with positive integers p and q, set $f \in \mathfrak{a}^{p/q}$ if and only if $f^q \in \overline{\mathfrak{a}^p}$, where the bar denotes the integral closure.

EXAMPLE 2.13. Let $\mathfrak{a} = (x^2, y^3)$. In this case we computed a log resolution with $F = 2E_1 + 3E_2 + 6E_3$. Using the basic formula (Proposition 2.4) for the relative canonical divisor of a blowup along a smooth center, one computes $K_{Y/X} = E_1 + 2E_2 + 4E_3$. Therefore,

$$\mathcal{J}(\mathfrak{a}^c) = \mu_* \big(\mathcal{O}_Y (E_1 + 2E_2 + 4E_3 - \lfloor c(2E_1 + 3E_2 + 6E_3) \rfloor) \big)$$
$$= \mu_* \big(\mathcal{O}_Y ((1 - \lfloor 2c \rfloor)E_1 + (2 - \lfloor 3c \rfloor)E_2 + (4 - \lfloor 6c \rfloor)E_3) \big).$$

This computation shows that for $c < \frac{5}{6}$ the multiplier ideal is trivial, that is, $\mathcal{J}(\mathfrak{a}^c) = \mathcal{O}_X$. Furthermore, $\mathcal{J}(\mathfrak{a}^{5/6}) = (x, y)$. The next coefficient for which the multiplier ideal changes is $c = 1$. This behavior of multiplier ideals to be piecewise constant with discrete jumps is true in general and will be discussed in more detail later.

EXERCISE 2.14 (SMOOTH IDEALS). Suppose that $\mathfrak{q} \subseteq k[X]$ is the ideal of a smooth subvariety $Z \subseteq X$ of pure codimension e. Then

$$\mathcal{J}(\mathfrak{q}^l) = \mathfrak{q}^{l+1-e}.$$

(Blowing up X along Z yields a log resolution of $\mathfrak{q}$.) The case of fractional exponents is similar.

2.3. Two basic properties. The definitions of the previous subsection are justified by the fact that they lead to two fundamental results. The first is that the ideal $\mathcal{J}(\mathfrak{a}^c)$ constructed in Definition 2.9 is actually independent of the choice of resolution.

THEOREM 2.15. *If* $X_1 \xrightarrow{\mu_1} X$ *and* $X_2 \xrightarrow{\mu_2} X$ *are log resolutions of the ideal* $\mathfrak{a} \subseteq \mathcal{O}_X$ *such that* $\mathfrak{a}\mathcal{O}_{X_i} = \mathcal{O}_{X_i}(-F_i)$, *then*

$$\mu_{1_*} \big(\mathcal{O}_{X_1}(K_{X_1/X} - \lfloor c \cdot F_1 \rfloor) \big) = \mu_{2_*} \big(\mathcal{O}_{X_2}(K_{X_2/X} - \lfloor c \cdot F_2 \rfloor) \big).$$

As one would expect, the proof involves dominating μ_1 and μ_2 by a third resolution. It is during this argument that it becomes important to know that F_1 and F_2 have normal crossing support. See [Lazarsfeld 2004, Chapter 9].

EXERCISE 2.16. By contrast, give an example to show that if c is nonintegral, the ideal $\mu_*(-\lfloor cF \rfloor)$ may indeed depend on the log resolution μ.

The second fundamental fact is a vanishing theorem for the sheaves computing multiplier ideals.

THEOREM 2.17 (LOCAL VANISHING THEOREM). *Consider an ideal* $\mathfrak{a} \subseteq k[X]$ *as above, and let* $\mu : Y \to X$ *be a log resolution of* $\mathfrak{a}$ *with* $\mathfrak{a} \cdot \mathcal{O}_Y = \mathcal{O}_Y(-F)$. *Then*

$$R^i \mu_* \mathcal{O}_Y (K_{Y/X} - \lfloor cF \rfloor) = 0$$

for all $i > 0$ *and* $c > 0$.

This leads one to expect that the multiplier ideal, being the zeroth derived image of $\mathscr{O}_Y(K_{Y/X} - \lfloor cF \rfloor)$ under μ_*, will display particularly good cohomological properties.

Theorem 2.17 is a special case of the Kawamata–Viehweg vanishing theorem for a mapping; see [Lazarsfeld 2004, Chapter 9]. It is the essential fact underlying all the applications of multiplier ideals appearing in this article. When c is a natural number, the result can be seen as a slight generalization of the classical Grauert–Riemenschneider Vanishing Theorem. However, as we shall see, it is precisely the possibility of working with nonintegral c that opens the door to applications of a nonclassical nature.

2.4. Analytic construction of multiplier ideals. We sketch briefly the analytic construction of multiplier ideals. Let X be a smooth complex affine variety, and $\mathfrak{a} \subseteq \mathbb{C}[X]$ an ideal. Choose generators $g_1, \ldots, g_p \in \mathfrak{a}$. Then

$$\mathscr{J}(\mathfrak{a}^c)^{\mathrm{an}} =_{\mathrm{locally}} \left\{ h \text{ holomorphic} \;\middle|\; \frac{|h|^2}{\left(\sum |g_i|^2 \right)^c} \text{ is locally integrable} \right\}.$$

In other words, the analytic ideal associated to $\mathscr{J}(\mathfrak{a}^c)$ arises as a sheaf of "multipliers". See [Demailly 1999, (5.9)] or [Lazarsfeld 2004, Chapter 9.3.D] for the proof. In brief the idea is to show that both the algebraic and the analytic definitions lead to ideals that transform the same way under birational maps. This reduces one to the situation where $\mathfrak{a}$ is the principal ideal generated by a single monomial in local coordinates. Here the stated equality can be checked by an explicit calculation.

2.5. Multiplier ideals via tight closure. As hinted at in the introduction, there is an intriguing parallel between effective results in local algebra obtained via multiplier ideals on the one hand and tight closure methods on the other. Almost all the results we will discuss in these notes are of this kind: there are tight closure versions of the Briançon–Skoda theorem, the uniform Artin–Rees lemma and even of the result on symbolic powers that we present as an application of the asymptotic multiplier ideals in Section 6.4. (For these tight closure analogues see [Hochster and Huneke 1990], [Huneke 1992] and [Hochster and Huneke 2002], respectively.) There is little understanding for why such different techniques (characteristic zero, analytic in origin vs. positive characteristic) seem to be tailor-made to prove the same results.

Recently, Hara and Yoshida [2003] and Takagi [Takagi 2004; Takagi and Watanabe 2004; Hara and Takagi 2002; Takagi 2003] strengthened this parallel by constructing multiplier-like ideals using methods modelled after tight closure theory. Their construction builds on earlier work of Smith [2000] and Hara [2001], who had established a connection between the multiplier ideal associated to the unit ideal (1) on certain singular varieties with the so-called test ideal in tight closure. The setting of the work of Hara and Yoshida is a regular[2] local ring R of positive characteristic p. For simplicity one might again assume R is the local ring of a point in $\mathbb{A}^n$. Just as with multiplier ideals, one assigns to an ideal $\mathfrak{a} \subseteq R$ and a rational parameter $c \geq 0$, the *test ideal*

$$\tau(\mathfrak{a}^c) = \{ h \in R \mid hI^{*\mathfrak{a}^c} \subseteq I \text{ for all ideals } I \}.$$

Here $I^{*\mathfrak{a}^c}$ denotes the $\mathfrak{a}^c$–tight closure of an ideal, specifically introduced for the purpose of constructing these test ideals $\tau(\mathfrak{a}^c)$.[3] The properties the test ideals enjoy are strikingly similar to those of the multiplier ideal in characteristic zero: For example the Restriction Theorem (Theorem 5.8) and Subadditivity (Theorem 5.10) hold. What makes the test ideal a true analog of the multiplier ideal is that under the process of reduction to positive characteristic the multiplier ideal $\mathcal{J}(\mathfrak{a}^c)$ corresponds to the test ideal $\tau(\mathfrak{a}^c)$, or more precisely to the test ideal of the reduction mod p of $\mathfrak{a}^c$ (for $p \gg 0$).

3. The Multiplier Ideal of Monomial Ideals

Although multiplier ideals enjoy excellent formal properties, they are hard to compute in general. An important exception is the class of monomial ideals, whose multiplier ideals are described by a simple combinatorial formula established by Howald [2001]. By way of illustration we discuss this result in detail.

To state the result let $\mathfrak{a} \subseteq k[x_1, \ldots, x_n]$ be a monomial ideal, that is, an ideal generated by monomials of the form $x^m = x_1^{m_1} \cdot \cdots \cdot x_n^{m_n}$ for $m \in \mathbb{Z}^n \subseteq \mathbb{R}^n$. In this way we can identify a monomial ideal $\mathfrak{a}$ of $k[x_1, \ldots, x_n]$ with the set of exponents (contained in $\mathbb{Z}^n$) of the monomials in $\mathfrak{a}$. The convex hull of this set in $\mathbb{R}^n = \mathbb{Z}^n \otimes \mathbb{R}$ is called the *Newton polytope* of $\mathfrak{a}$ and it is denoted by $\mathrm{Newt}(\mathfrak{a})$. Now Howald's result states:

THEOREM 3.1. *Let $\mathfrak{a} \subseteq k[x_1, \ldots, x_n]$ be a monomial ideal. Then for every $c > 0$,*

$$\mathcal{J}(\mathfrak{a}^c) = \langle x^m \mid m + (1, \ldots, 1) \in \text{interior of } c \cdot \mathrm{Newt}(\mathfrak{a}) \rangle$$

For example, the picture of the Newton polytope of the monomial ideal $\mathfrak{a} = (x^4, xy^2, y^4)$ in Figure 3 shows, using Howald's result, that $\mathcal{J}(\mathfrak{a}) = (x^2, xy, y^2)$. Note that even though $(0, 1) + (1, 1)$ lies in the Newton polytope $\mathrm{Newt}(\mathfrak{a})$ it does not lie in the interior. Therefore, the monomial y corresponding to $(0, 1)$ does *not* lie in the multiplier ideal $\mathcal{J}(\mathfrak{a})$. But for all $c < 1$, clearly $y \in \mathcal{J}(\mathfrak{a}^c)$.

To pave the way for clean proofs we need to formalize our setup slightly and recall some results from toric geometry.

---

[2]One feature of their theory is that there is no reference to resolutions of singularities. As a consequence no restriction on the singularity of R arises, whereas for multiplier ideals at least some sort of $\mathbb{Q}$–Gorenstein assumption is needed.

[3]Similarly as for tight closure, $x \in I^{*\mathfrak{a}^c}$ if there is a $h \neq 0$ such that for all $q = p^e$ one has $hx^q \mathfrak{a}^{\lceil qc \rceil} \subseteq I^{[q]}$. Note that $I^{[q]}$ denotes the ideal generated by all q-th powers of the elements of I, whereas $\mathfrak{a}^{\lceil qc \rceil}$ is the usual $\lceil qc \rceil$-th power of $\mathfrak{a}$.

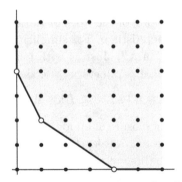

Figure 3. Newton polytope of (x^4, xy^2, y^4).

The ring $k[X] = k[x_1, \ldots, x_n]$ carries a natural $\mathbb{Z}^n$-grading that assigns to a monomial $x^m = x_1^{m_1} \cdot \cdots \cdot x_n^{m_n}$ the degree $m \in \mathbb{Z}^n$. Equivalently, the n-dimensional torus

$$T^n = \operatorname{Spec} k[x_1^{\pm 1}, \ldots, x_n^{\pm 1}] \cong (k^*)^n$$

acts on $k[X]$ via $\lambda \cdot x^m = \lambda^m x^m$ for $\lambda \in (k^*)^n$. In terms of the varieties this means that $X = \mathbb{A}^n$ contains the torus T^n as a dense open subset, and the action of T^n on itself naturally extends to an action of T^n on all of X. Under this action, the torus fixed ($= \mathbb{Z}^n$-graded) ideals are precisely the monomial ideals. We denote the lattice $\mathbb{Z}^n$ in which the grading takes place by M. It is just the lattice of the exponents of the Laurent monomials of $k[T^n]$.

As indicated above, the Newton polytope $\operatorname{Newt}(\mathfrak{a})$ of a monomial ideal $\mathfrak{a}$ is the convex hull in $M_{\mathbb{R}} = M \otimes_{\mathbb{Z}} \mathbb{R}$ of the set $\{m \in M \mid x^m \in \mathfrak{a}\}$. The Newton polytope of a principal ideal (x^v) is just the positive orthant in $M_{\mathbb{R}}$ shifted by v. In general, the Newton polytope of any ideal is an unbounded region contained in the first orthant. With every point v the Newton polytope also contains the first orthant shifted by v.

EXERCISE 3.2. Let $\mathfrak{a}$ be a monomial ideal in $k[x_1, \ldots, x_n]$. The lattice points (viewed as exponents) in the Newton polytope $\operatorname{Newt}(\mathfrak{a})$ of $\mathfrak{a}$ define an ideal $\bar{\mathfrak{a}} \supseteq \mathfrak{a}$. Show that $\bar{\mathfrak{a}}$ is the integral closure of $\mathfrak{a}$ (see [Fulton 1993]).

The fact that $X = \mathbb{A}^n$ contains the torus T^n as a dense open set such that the action of T^n on itself extends to an action on X as just described makes it a *toric variety*, by definition. The language of toric varieties is the most natural to phrase and prove Howald's result (and generalize it — see [Blickle 2004]). To set this up completely would take us far afield, so we choose a more direct approach using a bare minimum of toric geometry.

A first fact we have to take without proof from the theory of toric varieties is that log resolutions of torus fixed ideals of $k[X]$ exist in the category of toric varieties. (To be precise, a toric variety comes with the datum of the torus embedding $T^n \subseteq X$. Maps of toric varieties must preserve the torus action.)

THEOREM 3.3. *Let* $\mathfrak{a} \subseteq k[x_1, \ldots, x_n]$ *be a monomial ideal. Then there is a log resolution* $\mu : Y \to X$ *of* $\mathfrak{a}$ *such that* μ *is a map of toric varieties and consequently* $\mathfrak{a} \cdot \mathscr{O}_Y = \mathscr{O}_Y(-F)$ *is such that* F *is fixed by the torus action on* Y.

INDICATION OF PROOF. This follows from the theory of toric varieties. First one takes the normalized blowup of $\mathfrak{a}$, which is a (possibly singular) toric variety since $\mathfrak{a}$ was a torus-invariant ideal. Then one torically resolves the singularities of the resulting variety as described in [Fulton 1993]. This is a much easier task than resolution of singularities in general. It comes down to a purely combinatorial procedure.

An alternative proof could use Encinas and Villamayor's [2000] equivariant resolution of singularities. They give an algorithmic procedure of constructing a log resolution of $\mathfrak{a}$ such that the torus action is preserved — that is, by only blowing up along torus fixed centers.   □

Toric Divisors. A toric variety X has a finite set of torus-fixed prime (Weil) divisors. Indeed, since an arbitrary torus fixed prime divisor cannot meet the torus (T^n acts transitively on itself and is dense in X), it has to lie in the boundary $Y - T^n$, which is a variety of dimension at most $n - 1$ and thus can only contain finitely many components of dimension $n - 1$. Furthermore, these torus fixed prime divisors $E_1, \ldots, E_r$ generate the lattice of all torus fixed divisors, which we shall denote by L^X. We denote the sum of all torus-invariant prime divisors $E_1 + \cdots + E_r$ by 1_X.

The torus-invariant rational functions of a toric variety are just the Laurent monomials $x_1^{m_1} \cdot \cdots \cdot x_n^{m_n} \in k[T^n]$. For the toric variety $X = \mathbb{A}^n$ one clearly can identify M, the lattice of exponents, with L^X by sending m to $\operatorname{div} x^m$. In general this map will not be surjective and its image is precisely the set of torus-invariant Cartier divisors. We note the following easy lemma, which will nevertheless play an important role in our proof of Theorem 3.1. It makes precise the idea that a log resolution of a monomial ideal $\mathfrak{a}$ corresponds to turning its Newton polytope $\operatorname{Newt}(\mathfrak{a}) \subseteq M_{\mathbb{R}}$ into a translate of the first orthant in $L_{\mathbb{R}}^X$.

LEMMA 3.4. *Let* $\mu : Y \to X = \operatorname{Spec} k[x_1, \ldots, x_n]$ *be a toric resolution of the monomial ideal* $\mathfrak{a} \subseteq k[x_1, \ldots, x_n]$ *such that* $\mathfrak{a} \cdot \mathscr{O}_Y = \mathscr{O}_Y(-F)$. *For* $m \in M$ *we have*

$$c \cdot m \in c' \operatorname{Newt}(\mathfrak{a}) \iff c \cdot \mu^* \operatorname{div} x^m \geq c' \cdot F$$

for all rational $c, c' > 0$.

PROOF. We first show the case $c = c' = 1$. Assume that $m \in \operatorname{Newt}(\mathfrak{a})$. By Exercise 3.2, this is equivalent to $x^m \in \bar{\mathfrak{a}}$, the integral closure of $\mathfrak{a}$. Since, by Exercise 2.7, $\bar{\mathfrak{a}} = \mu_* \mathscr{O}_Y(-F)$ it follows that $x^m \in \bar{\mathfrak{a}}$ if and only if $\mu^* x^m \in \mathscr{O}_Y(-F)$. This, finally, is equivalent to $\mu^*(\operatorname{div} x^m) \geq F$.

For the general case, express c and c' as integer fractions. Then reduce to the previous case by clearing denominators and noticing that $a \operatorname{Newt}(\mathfrak{a}) = \operatorname{Newt}(\mathfrak{a}^a)$ if a is an integer.   □

The canonical divisor. As a further ingredient for computing the multiplier ideal we need an understanding of the canonical divisor (class) of a toric variety.

LEMMA 3.5. *Let X be a (smooth) toric variety and let $E_1, \ldots, E_r$ denote the collection of all torus-invariant prime Weil divisors. Then the canonical divisor is $K_Y = -\sum E_i = -1_X$.*

We leave the proof as an exercise or alternatively refer to [Fulton 1993] or [Danilov 1978] for this basic result. We verify it for $X = \mathbb{A}^n$. In this case $E_i = (x_i = 0)$ for $i = 1, \ldots, n$ are the torus-invariant divisors and K_X is represented by the divisor of the T^n-invariant rational n-form $dx_1/x_1 \wedge \cdots \wedge dx_n/x_n$, which is $-(E_1 + \cdots + E_n)$. As a consequence of the last lemma we get the following lemma.

LEMMA 3.6. *Let $\mu : Y \to X = \mathbb{A}^n$ be a birational map of (smooth) toric varieties. Then $K_{Y/X} = \mu^* 1_X - 1_Y$ and the support of $\mu^* 1_X$ is equal to the support of 1_Y.*

PROOF. As the strict transform of a torus-invariant divisor on X is a torus-invariant divisor on Y it follows that $\mu^* 1_X - 1_Y$ is supported on the exceptional locus of μ. Since -1_X represents the canonical class K_X and likewise for Y, the first assertion follows from the definition of $K_{Y/X}$. Since $\mu^* 1_X$ is torus-invariant, its support is included in 1_Y. Since μ is an isomorphism over the torus $T^n \subseteq X$ it follows that $\mu^{-1}(1_X) \supseteq 1_Y$, which implies the second assertion.   □

EXERCISE 3.7. This exercise shows how to avoid taking Lemma 3.5 on faith but instead using a result of Russel Goward [2002] that states that a log resolution of a monomial ideal can be obtained by a sequence of monomial blowups.

A *monomial blowup* $Y = Bl_Z(Y)$ of $\mathbb{A}^n$ is the blowing up of $\mathbb{A}^n$ at the intersection Z of some of the coordinate hyperplanes $E_i = (x_i = 0)$ of $\mathbb{A}^n$.

For such a monomial blowup $\mu : Y = Bl_Z(X) \to X \cong \mathbb{A}^n$, show that Y is a smooth toric variety canonically covered by $\mathrm{codim}(Z, X)$ many $\mathbb{A}^n$ patches. Show that $1_Y = E_1 + \cdots + E_n + E$, where E is the exceptional divisor of μ. Via a direct calculation verify the assertions of the last two lemmata for Y.

Since a monomial blowup is canonically covered by affine spaces, one can repeat the process in a *sequence of monomial blowups*. Using Goward's result, show directly that a monomial ideal has a toric log resolution $\mu : Y \to \mathbb{A}$ with the properties stated in Lemma 3.6.

We are now ready to wrap up the Proof of Theorem 3.1. By the existence of a toric (or equivariant) log resolution of a monomial ideal $\mathfrak{a}$, it follows immediately that the multiplier ideal $\mathcal{J}(\mathfrak{a}^c)$ is also generated by monomials. Thus, in order to determine $\mathcal{J}(\mathfrak{a}^c)$, it is enough to decide which monomials x^m lie in $\mathcal{J}(\mathfrak{a}^c)$. With our preparations this is now an easy task.

PROOF OF THEOREM 3.1. As usual we denote $\mathrm{Spec}\, k[x_1, \ldots, x_n]$ by X and let $\mu : Y \to X$ be a toric log resolution of $\mathfrak{a}$ such that $\mathfrak{a} \cdot \mathcal{O}_Y = \mathcal{O}_Y(-F)$.

Abusing notation by identifying $\operatorname{div}(x_1 \cdots x_n) = 1_X \in L^X$ with $(1, \ldots, 1) \in M$, the condition of the theorem that $m + 1_X$ is in the interior of the Newton polytope $c \cdot \operatorname{Newt}(\mathfrak{a})$ is equivalent to

$$m + 1_X - \varepsilon 1_X \in c \operatorname{Newt}(\mathfrak{a})$$

for small enough rational $\varepsilon > 0$. By Lemma 3.4 this holds if and only if

$$\mu^* \operatorname{div} g + \mu^* 1_X - \varepsilon \mu^* 1_X \geq cF.$$

Using the formula $K_{Y/X} = \mu^* 1_X - 1_Y$ from Lemma 3.5 this is equivalent to

$$\mu^* \operatorname{div} g + K_{Y/X} + \lfloor 1_Y - \varepsilon \mu^* 1_X - cF \rfloor \geq 0$$

for sufficiently small $\varepsilon > 0$. Since by Lemma 3.6, $\mu^* 1_X$ is effective with the same support as 1_Y it follows that all coefficients appearing in $1_Y - \varepsilon \mu^* 1_X$ are very close to but strictly smaller than 1 for small $\varepsilon > 0$. Therefore, $\lfloor 1_Y - \varepsilon \mu^* 1_X - cF \rfloor = \lceil -cF \rceil = -\lfloor cF \rfloor$. Thus we can finish our chain of equivalences with

$$\mu^* \operatorname{div} g \geq -K_{Y/X} + \lfloor cF \rfloor,$$

which says nothing but that $g \in \mathcal{J}(\mathfrak{a}^c)$.   □

This formula for the multiplier ideal of a monomial ideal is applied in the next section to concretely compute certain invariants arising from multiplier ideals.

4. Invariants Arising from Multiplier Ideals and Applications

We keep the notation of a smooth affine variety X over an algebraically closed field of characteristic zero, and an ideal $\mathfrak{a} \subseteq k[X]$. In this section we use multiplier ideals to attach some invariants to $\mathfrak{a}$, and we study their influence on some algebraic questions.

4.1. The log canonical threshold. If $c > 0$ is very small, then $\mathcal{J}(\mathfrak{a}^c) = k[X]$. For large c, on the other hand, the multiplier ideal $\mathcal{J}(\mathfrak{a}^c)$ is clearly nontrivial. This leads one to define:

DEFINITION 4.1. The *log canonical threshold* of $\mathfrak{a}$ is the number

$$\operatorname{lct}(\mathfrak{a}) = \operatorname{lct}(X, \mathfrak{a}) = \inf \left\{ c > 0 \mid \mathcal{J}(\mathfrak{a}^c) \neq \mathcal{O}_X \right\}.$$

The following exercise shows that $\operatorname{lct}(\mathfrak{a})$ is a rational number, and that the infimum appearing in the definition is actually a minimum. Consequently, the log canonical threshold is just the smallest $c > 0$ such that $\mathcal{J}(\mathfrak{a}^c)$ is nontrivial.

EXERCISE 4.2. As usual, fixing notation of a log resolution $\mu : Y \to X$ with $\mathfrak{a} \cdot \mathcal{O}_Y = \sum r_i E_i$ and $K_{Y/X} = \sum b_i E_i$, show that

$$\operatorname{lct}(X, \mathfrak{a}) = \min \left\{ \frac{b_i + 1}{r_i} \right\}.$$

Recall the notions from singularity theory [Kollár 1997] in which a pair $(X, \mathfrak{a}^c)$ is called *log terminal* if and only if $b_i - cr_i + 1 > 0$ for all i. It is called *log canonical* if and only if $b_i - cr_i + 1 \geq 0$ for all i. The last exercise also shows that $(X, \mathfrak{a}^c)$ is log terminal if and only if the multiplier ideal $\mathcal{J}(\mathfrak{a}^c)$ is trivial.

EXAMPLE 4.3. Continuing previous examples, we observe that $\mathrm{lct}((x^2, y^2)) = 1$ and $\mathrm{lct}((x^2, y^3)) = \frac{5}{6}$.

EXAMPLE 4.4 (THE LOG CANONICAL THRESHOLD OF A MONOMIAL IDEAL). The formula for the multiplier ideal of a monomial ideal $\mathfrak{a}$ on $X = \mathrm{Spec}\, k[x_1, \ldots, x_n]$ shows that $\mathcal{J}(\mathfrak{a}^c)$ is trivial if and only if $1_X = (1, \ldots, 1)$ is in the interior of the Newton polytope $c\,\mathrm{Newt}(\mathfrak{a})$. This allows to compute the log canonical threshold of $\mathfrak{a}$: $\mathrm{lct}(\mathfrak{a})$ is the largest $t > 0$ such that $1_X \in t \cdot \mathrm{Newt}(\mathfrak{a})$.

EXAMPLE 4.5. As a special case of the previous example, take

$$\mathfrak{a} = (x_1^{a_1}, \ldots, x_n^{a_n}).$$

Then the Newton polytope is the subset of the first orthant consisting of points $(v_1, \ldots, v_n)$ satisfying $\sum v_i/a_i \geq 1$. Therefore $1_X \in t \cdot \mathrm{Newt}(\mathfrak{a})$ if and only if $\sum 1/a_i \geq t$. In particular, $\mathrm{lct}(\mathfrak{a}) = \sum 1/a_i$.

4.2. Jumping numbers. The log canonical threshold measures the triviality or nontriviality of a multiplier ideal. By using the full algebraic structure of these ideals, it is natural to see this threshold as merely the first of a sequence of invariants. These so-called jumping numbers were first considered (at least implicitly) in [Libgober 1983] and [Loeser and Vaquié 1990]. They are studied more systematically in [Ein et al. 2004].

We start with a lemma:

LEMMA 4.6. *For $\mathfrak{a} \subseteq \mathcal{O}_X$, there is an increasing discrete sequence of rational numbers*

$$0 = \xi_0 < \xi_1 < \xi_2 < \cdots$$

such that $\mathcal{J}(\mathfrak{a}^c)$ is constant for $\xi_i \leq c < \xi_{i+1}$ and $\mathcal{J}(\mathfrak{a}^{\xi_i}) \supsetneq \mathcal{J}(\mathfrak{a}^{\xi_{i+1}})$.

We leave the (easy) proof to the reader.

The $\xi_i = \xi_i(\mathfrak{a})$ are called the *jumping numbers* or *jumping coefficients* of $\mathfrak{a}$. Referring to the log resolution μ appearing in Example 4.2, note that the only candidates for jumping numbers are those c such that cr_i is an integer for some i. Clearly the first jumping number $\xi_1(\mathfrak{a})$ is the log canonical threshold $\mathrm{lct}(\mathfrak{a})$.

EXAMPLE 4.7 (JUMPING NUMBERS OF MONOMIAL IDEALS). Consider a monomial ideal $\mathfrak{a} \subseteq k[x_1, \ldots, x_n]$. For the multiplier ideal $\mathcal{J}(\mathfrak{a}^c)$ to jump at $c = \xi$ is equivalent to the condition that some monomial, say x^v, is in $\mathcal{J}(\mathfrak{a}^\xi)$ but not in $\mathcal{J}(\mathfrak{a}^{\xi - \varepsilon})$ for all $\varepsilon > 0$. Thus, the largest $\xi > 0$ such that $v + (1, \ldots, 1) \in \xi\,\mathrm{Newt}(\mathfrak{a})$ is a jumping number. Performing this construction for all $v \in \mathbb{N}^n$ one obtains all jumping numbers of $\mathfrak{a}$ (this uses the fact that the multiplier ideal of a monomial ideal is a monomial ideal).

EXERCISE 4.8. Consider again $\mathfrak{a} = (x_1^{a_1}, \ldots, x_n^{a_n})$. The jumping numbers of $\mathfrak{a}$ are precisely the rational numbers of the form

$$\frac{v_1 + 1}{a_1} + \cdots + \frac{v_n + 1}{a_n},$$

where $(v_1, \ldots, v_n)$ ranges over $\mathbb{N}^n$. But different vectors $(v_1, \ldots, v_n)$ may give the same jumping number.

It is instructive to picture the jumping numbers of an ideal graphically. The figure below, taken from [Ein et al. 2004], shows the jumping numbers of the two ideals (x^9, y^{10}) and (x^3, y^{30}): the exponents are chosen so that the two ideals have the same Samuel multiplicity, and so that the pictured jumping coefficients occur "with multiplicity one" (in a sense whose meaning we leave to the reader).

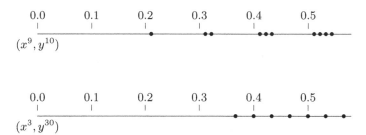

4.3. Jumping length.
Jumping numbers give rise to an additional invariant in the case of principal ideals.

LEMMA 4.9. *Let $f \in k[X]$ be a nonzero function. Then $\mathcal{J}(f) = (f)$ but $(f) \subsetneq \mathcal{J}(f^c)$ for $c < 1$. In other words, $\xi = 1$ is a jumping number of the principal ideal (f).*

Deferring the proof for a moment, we note that the lemma means that $\xi_l(f) = 1$ for some index l. We define $l = l(f)$ to be the *jumping length* of f. Thus $l(f)$ counts the number of jumping coefficients of (f) that are ≤ 1.

EXAMPLE 4.10. Let $f = x^4 + y^3 \in \mathbb{C}[x, y]$. One can show that f is sufficiently generic so that $\mathcal{J}(f^c) = \mathcal{J}((x^4, y^3)^c)$ provided that $c < 1$. Therefore the first few jumping numbers of f are

$$0 < \operatorname{lct}(f) = \tfrac{1}{4} + \tfrac{1}{3} < \tfrac{2}{4} + \tfrac{1}{3} < \tfrac{1}{4} + \tfrac{2}{3} < 1,$$

and $l(f) = 4$.

PROOF OF LEMMA 4.9. Let $\mu : Y \to X$ be a log resolution of (f) and denote the integral divisor $(f = 0)$ by $D = \sum a_i D_i$. Clearly, $\mathfrak{a} \cdot \mathcal{O}_Y = \mathcal{O}_Y(-\mu^* D)$ and $\mu^* D$ is also an integral divisor. Thus

$$\mathcal{J}(f) = \mu_* \mathcal{O}_Y(K_{Y/X} - \mu^* D) = \mu_*(\mathcal{O}_Y(K_{Y/X}) \otimes \mu^* \mathcal{O}_X(-D))$$
$$= \mathcal{O}_X \otimes \mathcal{O}_X(-D)$$
$$= (f).$$

On the other hand, choose a general point $x \in D_i$ on any of the components of $D = \operatorname{div}(f) = \sum a_i D_i$. Then μ is an isomorphism over x and consequently

$$\operatorname{ord}_{D_i}\big(\mathcal{J}(f^c)\big) < a_i \quad \text{for } 0 < c < 1.$$

Therefore $\mathcal{J}((f)^c) \subsetneq (f)$ whenever $c < 1$. $\qquad\qquad\square$

Finally, we note that the jumping length can be related to other invariants of the singularities of f:

PROPOSITION 4.11 [Ein et al. 2004]. *Assume the hypersurface defined by the vanishing of f has at worst an isolated singularity at $x \in X$. Then*

$$l(f) \le \tau(f, x) + 1,$$

where $\tau(f, x)$ is the Tjurina number of f at x, defined as the colength in $\mathcal{O}_{x,X}$ of $(f, \partial f/\partial z_1, \dots, \partial f/\partial z_n)$ for $z_1, \dots, z_n$ parameters around x.

4.4. Application to uniform Artin–Rees numbers. We next discuss a result relating jumping lengths to uniform Artin–Rees numbers of a principal ideal.

To set the stage, recall the statement of the Artin–Rees lemma in a simple setting:

THEOREM (ARTIN–REES). *Let $\mathfrak{b}$ be an ideal and f an element of $k[X]$. There exists an integer $k = k(f, \mathfrak{b})$ such that*

$$\mathfrak{b}^m \cap (f) \subseteq \mathfrak{b}^{m-k} \cdot (f)$$

for all $m \ge k$. In other words, if $fg \in \mathfrak{b}^m$ then $g \in \mathfrak{b}^{m-k}$.

Classically, k is allowed to depend both on $\mathfrak{b}$ and f. However, Huneke [1992] showed that in fact there is a single integer $k = k(f)$ that works simultaneously for all ideals $\mathfrak{b}$. Any such k is called a *uniform Artin–Rees number* of f. (Both the classical Artin–Rees Lemma and Huneke's theorem are valid in a much more general setting.)

The next result shows that the jumping length gives an effective estimate (of moderate size!) for uniform Artin–Rees numbers.

THEOREM 4.12 [Ein et al. 2004]. *As above, write $l(f)$ for the jumping length of f. Then the integer $k = l(f) \cdot \dim X$ is a uniform Artin–Rees number of f.*

If f defines a smooth hypersurface, its jumping length is 1 and it follows that $n = \dim X$ is a uniform Artin–Rees number in this case. (In fact, Huneke showed that $n - 1$ also works in this case.)

If f defines a hypersurface with only an isolated singular point $x \in X$, it follows from Proposition 4.11 and the theorem that $k = n \cdot (\tau(f, x) + 1)$ is a uniform Artin–Rees number. (One can show using the next lemma and observations of Huneke that $k = \tau(f, x) + n$ also works: see [Ein et al. 2004, § 3].)

The essential input to Theorem 4.12 is a statement involving consecutive jumping coefficients:

LEMMA 4.13. *Consider two consecutive jumping numbers*

$$\xi = \xi_i(f) < \xi_{i+1}(f) = \xi'$$

of f, and let $\mathfrak{b} \subseteq k[X]$ be any ideal. Then given a natural number $m > n = \dim X$, one has

$$\mathfrak{b}^m \cdot \mathcal{J}(f^\xi) \cap \mathcal{J}(f^{\xi'}) \subseteq \mathfrak{b}^{m-n} \cdot \mathcal{J}(f^{\xi'}).$$

We will deduce this from Skoda's theorem in the next section. In the meantime, we observe an immediate application:

PROOF OF THEOREM 4.12. We apply Lemma 4.13 repeatedly to successive jumping numbers in the chain of multiplier ideals:

$$k[X] = \mathcal{J}(f^0) \supsetneq \mathcal{J}(f^{\xi_1}) \supsetneq \mathcal{J}(f^{\xi_2}) \supsetneq \cdots \supsetneq \mathcal{J}(f^{\xi_l}) = \mathcal{J}(f) = (f).$$

After further intersection with (f) one finds

$$\begin{aligned}
\mathfrak{b}^m \cap (f) &\subseteq \mathfrak{b}^{m-n} \cdot \mathcal{J}(f^{\xi_1}) \cap (f) \\
&\subseteq \mathfrak{b}^{m-2n} \cdot \mathcal{J}(f^{\xi_2}) \cap (f) \subseteq \cdots \subseteq \mathfrak{b}^{m-ln}(f),
\end{aligned}$$

as required.   □

REMARK 4.14. When $\mathfrak{a} = (f)$ is a principal ideal, the jumping numbers of f are related to other invariants appearing in the literature. In particular, if f has an isolated singularity, suitable translates of the jumping coefficients appear in the Hodge-theoretically defined *spectrum* of f. See [Ein et al. 2004, § 5] for precise statements and references.

5. Further Local Properties of Multiplier Ideals

In this section we discuss some results involving the local behavior of multiplier ideals. We start with Skoda's theorem and some variants. Then we discuss the restriction and subadditivity theorems, which will be used in the next section.

5.1. Skoda's theorem. An important and early example of a uniform result in local algebra was established by Briançon and Skoda [1974] using analytic results of Skoda [1972]. In our language, Skoda's result is this:

THEOREM 5.1 (SKODA'S THEOREM, I). *Consider any ideal $\mathfrak{b} \subseteq k[X]$ with X smooth of dimension n. Then, for all $m \geq n$,*

$$\mathcal{J}(\mathfrak{b}^m) = \mathfrak{b} \cdot \mathcal{J}(\mathfrak{b}^{m-1}) = \cdots = \mathfrak{b}^{m+1-n} \cdot \mathcal{J}(\mathfrak{b}^{n-1}).$$

REMARK 5.2. The statement and proof in [Skoda 1972] have a more analytic flavor (see [Hochster 2004, pp. 125 and 126] in this volume for some more on this). In fact, using the analytic interpretation of multiplier ideals (Section 2.4) one sees that the analytic analogue of Theorem 5.1 is essentially equivalent to the following statement.

Suppose that $\mathfrak{b}$ is generated by $(g_1, \ldots, g_t)$, and that f is a holomorphic function such that

$$\int \frac{|f|^2}{(\sum |g_i|^2)^m} < \infty$$

for some $m \geq n = \dim X$. Then locally there exist holomorphic functions h_i such that $f = \sum h_i g_i$, and moreover each of the h_i satisfies the local integrability condition

$$\int \frac{|h_i|^2}{(\sum |g_i|^2)^{m-1}} < \infty.$$

(The hypothesis expresses the membership of f in $\mathcal{J}(\mathfrak{b}^m)^{\mathrm{an}}$ and the conclusion writes f as belonging to $\mathfrak{b}^{\mathrm{an}} \cdot \mathcal{J}(\mathfrak{b}^{m-1})^{\mathrm{an}}$.)

As a corollary of Skoda's theorem, one obtains the classical theorem of Briançon–Skoda.

COROLLARY 5.3 (BRIANÇON–SKODA). *With the notation as before,*

$$\overline{\mathfrak{b}^m} \subseteq \mathcal{J}(\mathfrak{b}^m) \subseteq \mathfrak{b}^{m+1-n}$$

where $^-$ denotes the integral closure and $n = \dim X$.

SKETCH OF PROOF OF THEOREM 5.1. The argument follows ideas of Teissier and Lipman. We choose generators $g_1, \ldots, g_k$ for the ideal $\mathfrak{b}$ and fix a log resolution $\mu : Y \to X$ of $\mathfrak{b}$ with $\mathfrak{b} \cdot \mathcal{O}_Y = \mathcal{O}_Y(-F)$. Write $g_i' = \mu^*(g_i) \in \Gamma(Y, \mathcal{O}_Y(-F))$ to define the surjective map

(5–1) $$\bigoplus_{i=0}^k \mathcal{O}_Y \to \mathcal{O}_Y(-F)$$

by sending $(x_1, \ldots, x_k)$ to $\sum x_i g_i'$. Tensoring this map with $\mathcal{O}_Y(K_{Y/X} - (m-1)F)$ yields the surjection

$$\bigoplus_{i=1}^k \mathcal{O}_Y(K_{Y/X} - (m-1)F) \xrightarrow{\varphi} \mathcal{O}_Y(K_{Y/X} - mF).$$

Further applying μ_* we get the map $\bigoplus_{i=0}^k \mathcal{J}(\mathfrak{b}^{m-1}) \xrightarrow{\mu_*\varphi} \mathcal{J}(\mathfrak{b}^m)$, which again sends a tuple $(y_1, \ldots, y_k)$ to $\sum y_i g_i$. Therefore, the image of $\mu_*(\varphi)$ is

$$\mathrm{Image}(\mu_*\varphi) = \mathfrak{b}\mathcal{J}(\mathfrak{b}^{m-1}) \subseteq \mathcal{J}(\mathfrak{b}^m).$$

What remains to show is that $\mu_*\varphi$ is surjective. For this consider the Koszul complex on the g_i' on Y that resolves the map in (5–1):

$$0 \to \mathcal{O}_Y((k-1)F) \to \bigoplus^k \mathcal{O}_Y((k-2)F) \to \cdots$$
$$\cdots \to \bigoplus^{\binom{k}{2}} \mathcal{O}_Y(F) \to \bigoplus^k \mathcal{O}_Y \to \mathcal{O}_Y(-F) \to 0.$$

As above, tensor through by $\mathscr{O}_Y(K_{Y/X} - (m{-}1)F)$ to get a resolution of φ. Local vanishing (Theorem 2.17) applies to the $m \geq n = \dim X$ terms on the right. Chasing through the sequence while taking direct images then gives the required surjectivity. See [Lazarsfeld 2004, Chapter 9] or [Ein and Lazarsfeld 1999] for details. $\square$

It will be useful to have a variant involving several ideals and fractional coefficients. For this we extend slightly the definition of multiplier ideals. Fix a sequence of ideals $\mathfrak{a}_1, \ldots, \mathfrak{a}_t$ and positive rational numbers $c_1, \ldots, c_t$. Then we define the multiplier ideal

$$\mathcal{J}(\mathfrak{a}_1^{c_1} \cdot \cdots \cdot \mathfrak{a}_t^{c_t})$$

starting with a log resolution $\mu : Y \to X$ of the product $\mathfrak{a}_1 \cdot \cdots \cdot \mathfrak{a}_t$. Since this is at the same time also a log resolution of each $\mathfrak{a}_i$ write $\mathfrak{a}_i \cdot \mathscr{O}_Y = \mathscr{O}_Y(-F_i)$ for simple normal crossing divisors F_i.

DEFINITION 5.4. With the notation as indicated, the *mixed multiplier ideal* is

$$\mathcal{J}(\mathfrak{a}_1^{c_1} \cdot \cdots \cdot \mathfrak{a}_t^{c_t}) = \mu_*(\mathscr{O}_Y(K_{Y/X} - \lfloor c_1 F_1 + \cdots + c_t F_t \rfloor)).$$

As before, this definition is independent of the chosen log resolution.

Once again we do not attempt to assign any meaning to the expression $\mathfrak{a}_1^{c_1} \cdots \mathfrak{a}_t^{c_t}$ in the argument of $\mathcal{J}$. This expression is meaningful a priori whenever all the c_i are positive integers and our definition is consistent with this prior meaning.

With this generalization of the concept of multiplier ideals we get the following variant of Skoda's theorem.

THEOREM 5.5 (SKODA'S THEOREM, II). *For every integer $c \geq n = \dim X$ and any $d > 0$ one has*

$$\mathcal{J}(\mathfrak{a}_1^c \cdot \mathfrak{a}_2^d) = \mathfrak{a}_1^{c-(n-1)} \mathcal{J}(\mathfrak{a}_1^{n-1} \cdot \mathfrak{a}_2^d).$$

The proof of this result is only a technical complication of the proof of the first version, Theorem 5.1. See [Lazarsfeld 2004, Chapter 9] for details.

We conclude by using Skoda's Theorem to prove (a slight generalization of) the Lemma 4.13 underlying the results on uniform Artin–Rees numbers in the previous section.

LEMMA 5.6. *Let $\mathfrak{a} \subseteq k[X]$ be an ideal and let $\xi < \xi'$ be consecutive jumping numbers of $\mathfrak{a}$. Then for $m > n$ we have*

$$\mathfrak{b}^m \cdot \mathcal{J}(\mathfrak{a}^\xi) \cap \mathcal{J}(\mathfrak{a}^{\xi'}) \subseteq \mathfrak{b}^{m-n} \cdot \mathcal{J}(\mathfrak{a}^{\xi'})$$

for all ideals $\mathfrak{b} \subseteq k[X]$.

PROOF. We first claim that

$$\mathfrak{b}^m \mathcal{J}(\mathfrak{a}^\xi) \cap \mathcal{J}(\mathfrak{a}^{\xi'}) \subseteq \mathcal{J}(\mathfrak{b}^{m-1} \cdot \mathfrak{a}^{\xi'}).$$

This is shown via a simple computation. In fact, to begin with one can replace ξ by $c \in [\xi, \xi')$ arbitrarily close to ξ' since this does not change the statement. Let

$\mu : Y \to X$ be a common log resolution of $\mathfrak{a}$ and $\mathfrak{b}$ such that $\mathfrak{a} \cdot \mathscr{O}_Y = \mathscr{O}_Y(-A)$ and $\mathfrak{b} \cdot \mathscr{O}_Y = \mathscr{O}_Y(-B)$. Let E be a prime divisor on Y and denote by a, b and e the coefficient of E in A, B and $K_{Y/X}$, respectively. Then f is in the left-hand side if and only if

$$\mathrm{ord}_E\, f \geq \max\big(-e + mb + \lfloor ca \rfloor, -e + \lfloor \xi' a \rfloor\big).$$

If $b = 0$ this implies that $\mathrm{ord}_E\, f \geq -e + (m-1)b + \lfloor \xi' a \rfloor$. If $b \neq 0$ then b is a positive integer ≥ 1. Since c is arbitrarily close to ξ' we get

$$\lfloor \xi' a \rfloor - b \leq \lfloor \xi' a \rfloor - 1 \leq \lfloor ca \rfloor.$$

Adding $-e + mb$ it follows that also in this case $\mathrm{ord}_E\, f \geq -e + (m-1)b + \lfloor \xi' a \rfloor$. Since this holds for all E it follows that $f \in \mathcal{J}(\mathfrak{b}^{m-1} \cdot \mathfrak{a}^{\xi'})$.

Now, using Theorem 5.5 we deduce

$$\mathcal{J}(\mathfrak{b}^{m-1} \cdot \mathfrak{a}^{\xi'}) \subseteq \mathfrak{b}^{m-n} \mathcal{J}(\mathfrak{b}^{n-1} \cdot \mathfrak{a}^{\xi'}) \subseteq \mathfrak{b}^{m-n} \mathcal{J}(\mathfrak{a}^{\xi'}).$$

Putting all the inclusions together, the lemma follows. □

EXERCISE 5.7. Let $\mathfrak{a} \subseteq k[X]$ be an ideal. Starting at $\dim X - 1$, the jumping numbers are periodic with period 1. That is, $\xi \geq \dim X - 1$ is a jumping number if and only if $\xi + 1$ is a jumping number.

5.2. Restriction theorem. The next result deals with restrictions of multiplier ideals. Consider a smooth subvariety $Y \subseteq X$ and an ideal $\mathfrak{b} \subseteq k[X]$ that does not vanish on Y. There are then two ways to get an ideal on Y. First, one can compute the multiplier ideal $\mathcal{J}(X, \mathfrak{b}^c)$ on X and then restrict it to Y. Or one can restrict $\mathfrak{b}$ to Y and then compute the multiplier ideal on Y of this restricted ideal. The Restriction Theorem — arguably the most important local property of multiplier ideals — says there is always an inclusion among these ideals on Y.

THEOREM 5.8 (RESTRICTION THEOREM). *Let $Y \subseteq X$ be a smooth subvariety of X and $\mathfrak{b}$ an ideal of $k[X]$ such that Y is not contained in the zero locus of $\mathfrak{b}$. Then*

$$\mathcal{J}\big(Y, (\mathfrak{b} \cdot k[Y])^c\big) \subseteq \mathcal{J}(X, \mathfrak{b}^c) \cdot k[Y].$$

One can think of the theorem as reflecting the principle that singularities can only get worse under restriction.

In the present setting, the result is due to Esnault and Viehweg [1992, Proposition 7.5]. When Y is a hypersurface, the statement is proved using the Local Vanishing Theorem, page 93. Since in any event a smooth subvariety is a local complete intersection, the general case then follows from this.

EXERCISE 5.9. Give an example where strict inclusion holds in the theorem.

5.3. Subadditivity theorem. We conclude with a result due to Demailly, Ein and Lazarsfeld [Demailly et al. 2000] concerning the multiplicative behavior of multiplier ideals. This subadditivity theorem will be used in the next section to obtain uniform bounds on symbolic powers of ideals.

THEOREM 5.10 (SUBADDITIVITY). *Let $\mathfrak{a}$ and $\mathfrak{b}$ be ideals in $k[X]$. Then, for all $c, d > 0$,*

$$\mathcal{J}(\mathfrak{a}^c \cdot \mathfrak{b}^d) \subseteq \mathcal{J}(\mathfrak{a}^c) \cdot \mathcal{J}(\mathfrak{b}^d).$$

In particular, for every positive integer m, $\mathcal{J}(\mathfrak{a}^{cm}) \subseteq \mathcal{J}(\mathfrak{a}^c)^m$.

SKETCH OF PROOF. The idea is to pull back the data to the product $X \times X$ and then to restrict to the diagonal Δ. Specifically, assume for simplicity that $c = d = 1$, and consider the product

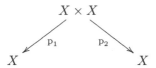

along with its projections as indicated. For log resolutions μ_1 and μ_2 of $\mathfrak{a}$ and $\mathfrak{b}$ respectively one can verify that $\mu_1 \times \mu_2$ is a log resolution of the ideal $p_1^{-1}(\mathfrak{a}) \cdot p_2^{-1}(\mathfrak{b})$ on $X \times X$. Using this one shows that

$$\mathcal{J}(X \times X, \, p_1^{-1}(\mathfrak{a}) \cdot p_2^{-1}(\mathfrak{b})) = p_1^{-1}\, \mathcal{J}(X, \mathfrak{a}) \cdot p_2^{-1}\, \mathcal{J}(X, \mathfrak{b}).$$

Now let $\Delta \subseteq X \times X$ be the diagonal. Apply the Restriction Theorem 5.8 with $Y = \Delta$ to conclude that, as required,

$$
\begin{aligned}
\mathcal{J}(X, \mathfrak{a} \cdot \mathfrak{b}) &= \mathcal{J}(\Delta, \, p_1^{-1}(\mathfrak{a}) \cdot p_2^{-1}(\mathfrak{b}) \cdot \mathscr{O}_\Delta) \\
&\subseteq \mathcal{J}(X \times X, \, p_1^{-1}(\mathfrak{a}) \cdot p_2^{-1}(\mathfrak{b})) \cdot \mathscr{O}_\Delta \\
&= \mathcal{J}(X, \mathfrak{a}) \cdot \mathcal{J}(X, \mathfrak{b}). \qquad \square
\end{aligned}
$$

6. Asymptotic Constructions

In many natural situations in geometry and algebra, one must confront rings or algebras that fail to be finitely generated. For example, if D is a nonample divisor on a projective variety V, the section ring $R(V, D) = \bigoplus \Gamma(V, \mathscr{O}_V(mD))$ is typically not finitely generated. Likewise, if $\mathfrak{q}$ is a radical ideal in some ring, the symbolic blow-up algebra $\bigoplus \mathfrak{q}^{(m)}$ likewise fails to be finitely generated in general. It is nonetheless possible to extend the theory of multiplier ideals to such settings. It turns out that there is finiteness built into the resulting multiplier ideals that may not be present in the underlying geometry or algebra. This has led to some of the most interesting applications of the theory.

In the geometric setting, the asymptotic constructions have been known for some time, but it was only with Siu's work [1998] on deformation invariance of plurigenera that their power became clear. Here we focus on an algebraic formulation of the theory from [Ein et al. 2001]. As before, we work with a smooth affine variety X defined over an algebraically closed field k of characteristic zero.

6.1. Graded systems of ideals. We start by defining certain collections of ideals, to which we will later attach multiplier ideals.

DEFINITION 6.1. A *graded system* or *graded family of ideals* is a family $\mathfrak{a}_\bullet = \{\mathfrak{a}_k\}_{k \in \mathbb{N}}$ of ideals in $k[X]$ such that

$$\mathfrak{a}_l \cdot \mathfrak{a}_m \subseteq \mathfrak{a}_{l+m}$$

for all $l, m \geq 1$. To avoid trivialities, we also assume that $\mathfrak{a}_k \neq (0)$ for $k \gg 1$.

The condition in the definition means that the direct sum

$$R(\mathfrak{a}_\bullet) \overset{\text{def}}{=} k[X] \oplus \mathfrak{a}_1 \oplus \mathfrak{a}_2 \oplus \cdots$$

naturally carries a graded $k[X]$-algebra structure and $R(\mathfrak{a}_\bullet)$ is called the *Rees algebra* of $\mathfrak{a}_\bullet$. In the interesting situations $R(\mathfrak{a}_\bullet)$ is not finitely generated, and it is here that the constructions of the present section give something new. One can view graded systems as local objects displaying complexities similar to those that arise from linear series on a projective variety V. If D is an effective divisor on V, the base ideals $\mathfrak{b}_k = \mathfrak{b}(|kD|) \subseteq \mathcal{O}_V$ form a graded family of ideal sheaves on V: this is the prototypical example.

EXAMPLE 6.2. We give several examples of graded systems.

(i) Let $\mathfrak{b} \subseteq k[X]$ be a fixed ideal, and set $\mathfrak{a}_k = \mathfrak{b}^k$. One should view the resulting graded system as a trivial example.

(ii) Let $Z \subseteq X$ be a reduced subvariety defined by the radical ideal $\mathfrak{q}$. The symbolic powers

$$\mathfrak{q}^{(k)} \overset{\text{def}}{=} \{f \in k[X] \mid \operatorname{ord}_z f \geq k \text{ for } z \in Z \text{ generic}\}$$

form a graded system.[4]

(iii) Let $<$ be a term order on $k[x_1, \ldots, x_n]$ and $\mathfrak{b}$ be an ideal. Then

$$\mathfrak{a}_k \overset{\text{def}}{=} \operatorname{in}_<(\mathfrak{b}^k)$$

defines a graded system of monomial ideals, where $\operatorname{in}_<(\mathfrak{b}^k)$ denotes the initial ideal with respect to the given term order.

EXAMPLE 6.3 (VALUATION IDEALS). Let ν be a $\mathbb{R}$-valued valuation centered on $k[X]$. Then the valuation ideals

$$\mathfrak{a}_k \overset{\text{def}}{=} \{f \in k[X] \mid \nu(f) \geq k\}$$

---

[4]When Z is reducible, we ask that the condition hold at a general point of each component. That this is equivalent to the usual algebraic definition is a theorem of Zariski and Nagata: see [Eisenbud 1995, Chapter 3].

form a graded family. Special cases of this construction are interesting even when $X = \mathbb{A}^2_{\mathbb{C}}$.

(i) Let $\eta : Y \to \mathbb{A}^2$ be a birational map with Y also smooth and let $E \subseteq Y$ be a prime divisor. Define the valuation $\nu(f) \stackrel{\text{def}}{=} \operatorname{ord}_E(f)$. Then

$$\mathfrak{a}_k \stackrel{\text{def}}{=} \mu_* \mathcal{O}_Y(-kE) = \{ f \in \mathcal{O}_X \mid \nu(f) = \operatorname{ord}_E(f) \geq k \}.$$

(ii) In $\mathbb{C}[x, y]$ put $\nu(x) = 1$ and $\nu(y) = 1/\sqrt{2}$. Then one gets a valuation by weighted degree. Here $\mathfrak{a}_k$ is the monomial ideal generated by the monomials $x^i y^j$ such that $i + j/\sqrt{2} \geq k$.

(iii) Given $f \in \mathbb{C}[x, y]$ define $\nu(f) = \operatorname{ord}_z(f(z, e^z - 1))$. This yields a valuation giving rise to the graded system

$$\mathfrak{a}_k \stackrel{\text{def}}{=} (x^k, \, y - P_{k-1}(x)),$$

where $P_{k-1}(x)$ is the $(k-1)$-st Taylor polynomial of $e^x - 1$. Note that the general element in $\mathfrak{a}_k$ defines a smooth curve in the plane.

REMARK 6.4. Except for Example 6.2(i), all these constructions give graded families $\mathfrak{a}_\bullet$ whose corresponding Rees algebra need not be finitely generated.

6.2. Asymptotic multiplier ideals. We now attach multiplier ideals $\mathcal{J}(\mathfrak{a}_\bullet^c)$ to a graded family $\mathfrak{a}_\bullet$ of ideals. The starting point is:

LEMMA 6.5. *Let $\mathfrak{a}_\bullet$ be a graded system of ideals on X, and fix a rational number $c > 0$. Then for $p \gg 0$ the multiplier ideals $\mathcal{J}(\mathfrak{a}_p^{c/p})$ all coincide.*

DEFINITION 6.6. Let $\mathfrak{a}_\bullet = \{\mathfrak{a}_k\}_{k \in \mathbb{N}}$ be a graded system of ideals on X. Given $c > 0$ we define the *asymptotic multiplier ideal* of $\mathfrak{a}_\bullet$ with exponent c to be the common ideal

$$\mathcal{J}(\mathfrak{a}_\bullet^c) \stackrel{\text{def}}{=} \mathcal{J}(\mathfrak{a}_p^{c/p})$$

for any sufficiently big $p \gg 0$.[5]

INDICATION OF PROOF OF LEMMA 6.5. We first claim that one has an inclusion of multiplier ideals $\mathcal{J}(\mathfrak{a}_p^{c/p}) \subseteq \mathcal{J}(\mathfrak{a}_{pq}^{c/pq})$ for all $p, q \geq 0$. Granting this, it follows from the Noetherian condition that the collection of ideals $\{\mathcal{J}(\mathfrak{a}_p^{c/p})\}_{p \geq 0}$ has a unique maximal element. This proves the lemma at least for sufficiently divisible p. (The statement for all $p \gg 0$ requires a little more work; see [Lazarsfeld 2004, Chapter 11].)

To verify the claim let $\mu : X' \longrightarrow X$ be a common log resolution of $\mathfrak{a}_p$ and $\mathfrak{a}_{pq}$ with $\mathfrak{a}_p \cdot \mathcal{O}_Y = \mathcal{O}_Y(-F_p)$ and $\mathfrak{a}_{pq} \cdot \mathcal{O}_Y = \mathcal{O}_Y(-F_{pq})$. Since the $\mathfrak{a}_k$ form a graded system one has $\mathfrak{a}_p^q \subseteq \mathfrak{a}_{pq}$ and therefore $-cqF_p \leq -cF_{pq}$. Thus, as claimed,

$$\mu_* \mathcal{O}_Y(K_{Y/X} - \lfloor \tfrac{cq}{pq} F_p \rfloor) \subseteq \mu_* \mathcal{O}_Y(K_{Y/X} - \lfloor \tfrac{c}{pq} F_{pq} \rfloor). \qquad \square$$

---

[5]In [Ein et al. 2001] and early versions of [Lazarsfeld 2004], one only dealt with the ideals $\mathcal{J}(\mathfrak{a}_\bullet^l)$ for integral l, which were written $\mathcal{J}(\|\mathfrak{a}_l\|)$.

REMARK 6.7. Lemma 6.5 shows that any information captured by the multiplier ideals $\mathcal{J}(\mathfrak{a}_p^{c/p})$ is present already for any one sufficiently large index p. It is in this sense that multiplier ideals have some finiteness built in that may not be present in the underlying graded system $\mathfrak{a}_\bullet$.

EXERCISE 6.8. We return to the graded systems in Example 6.3 coming from valuations on $\mathbb{A}^2$.

(ii) Here $\mathfrak{a}_k$ is the monomial ideal generated by $x^i y^j$ with $i + j/\sqrt{2} \geq k$, and $\mathcal{J}(\mathfrak{a}_\bullet^c)$ is the monomial ideal generated by all $x^i y^j$ with

$$(i+1) + \frac{(j+1)}{\sqrt{2}} > c.$$

(Compare with Theorem 3.1.)

(iii) Now take the valuation $\nu(f) = \text{ord}_z f(z, e^z - 1)$. Then

$$\mathcal{J}(\mathfrak{a}_\bullet^c) = \mathbb{C}[x, y]$$

for all $c > 0$. (Use the fact that each $\mathfrak{a}_k$ contains a smooth curve.)

6.3. Growth of graded systems. We now use the Subadditivity Theorem 5.10 to prove a result from [Ein et al. 2001] concerning the multiplicative behavior of graded families of ideals:

THEOREM 6.9. *Let $\mathfrak{a}_\bullet$ be a graded system of ideals and fix any $l \in \mathbb{N}$. Then*

$$\mathcal{J}(\mathfrak{a}_\bullet^l) = \mathcal{J}(\mathfrak{a}_{lp}^{1/p}) \quad \text{for } p \gg 0.$$

Moreover, for every $m \in \mathbb{N}$,

$$\mathfrak{a}_l^m \subseteq \mathfrak{a}_{lm} \subseteq \mathcal{J}(\mathfrak{a}_\bullet^{lm}) \subseteq \mathcal{J}(\mathfrak{a}_\bullet^l)^m.$$

In particular, if $\mathcal{J}(\mathfrak{a}_\bullet^l) \subseteq \mathfrak{b}$ for some natural number l and ideal $\mathfrak{b}$, then $\mathfrak{a}_{lm} \subseteq \mathfrak{b}^m$ for all m.

REMARK 6.10. The crucial point here is the containment $\mathcal{J}(\mathfrak{a}_\bullet^{lm}) \subseteq \mathcal{J}(\mathfrak{a}_\bullet^l)^m$: it shows that passing to multiplier ideals "reverses" the inclusion $\mathfrak{a}_l^m \subseteq \mathfrak{a}_{lm}$.

PROOF OF THEOREM 6.9. For the first statement, observe that if $p \gg 0$ then

$$\mathcal{J}(\mathfrak{a}_\bullet^l) = \mathcal{J}(\mathfrak{a}_p^{l/p}) = \mathcal{J}(\mathfrak{a}_{lp}^{l/lp}) = \mathcal{J}(\mathfrak{a}_{lp}^{1/p}),$$

where the second equality is obtained by taking lp in place of p as the large index in Lemma 6.5. For the containment $\mathfrak{a}_{lm} \subseteq \mathcal{J}(\mathfrak{a}_\bullet^{lm})$ it is then enough to prove that $\mathfrak{a}_{lm} \subseteq \mathcal{J}(\mathfrak{a}_{lmp}^{1/p})$. But we have $\mathfrak{a}_{lm} \subseteq \mathcal{J}(\mathfrak{a}_{lm})$ thanks to Exercise 2.7, while the inclusion $\mathcal{J}(\mathfrak{a}_{lm}) \subseteq \mathcal{J}(\mathfrak{a}_{lmp}^{1/p})$ was established during the proof of 6.5.

It remains only to prove that $\mathcal{J}(\mathfrak{a}_\bullet^{lm}) \subseteq \mathcal{J}(\mathfrak{a}_\bullet^l)^m$. To this end, fix $p \gg 0$. Then by the definition of asymptotic multiplier ideals and the Subadditivity Theorem one has, as required,

$$\mathcal{J}(\mathfrak{a}_\bullet^{lm}) = \mathcal{J}(\mathfrak{a}_p^{lm/p}) \subseteq \mathcal{J}(\mathfrak{a}_p^{l/p})^m = \mathcal{J}(\mathfrak{a}_\bullet^l)^m \qquad \square$$

EXAMPLE 6.11. The Theorem gives another explanation of the fact that the multiplier ideals associated to the graded system $\mathfrak{a}_\bullet$ from Example 6.3.(iii) are trivial. In fact, in this example the colength of $\mathfrak{a}_k$ in $\mathbb{C}[X]$ grows linearly in k. It follows from Theorem 6.9 that then $\mathcal{J}(\mathfrak{a}_\bullet^l) = (1)$ for all l.

EXERCISE 6.12. Let $\mathfrak{a}_k = \mathfrak{b}^k$ be the trivial graded family consisting of powers of a fixed ideal. Then $\mathcal{J}(\mathfrak{a}_\bullet^c) = \mathcal{J}(\mathfrak{b}^c)$ for all $c > 0$. So we do not get anything new in this case.

6.4. A comparison theorem for symbolic powers.

As a quick but surprising application of Theorem 6.9 we discuss a result due to Ein, Smith and Lazarsfeld [Ein et al. 2001] concerning symbolic powers of radical ideals.

Consider a reduced subvariety $Z \subseteq X$ defined by a radical ideal $\mathfrak{q} \subseteq k[X]$. Recall from Example 6.2(ii) that one can define the symbolic powers $\mathfrak{q}^{(k)}$ of $\mathfrak{q}$ as

$$\mathfrak{q}^{(k)} \overset{\text{def}}{=} \{ f \in \mathscr{O}_X \mid \operatorname{ord}_z f \geq k \text{ for } z \in Z \}.$$

Thus evidently $\mathfrak{q}^k \subseteq \mathfrak{q}^{(k)}$, and equality holds if Z is smooth. However, if Z is singular, the inclusion is strict in general:

EXAMPLE 6.13. Take $Z \subseteq \mathbb{C}^3$ to be the union of the three coordinate axes, defined by the ideal

$$\mathfrak{q} = (xy, yz, xz) \subseteq \mathbb{C}[x, y, z].$$

Then $xyz \in \mathfrak{q}^{(2)}$, since the union of the three coordinate planes has multiplicity 2 at a general point of Z. But $\mathfrak{q}^2$ is generated by monomials of degree 4, and so cannot contain xyz.

Swanson [2000] proved (in a much more general setting) that there exists an integer $k = k(Z)$ such that

$$\mathfrak{q}^{(km)} \subseteq \mathfrak{q}^m$$

for all $m \geq 0$. At first glance, one might be tempted to suppose that for very singular Z the coefficient $k(Z)$ will have to become quite large. The main result of [Ein et al. 2001] shows that this isn't the case, and that in fact one can take $k(Z) = \operatorname{codim} Z$:

THEOREM 6.14. *Assume that every irreducible component of Z has codimension at most e in X. Then*

$$\mathfrak{q}^{(em)} \subseteq \mathfrak{q}^m \quad \text{for all } m \geq 0.$$

In particular, $\mathfrak{q}^{(m \cdot \dim X)} \subseteq \mathfrak{q}^m$ for all radical ideals $\mathfrak{q} \subseteq k[X]$ and all $m \geq 0$.

EXAMPLE 6.15 (POINTS IN THE PLANE). Let $T \subseteq \mathbb{P}^2$ be a finite set, considered as a reduced scheme, and let $I \subseteq S = \mathbb{C}[x, y, z]$ be the homogeneous ideal of T. Suppose that $f \in S$ is a homogeneous form having multiplicity at least $2m$ at each of the points of T. Then $f \in I^m$. (Apply Theorem 6.14 to the homogeneous ideal I of T.) In spite of the classical nature of this statement, we do not know a direct elementary proof.

PROOF OF THEOREM 6.14. Applying Theorem 6.9 to the graded system $\mathfrak{a}_k = \mathfrak{q}^{(k)}$, it suffices to show that $\mathcal{J}(\mathfrak{a}_\bullet^e) \subseteq \mathfrak{q}$. Since $\mathfrak{q}$ is radical, it suffices to test this inclusion at a general point of Z. Therefore we can assume that Z is smooth, in which case $\mathfrak{q}^{(k)} = \mathfrak{q}^k$. Now Exercises 2.14 and 6.12 apply. $\square$

REMARK 6.16. Using their theory of tight closure, Hochster and Huneke [2002] have extended Theorem 6.14 to arbitrary regular Noetherian rings containing a field.

REMARK 6.17. Theorem 6.9 is applied in [Ein et al. 2003] to study the multiplicative behavior of Abhyankar valuations centered at a smooth point of a complex variety.

REMARK 6.18. Working with the asymptotic multiplier ideals $\mathcal{J}(\mathfrak{a}_\bullet^c)$ one can define the log canonical threshold and jumping coefficients of a graded system $\mathfrak{a}_\bullet$, much as in Section 4. However now these numbers need no longer be rational, the periodicity of jumping numbers (Exercise 5.7) may fail, and the set of jumping coefficients of $\mathfrak{a}_\bullet$ can contain accumulation points. See [Ein et al. 2004, § 5].

References

[Angehrn and Siu 1995] U. Angehrn and Y. T. Siu, "Effective freeness and point separation for adjoint bundles", *Invent. Math.* **122**:2 (1995), 291–308.

[Bierstone and Milman 1997] E. Bierstone and P. D. Milman, "Canonical desingularization in characteristic zero by blowing up the maximum strata of a local invariant", *Invent. Math.* **128**:2 (1997), 207–302.

[Blickle 2004] M. Blickle, "Multiplier ideal and module on toric varieties", *Math. Zei.* **248**:1 (2004), 113–121.

[Danilov 1978] V. I. Danilov, "The geometry of toric varieties", *Russian Math. Surveys* **33** (1978), 97–154.

[Demailly 1993] J.-P. Demailly, "A numerical criterion for very ample line bundles", *J. Differential Geom.* **37**:2 (1993), 323–374.

[Demailly 1999] J.-P. Demailly, "Méthodes L^2 et résultats effectifs en géométrie algébrique", pp. 59–90 (Exp. 852) in *Séminaire Bourbaki* (1998/99), Astérisque **266**, 1999.

[Demailly 2001] J.-P. Demailly, "Multiplier ideal sheaves and analytic methods in algebraic geometry", pp. 1–148 in *School on Vanishing Theorems and Effective Results in Algebraic Geometry* (Trieste, 2000), edited by J. P. Demailly et al., ICTP Lect. Notes **6**, Abdus Salam Int. Cent. Theoret. Phys., Trieste, 2001.

[Demailly et al. 2000] J.-P. Demailly, L. Ein, and R. Lazarsfeld, "A subadditivity property of multiplier ideals", *Michigan Math. J.* **48** (2000), 137–156.

[Ein and Lazarsfeld 1997] L. Ein and R. Lazarsfeld, "Singularities of theta divisors and the birational geometry of irregular varieties", *J. Amer. Math. Soc.* **10**:1 (1997), 243–258.

[Ein and Lazarsfeld 1999] L. Ein and R. Lazarsfeld, "A geometric effective Nullstellensatz", *Invent. Math.* **137**:2 (1999), 427–448.

[Ein et al. 2001] L. Ein, R. Lazarsfeld, and K. E. Smith, "Uniform bounds and symbolic powers on smooth varieties", *Invent. Math.* **144**:2 (2001), 241–252.

[Ein et al. 2003] L. Ein, R. Lazarsfeld, and K. E. Smith, "Uniform approximation of Abhyankar valuation ideals in smooth function fields", *Amer. J. Math.* **125**:2 (2003), 409–440.

[Ein et al. 2004] L. Ein, R. Lazarsfeld, K. E. Smith, and D. Varolin, "Jumping coefficients of multiplier ideals", *Duke Math. J.* **123**:3 (2004), 469–506.

[Eisenbud 1995] D. Eisenbud, *Commutative algebra, with a view toward algebraic geometry*, Graduate Texts in Mathematics **150**, Springer, New York, 1995.

[Encinas and Villamayor 2000] S. Encinas and O. Villamayor, "A course on constructive desingularization and equivariance", pp. 147–227 in *Resolution of singularities* (Obergurgl, 1997), edited by H. Hauser et al., Progr. Math. **181**, Birkhäuser, Basel, 2000.

[Esnault and Viehweg 1992] H. Esnault and E. Viehweg, *Lectures on vanishing theorems*, DMV Seminar **20**, Birkhäuser, Basel, 1992.

[Fulton 1993] W. Fulton, *Introduction to toric varieties*, vol. 131, Annals of Mathematics Studies, Princeton University Press, Princeton, NJ, 1993.

[Goward 2002] R. A. J. Goward, "A simple algorithm for principalization of monomial ideals", 2002. preprint.

[Hara 2001] N. Hara, "Geometric interpretation of tight closure and test ideals", *Trans. Am. Math. Soc.* **353**:5 (2001), 1885–1906.

[Hara and Takagi 2002] N. Hara and S. Takagi, "Some remarks on a generalization of test ideals", 2002. Available at arXiv:math.AC/0210131.

[Hara and Yoshida 2003] N. Hara and K.-I. Yoshida, "A generalization of tight closure and multiplier ideals", *Trans. Amer. Math. Soc.* **355**:8 (2003), 3143–3174.

[Hartshorne 1977] R. Hartshorne, *Algebraic geometry*, Graduate Texts in Mathematics **52**, Springer, New York, 1977.

[Hironaka 1964] H. Hironaka, "Resolution of singularities of an algebraic variety over a field of characteristic zero. I, II", *Ann. of Math.* (2) **79**:1–2 (1964), 109–203, 205–326.

[Hochster 2004] M. Hochster, "Tight closure theory and characteristic p methods", pp. 181–210 in *Trends in algebraic geometry*, edited by L. Avramov et al., Math. Sci. Res. Inst. Publ. **51**, Cambridge University Press, New York, 2004.

[Hochster and Huneke 1990] M. Hochster and C. Huneke, "Tight closure, invariant theory, and the Briançon–Skoda theorem", *J. Amer. Math. Soc.* **3**:1 (1990), 31–116.

[Hochster and Huneke 2002] M. Hochster and C. Huneke, "Comparison of symbolic and ordinary powers of ideals", *Invent. Math.* **147**:2 (2002), 349–369.

[Howald 2001] J. A. Howald, "Multiplier ideals of monomial ideals", *Trans. Amer. Math. Soc.* **353**:7 (2001), 2665–2671.

[Huneke 1992] C. Huneke, "Uniform bounds in Noetherian rings", *Invent. Math.* **107**:1 (1992), 203–223.

[Kollár 1997] J. Kollár, "Singularities of pairs", pp. 221–287 in *Algebraic geometry* (Santa Cruz, 1995), vol. 1, edited by J. Kollár et al., Proc. Sympos. Pure Math. **62**, Amer. Math. Soc., Providence, RI, 1997.

[Lazarsfeld 2004] R. Lazarsfeld, *Positivity in algebraic geometry*, Ergebnisse der Math. (3.F.) **48/49**, Springer, Berlin, 2004.

[Libgober 1983] A. Libgober, "Alexander invariants of plane algebraic curves", pp. 135–143 in *Singularities* (Arcata, CA, 1981), vol. 2, edited by P. Orlik, Proc. Sympos. Pure Math. **40**, Amer. Math. Soc., Providence, RI, 1983.

[Lipman 1993] J. Lipman, "Adjoints and polars of simple complete ideals in two-dimensional regular local rings", *Bull. Soc. Math. Belg. Sér. A* **45**:1-2 (1993), 223–244.

[Loeser and Vaquié 1990] F. Loeser and M. Vaquié, "Le polynôme d'Alexander d'une courbe plane projective", *Topology* **29**:2 (1990), 163–173.

[Nadel 1990] A. M. Nadel, "Multiplier ideal sheaves and Kähler–Einstein metrics of positive scalar curvature", *Ann. of Math.* (2) **132**:3 (1990), 549–596.

[Paranjape 1999] K. H. Paranjape, "The Bogomolov–Pantev resolution, an expository account", pp. 347–358 in *New trends in algebraic geometry* (Warwick, 1996), London Math. Soc. Lecture Note Ser. **264**, Cambridge Univ. Press, Cambridge, 1999.

[Siu 1998] Y.-T. Siu, "Invariance of plurigenera", *Invent. Math.* **134**:3 (1998), 661–673.

[Skoda 1972] H. Skoda, "Application des techniques L^2 à la théorie des idéaux d'une algèbre de fonctions holomorphes avec poids", *Ann. Sci. École Norm. Sup.* (4) **5** (1972), 545–579.

[Skoda and Briançon 1974] H. Skoda and J. Briançon, "Sur la clôture intégrale d'un idéal de germes de fonctions holomorphes en un point de $\mathbf{C}^n$", *C. R. Acad. Sci. Paris Sér. A* **278** (1974), 949–951.

[Smith 2000] K. E. Smith, "The multiplier ideal is a universal test ideal", *Commun. Algebra* **28**:12 (2000), 5915–5929.

[Swanson 2000] I. Swanson, "Linear equivalence of ideal topologies", *Math. Z.* **234**:4 (2000), 755–775.

[Takagi 2003] S. Takagi, "F-singularities of pairs and Inversion of Adjunction of arbitrary codimension", 2003.

[Takagi 2004] S. Takagi, "An interpretation of multiplier ideals via tight closure", *J. Algebraic Geom.* **13**:2 (2004), 393–415.

[Takagi and Watanabe 2004] S. Takagi and K. Watanabe, "When does the subadditivity theorem for multiplier ideals hold?", *Trans. Amer. Math. Soc.* **356**:10 (2004), 3951–3961.

MANUEL BLICKLE
UNIVERSITÄT ESSEN
FB6 MATHEMATIK
45117 ESSEN
GERMANY
manuel.blickle@uni-essen.de

ROBERT LAZARSFELD
DEPARTMENT OF MATHEMATICS
UNIVERSITY OF MICHIGAN
ANN ARBOR, MI 48109
UNITED STATES
rlaz@umich.edu

Trends in Commutative Algebra
MSRI Publications
Volume **51**, 2004

Lectures on the Geometry of Syzygies

DAVID EISENBUD

WITH A CHAPTER BY JESSICA SIDMAN

ABSTRACT. The theory of syzygies connects the qualitative study of algebraic varieties and commutative rings with the study of their defining equations. It started with Hilbert's work on what we now call the Hilbert function and polynomial, and is important in our day in many new ways, from the high abstractions of derived equivalences to the explicit computations made possible by Gröbner bases. These lectures present some highlights of these interactions, with a focus on concrete invariants of syzygies that reflect basic invariants of algebraic sets.

CONTENTS

These notes illustrate a few of the ways in which properties of syzygies reflect qualitative geometric properties of algebraic varieties. Chapters 1, 3 and 4 were written by David Eisenbud, and closely follow his lectures at the introductory workshop. Chapter 2 was written by Jessica Sidman, from the lecture she gave enlarging on the themes of the first lecture and providing examples. The lectures may serve as an introduction to the book *The Geometry of Syzygies* [Eisenbud 1995]; in particular, the book contains proofs for the many of the unproved assertions here.

Both authors are grateful for the hospitality of MSRI during the preparation of these lectures. Eisenbud was also supported by NSF grant DMS-9810361, and Sidman was supported by an NSF Postdoctoral Fellowship.

1. Hilbert Functions and Syzygies

1.1. Counting functions that vanish on a set. Let $\mathbb{K}$ be a field and let $S = \mathbb{K}[x_0, \ldots, x_r]$ be a ring of polynomials over $\mathbb{K}$. If $X \subset \mathbb{P}^r$ is a projective variety, the dimension of the space of forms (homogeneous polynomials) of each degree d vanishing on X is an invariant of X, called the Hilbert function of the ideal I_X of X. More generally, any finitely generated graded S-module $M = \bigoplus M_d$ has a *Hilbert function* $H_M(d) = \dim_{\mathbb{K}} M_d$. The *minimal free resolution* of a finitely generated graded S-module M provides invariants that refine the information in the Hilbert function. We begin by reviewing the origin and significance of Hilbert functions and polynomials and the way in which they can be computed from free resolutions.

Hilbert's interest in what is now known as the Hilbert function came from invariant theory. Given a group G acting on a vector space with basis $z_1, \ldots, z_n$, it was a central problem of nineteenth century algebra to determine the set of polynomial functions $p(z_1, \ldots, z_n)$ that are invariant under G in the sense that $p(g(z_1, \ldots, z_n)) = p(z_1, \ldots, z_n)$. The invariant functions form a graded subring, denoted T^G, of the ring $T = \mathbb{K}[z_1, \ldots, z_n]$ of all polynomials; the problem of invariant theory was to find generators for this subring.

For example, if G is the full symmetric group on $z_1, \ldots, z_n$, then T^G is the polynomial ring generated by the elementary symmetric functions $\sigma_1, \ldots, \sigma_n$, where

$$\sigma_i = \sum_{j_1 < \cdots < j_i} \prod_{t=1}^{i} z_{j_t};$$

see [Lang 2002, V.9] or [Eisenbud 1995, Example 1.1 and Exercise 1.6]. The result that first made Hilbert famous [1890] was that over the complex numbers ($\mathbb{K} = \mathbb{C}$), if G is either a finite group or a classical group of matrices (such as GL_n) acting algebraically—that is, via matrices whose entries are rational functions of the entries of the matrix representing an element of G—then the ring T^G is a finitely generated $\mathbb{K}$-algebra.

The homogeneous components of any invariant function are again invariant, so the ring T^G is naturally graded by (nonnegative) degree. For each integer d the homogeneous component $(T^G)_d$ of degree d is contained in T_d, a finite-dimensional vector space, so it too has finite dimension.

How does the number of independent invariant functions of degree d, say $h_d = \dim_{\mathbb{K}}(T^G)_d$, change with d? Hilbert's argument, reproduced in a similar case below, shows that the generating function of these numbers, $\sum_0^{\infty} h_d t^d$, is a rational function of a particularly simple form:

$$\sum_0^{\infty} h_d t^d = \frac{p(t)}{\prod_0^s (1 - t_i^{\alpha})},$$

for a polynomial p and positive integers α_i.

A similar problem, which will motivate these lectures, arises in projective geometry: Let $X \subset \mathbb{P}^r = \mathbb{P}^r_{\mathbb{K}}$ be a projective algebraic variety (or more generally a projective scheme) and let $I = I_X \subset S = \mathbb{K}[x_0, \ldots, x_r]$ be the homogeneous ideal of forms vanishing on X. An easy discrete invariant of X is given by the vector-space dimension $\dim_{\mathbb{K}} I_d$ of the degree d component of I. Again, we may ask how this "number of forms of degree d vanishing on X" changes with d. This number is usually expressed in terms of its complement in $\dim S_d$. We write $S_X := S/I$ for the homogeneous coordinate ring of X and we set $H_X(d) = \dim_{\mathbb{K}}(S_X)_d = \dim_{\mathbb{K}} S_d - \dim_{\mathbb{K}} I_d = \binom{r+d}{r} - \dim_{\mathbb{K}} I_d$. We call $H_X(d)$ the *Hilbert function* of X. Using Hilbert's ideas we will see that $H_X(d)$ agrees with a polynomial $P_X(d)$, called the Hilbert polynomial of X, when d is sufficiently large. Further, its generating function $\sum_d H_X(d) t^d$ can be written as a rational function in t, t^{-1} as above with denominator $(1 - t)^{r+1}$. Hilbert proved both the Hilbert Basis Theorem (polynomial rings are Noetherian) and the Hilbert Syzygy Theorem (modules over polynomial rings have finite free resolutions) in order to deduce this. As a first illustration of the usefulness of syzygies we shall see how these results fit together.

This situation of projective geometry is a little simpler than that of invariant theory because the generators x_i of S have degree 1, whereas in the case of invariants we have to deal with graded rings generated by elements of different degrees (the α_i). For simplicity we will henceforward stick to the case of degree-1 generators. See [Goto and Watanabe 1978a; 1978b] for more information.

Hilbert's argument *requires* us to generalize to the case of modules. If M is any finitely generated graded S-module (such as the ideal I or the homogeneous coordinate ring S_X), then the d-th homogeneous component M_d of M is a finite-dimensional vector space. We set $H_M(d) := \dim_{\mathbb{K}} M_d$. The function H_M is called the *Hilbert function* of M.

THEOREM 1.1. *Let $S = \mathbb{K}[x_0, \ldots, x_r]$ be the polynomial ring in $r + 1$ variables over a field $\mathbb{K}$. Let M be a finitely generated graded S-module.*

(i) *$H_M(d)$ is equal to a finite sum of the form $\sum_i \pm \binom{r+d-e_i}{r}$, and thus $H_M(d)$ agrees with a polynomial function $P_M(d)$ for $d \geq \max_i e_i - r$.*

(ii) *The generating function $\sum_d H_M(d) t^d$ can be expressed as a rational function of the form*

$$\frac{p(t, t^{-1})}{(1 - t)^{r+1}}$$

for some polynomial $p(t, t^{-1})$.

PROOF. First consider the case $M = S$. The dimension of the d-th graded component is $\dim_{\mathbb{K}} S_d = \binom{r+d}{r}$, which agrees with the polynomial in d

$$\frac{(r + d) \cdots (1 + d)}{r \cdots 1} = \frac{d^r}{r!} + \cdots + 1$$

for $d \geq -r$. Further,

$$\sum_0^\infty H_S(d)t^d = \sum_0^\infty \binom{r+d}{r} t^d = \frac{1}{(1-t)^{r+1}}$$

proving the theorem in this case.

At this point it is useful to introduce some notation: If M is any module we write $M(e)$ for the module obtained by "shifting" M by e positions, so that $M(e)_d = M_{e+d}$. Thus for example $S(-e)$ is the free module of rank 1 generated in degree e (note the change of signs!) Shifting the formula above we see that $H_{S(-e)}(d) = \binom{r+d-e}{r}$.

We immediately deduce the theorem in case $M = \bigoplus_i S(-e_i)$ is a free graded module, since then

$$H_M(d) = \sum_i H_{S(-e_i)}(d) = \sum_i \binom{r+d-e_i}{r}.$$

This expression is equal to a polynomial for $d \geq \max_i e_i - r$, and

$$\sum_{d=-\infty}^\infty H_M(d) = \frac{\sum_i t^{e_i}}{(1-t)^{r+1}}.$$

Hilbert's strategy for the general case was to compare an arbitrary module M to a free module. For this purpose, we choose a finite set of homogeneous generators m_i in M. Suppose $\deg m_i = e_i$. We can define a map (all maps are assumed homogeneous of degree 0) from a free graded module $F_0 = \bigoplus S(-e_i)$ onto M by sending the i-th generator to m_i. Let $M_1 := \ker F_0 \to M$ be the kernel of this map. Since $H_M(d) = H_{F_0}(d) - H_{M_1}(d)$, it suffices to prove the desired assertions for M_1 in place of M.

To use this strategy, Hilbert needed to know that M_1 would again be finitely generated, and that M_1 was in some way closer to being a free module than was M. The following two results yield exactly this information.

THEOREM 1.2 (HILBERT'S BASIS THEOREM). *Let S be the polynomial ring in $r+1$ variables over a field $\mathbb{K}$. Any submodule of a finitely generated S-module is finitely generated.*

Thus the module $M_1 = \ker F_0 \to M$, as a submodule of F_0, is finitely generated. To define the sense in which M_1 might be "more nearly free" than M, we need the following result:

THEOREM 1.3 (HILBERT'S SYZYGY THEOREM). *Let S be the polynomial ring in $r+1$ variables over a field $\mathbb{K}$. Any finitely generated graded S-module M has a finite free resolution of length at most $r+1$, that is, an exact sequence*

$$0 \longrightarrow F_n \xrightarrow{\phi_n} F_{n-1} \longrightarrow \cdots \longrightarrow F_1 \xrightarrow{\phi_1} F_0 \longrightarrow M \longrightarrow 0,$$

where the modules F_i are free and $n \leq r+1$.

We will not prove the Basis Theorem and the Syzygy Theorem here; see the very readable [Hilbert 1890], or [Eisenbud 1995, Corollary 19.7], for example. The Syzygy Theorem is true without the hypotheses that M is finitely generated and graded (see [Rotman 1979, Theorem 9.12] or [Eisenbud 1995, Theorem 19.1]), but we shall not need this.

If we take

$$\boldsymbol{F}: \quad 0 \longrightarrow F_n \xrightarrow{\phi_n} F_{n-1} \longrightarrow \cdots \longrightarrow F_1 \xrightarrow{\phi_1} F_0 \longrightarrow M \longrightarrow 0$$

to be a free resolution of M with the smallest possible n, then n is called the *projective dimension* of M. Thus the projective dimension of M is zero if and only if M is free. If M is not free, and we take $M_1 = \operatorname{im}\phi_1$ in such a minimal resolution, we see that the projective dimension of M_1 is strictly less than that of M. Thus we could complete the proof of Theorem 1.1 by induction.

However, given a finite free resolution of M we can compute the Hilbert function of M, and its generating function, directly. To see this, notice that if we take the degree d part of each module we get an exact sequence of vector spaces. In such a sequence the alternating sum of the dimensions is zero. With notation as above we have $H_M(d) = \sum (-1)^i H_{F_i}(d)$. If we decompose each F_i as $F_i = \sum_j S(-j)^{\beta_{i,j}}$ we may write this more explicitly as

$$H_M(d) = \sum_i (-1)^i \sum_j \beta_{i,j} \binom{r+d-j}{r}.$$

The sums are finite, so this function agrees with a polynomial in d for $d \geq \max\{j - r \mid \beta_{i,j} \neq 0 \text{ for some } i\}$. Further,

$$\sum_d H_M(d)\, t^d = \frac{\sum_i (-1)^i \sum_j \beta_{i,j} t^j}{(1-t)^{r+1}}$$

as required for Theorem 1.1.

Conversely, given the Hilbert function of a finitely generated module, one can recover some information about the $\beta_{i,j}$ in any finite free resolution $\boldsymbol{F}$. For this we use the fact that $\binom{r+d-j}{r} = 0$ for all $d < j$. We have

$$H_M(d) = \sum_i (-1)^i \sum_j \beta_{i,j} \binom{r+d-j}{r} = \sum_j \left(\sum_i (-1)^i \beta_{i,j} \right) \binom{r+d-j}{r}.$$

Since $\boldsymbol{F}$ is finite there is an integer d_0 such that $\beta_{i,j} = 0$ for $j < d_0$. If we put $d = d_0$ in the expression for $H_M(d)$ then all the $\binom{r+d-j}{r}$ vanish except for $j = d_0$, and because $\binom{r+d_0-d_0}{r} = 1$ we get $\sum_i (-1)^i \beta_{i,d_0} = H_M(d_0)$. Proceeding inductively we arrive at the proof of:

PROPOSITION 1.4. *Let M be a finitely generated graded module over $S = \mathbb{K}[x_0, \ldots, x_r]$, and suppose that $\boldsymbol{F}$ is a finite free resolution of M with graded*

Betti numbers $\beta_{i,j}$. *If* $\beta_{i,j} = 0$ *for all* $j < d_0$, *then the numbers* $B_j = \sum_i (-1)^i \beta_{i,j}$ *are inductively determined by the formulas*

$$B_{d_0} = H_M(d_0)$$

and

$$B_j = H_M(j) - \sum_{k<j} B_k \binom{j+d-k}{r}.$$

1.2. Meaning of the Hilbert function and polynomial.

The Hilbert function and polynomial are easy invariants to define, so it is perhaps surprising that they should be so important. For example, consider a variety $X \subset \mathbb{P}^r$ with homogeneous coordinate ring S_X. The restriction map to X gives an exact sequence of sheaves $0 \to \mathscr{I}_X \to \mathscr{O}_{\mathbb{P}^r} \to \mathscr{O}_X \to 0$. Tensoring with the line bundle $\mathscr{O}_{\mathbb{P}^r}(d)$ and taking cohomology we get a long exact sequence beginning

$$0 \to \mathrm{H}^0 \mathscr{I}_X(d) \to \mathrm{H}^0 \mathscr{O}_{\mathbb{P}^r}(d) \to \mathrm{H}^0 \mathscr{O}_X(d) \to \mathrm{H}^1 \mathscr{I}_X(d) \to \cdots.$$

The term $\mathrm{H}^0 \mathscr{O}_{\mathbb{P}^r}(d)$ may be identified with the vector space S_d of forms of degree d in S. The space $\mathrm{H}^0 \mathscr{I}_X(d)$ is thus the space of forms of degree d that induce 0 on X, that is $(I_X)_d$. Further, by Serre's vanishing theorem [Hartshorne 1977, Ch. III, Theorem 5.2], $\mathrm{H}^1 \mathscr{I}_X(d) = 0$ for large d. Thus for large d

$$(S_X)_d = S_d/(I_X)_d = \mathrm{H}^0 \mathscr{O}_X(d).$$

Applying Serre's theorem again, we see that all the higher cohomology of $\mathscr{O}_X(d)$ is zero for large d. Taking dimensions, we see that for large d the Hilbert function of S_X equals the Euler characteristic

$$\chi(\mathscr{O}_X(d)) := \sum_i (-1)^i \dim_{\mathbb{K}} \mathrm{H}^i(\mathscr{O}_X(d)).$$

The Hilbert function equals the Hilbert polynomial for large d; and the Euler characteristic is a polynomial for all d. Thus we may interpret the Hilbert polynomial as the Euler characteristic, and the difference from the Hilbert function (for small d) as an effect of the nonvanishing of higher cohomology.

For a trivial case, take X to be a set of points. Then $\mathscr{O}_X(d)$ is isomorphic to $\mathscr{O}_X$ whatever the value of d, and its global sections are spanned by the characteristic functions of the individual points. Thus $\chi(\mathscr{O}_X(d)) = P_X(d)$ is a constant function of d, equal to the number of points in X.

In general the Riemann–Roch Theorem gives a formula for the Euler characteristic, and thus the Hilbert polynomial, in terms of geometric data on X. In the simplest interesting case, where X is a smooth curve, the Riemann–Roch theorem says that

$$\chi(\mathscr{O}_X(d)) = P_X(d) = d + 1 - g,$$

where g is the genus of X.

These examples only suggest the strength of the invariants $P_X(d)$ and $H_X(d)$. To explain their real role, we recall some basic definitions. A *family of algebraic* *sets parametrized by a variety* T is simply a map of algebraic sets $\pi : \mathscr{X} \to T$. The subschemes $X_t = \pi^{-1}(t)$ for $t \in T$ are called the fibers of the family. Of course we are most interested in families where the fibers vary continuously in some reasonable sense! Of the various conditions we might put on the family to ensure this, the most general and the most important is the notion of flatness, due to Serre: the family $\pi : \mathscr{X} \to T$ is said to be flat if, for each point $p \in T$ and each point $x \in \mathscr{X}$ mapping to t, the pullback map on functions $\pi^* : \mathscr{O}_{T,t} \to \mathscr{O}_{\mathscr{X},x}$ is flat. This means simply that $\mathscr{O}_{\mathscr{X},x}$ is a flat $\mathscr{O}_{T,t}$-module; tensoring it with short exact sequences of $\mathscr{O}_{T,t}$-modules preserves exactness. More generally, a sheaf $\mathscr{F}$ on $\mathscr{X}$ is said to be flat if the $\mathscr{O}_{T,t}$-module $\mathscr{F}_x$ (the stalk of $\mathscr{F}$ at x) is flat for all x mapping to t. The same definitions work for the case of maps of schemes.

The condition of flatness for a family $\mathscr{X} \to T$ has many technical advantages. It includes the important case where $\mathscr{X}$, T, and all the fibers X_t are smooth and of the same dimension. It also includes the example of a family from which algebraic geometry started, the family of curves of degree d in the projective plane, even though the geometry and topology of such curves varies considerably as they acquire singularities. But the geometric meaning of flatness in general could well be called obscure.

In some cases flatness is nonetheless easy to understand. Suppose that $\mathscr{X} \subset \mathbb{P}^r \times T$ and the map $\pi : \mathscr{X} \to T$ is the inclusion followed by the projection onto T (this is not a very restrictive condition: any map of projective varieties, for example, has this form). In this case each fiber X_t is naturally contained as an algebraic set in $\mathbb{P}^r$.

We say in this case that $\pi : \mathscr{X} \to T$ is a *projective family*. Corresponding to a projective family $\mathscr{X} \to T$ we can look at the family of cones

$$\tilde{\mathscr{X}} \subset \mathbb{A}^{r+1} \times T \to T$$

obtained as the affine set corresponding to the (homogeneous) defining ideal of $\mathscr{X}$. The fibers $\tilde{X}_t$ are then all affine cones.

THEOREM 1.5. *Let* $\pi : \mathscr{X} \to T$ *be a projective family, as above. If* T *is a reduced algebraic set then* $\pi : \mathscr{X} \to T$ *is flat if and only if all the fibers* X_t *of* $\mathscr{X}$ *have the same Hilbert polynomial. The family of affine cones over* X_t *is flat if and only if all the* $\tilde{X}_t$ *have the same Hilbert function.*

These ideas can be generalized to the flatness of families of sheaves, giving an interpretation of the Hilbert function and polynomial of modules.

1.3. Minimal free resolutions. As we have defined it, a free resolution F of M does not seem to offer any easy invariant of M beyond the Hilbert function, since F depends on the choice of generators for M, the choice of generators for $M_1 = \ker F_0 \to M$, and so on. But this dependence on choices turns out to be very weak. We will say that F is a *minimal free resolution* of M if at each stage we choose the minimal number of generators.

PROPOSITION 1.6. *Let S be the polynomial ring in $r+1$ variables over a field $\mathbb{K}$, and M a finitely generated graded S-module. Any two minimal free resolutions of M are isomorphic. Moreover, any free resolution of M can be obtained from a minimal one by adding "trivial complexes" of the form*

$$G_i = S(-a) \xrightarrow{\ 1\ } S(-a) = G_{i-1}$$

for various integers i and a.

The proof is an exercise in the use of Nakayama's Lemma; see for example [Eisenbud 1995, Theorem 20.2].

Thus the ranks of the modules in the minimal free resolution, and even the numbers $\beta_{i,j}$ of generators of degree j in F_i, are invariants of M. Theorem 1.1 shows that these invariants are at least as strong as the Hilbert function of M, and we will soon see that they contain interesting additional information.

The numerical invariants in the minimal free resolution of a module in non-negative degrees can be described conveniently using a piece of notation introduced by Bayer and Stillman: the Betti diagram. This is a table displaying the numbers $\beta_{i,j}$ in the pattern

$$
\begin{array}{cccc}
\cdots & & & \\
\beta_{0,0} & \beta_{1,1} & \cdots & \beta_{i,i} \\
\beta_{0,1} & \beta_{1,2} & \cdots & \beta_{i,i+1} \\
\cdots & & &
\end{array}
$$

with $\beta_{i,j}$ in the i-th column and $(j-i)$-th row. Thus the i-th column corresponds to the i-th free module in the resolution, $F_i = \bigoplus_j S(-j)^{\beta_{i,j}}$. The utility of this pattern will become clearer later in these notes, but it was introduced partly to save space. For example, suppose that a module M has all its minimal generators in degree j, so that $\beta_{0,j} \neq 0$ but $\beta_{0,m} = 0$ for $m < j$. The minimality of $\boldsymbol{F}$ then implies that $\beta_{1,j} = 0$; otherwise, there would be a generator of F_1 of degree j, and it would map to a nonzero scalar linear combination of the generators of F_0. Since this combination would go to 0 in M, one of the generators of M would be superfluous, contradicting minimality. Thus there is no reason to leave a space for $\beta_{1,j}$ in the diagram. Arguing in a similar way we can show that $\beta_{i,m} = 0$ for all $m < i + j$. Thus if we arrange the $\beta_{i,j}$ in a Betti diagram as above we will be able to start with the j-th row, simply leaving out the rest.

To avoid confusion, we will label the rows, and sometimes the columns of the Betti diagram. The column containing $\beta_{i,j}$ (for all j) will be labeled i while the row with $\beta_{0,j}$ will be labeled j. For readability we often replace entries that are zero with $-$, and unknown entries with $*$, and we suppress rows in the region where all entries are 0. Thus for example if I is an ideal with 2 generators of degree 4 and one of degree 5, and relations of degrees 6 and 7, then the free resolution of S/I has the form

$$0 \to S(-6) \oplus S(-7) \to S^2(-4) \oplus S(-5) \to S$$

and Betti diagram

	0	1	2
0	1	–	–
1	–	–	–
2	–	–	–
3	–	2	–
4	–	1	1
5	–	–	1

An example that makes the space-saving nature of the notation clearer is the Koszul complex (the minimal free resolution of $S/(x_0,\ldots,x_r)$ — see [Eisenbud 1995, Ch. 17]), which has Betti diagram

	0	1	$\cdots$	i	$\cdots$	$r+1$
0	1	$r+1$	$\cdots$	$\binom{r+1}{i}$	$\cdots$	1

1.4. Four points in $\mathbb{P}^2$. We illustrate what has gone before by describing the Hilbert functions, polynomials, and Betti diagrams of each possible configuration $X \subset \mathbb{P}^2$ of four distinct points in the plane. We let $S = \mathbb{K}[x_0, x_1, x_2]$ be the homogeneous coordinate ring of the plane. We already know that the Hilbert polynomial of a set of four points, no matter what the configuration, is the constant polynomial $P_X(d) \equiv 4$. In particular, the family of 4-tuples of points is flat over the natural parameter variety

$$T = \mathbb{P}^2 \times \mathbb{P}^2 \times \mathbb{P}^2 \times \mathbb{P}^2 \setminus \text{diagonals}.$$

We shall see that the Hilbert function of X depends only on whether all four points lie on a line. The graded Betti numbers of the minimal resolution, in contrast, capture all the remaining geometry: they tell us whether any three of the points are collinear as well.

PROPOSITION 1.7. (i) *If X consists of four collinear points, $H_{S_X}(d)$ has the values $1, 2, 3, 4, 4, \ldots$ at $d = 0, 1, 2, 3, 4, \ldots$*

(ii) *If $X \subset \mathbb{P}^2$ consists of four points not all on a line, $H_{S_X}(d)$ has the values $1, 3, 4, 4, \ldots$ at $d = 0, 1, 2, 3, \ldots$. In classical language: X imposes 4 conditions on degree d curves for $d \geq 2$.*

PROOF. Let $H_X := H_{S_X}(d)$. In case (i), H_X has the same values that it would if we considered X to be a subset of $\mathbb{P}^1$. But in $\mathbb{P}^1$ the ideal of any d points is generated by one form of degree d, so the Hilbert function $H_X(d)$ for four collinear points X takes the values $1, 2, 3, 4, 4, \ldots$ at $d = 0, 1, 2, 3, 4, \ldots$.

In case (ii) there are no equations of degree $d \leq 1$, so for $d = 0, 1$ we get the claimed values for $H_X(d)$. In general, $H_X(d)$ is the number of independent functions induced on X by ratios of forms of degree d (see the next lecture) so $H_X(d) \leq 4$ for any value of d.

To see that $H_X(2) = 4$ it suffices to produce forms of degree 2 vanishing at all $X \setminus p$ for each of the four points p in X, since these forms must be linearly

independent modulo the forms vanishing on X. But it is easy to draw two lines
going through the three points of $X \setminus p$ but not through p:

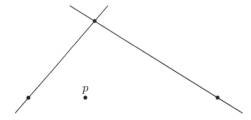

and the union of two lines has as equation the quadric given by the product
of the corresponding pair of linear forms. A similar argument works in higher
degree: just add lines to the quadric that do not pass through any of the points
to get curves of the desired higher degree.   □

In particular we see that the set of lines through a point in affine 3-space (the
cones over the sets of four points) do not form a flat family; but the ones where
not all the lines are coplanar do form a flat family. (For those who know about
schemes: the limit of a set of four noncoplanar lines as they become coplanar
has an embedded point at the vertex.)

    When all four points are collinear it is easy to compute the free resolution:
The ideal of X contains the linear form L that vanishes on the line containing
the points. But S/L is the homogeneous coordinate ring of the line, and in the
line the ideal of four points is a single form of degree 4. Lifting this back (in
any way) to S we see that I_X is generated by L and a quartic form, say f.
Since L does not divide f the two are relatively prime, so the free resolution of
$S_X = S/(L, f)$ has the form

$$0 \to S(-5) \xrightarrow{\begin{pmatrix} f \\ -L \end{pmatrix}} S(-1) \oplus S(-4) \xrightarrow{(L, f)} S,$$

with Betti diagram

	0	1	2
0	1	1	–
1	–	–	–
2	–	–	–
3	–	1	1

    We now suppose that the points of X are not all collinear, and we want to
see that the minimal free resolutions determine whether three are on a line. In
fact, this information is already present in the number of generators required by
I_X. If three points of X lie on a line $L = 0$, then by Bézout's theorem any conic
vanishing on X must contain this line, so the ideal of X requires at least one
cubic generator.

    On the other hand, any four noncollinear points lie on an irreducible conic
(to see this, note that any four noncollinear points can be transformed into any

other four noncollinear points by an invertible linear transformation of $\mathbb{P}^2$; and we can choose four noncollinear points on an irreducible conic.) From the Hilbert function we see that there is a two dimensional family of conics through the points, and since one is irreducible, any two distinct quadrics Q_1, Q_2 vanishing on X are relatively prime. It is easy to see from relative primeness that the syzygy $(Q_2, -Q_1)$ generates all the syzygies on (Q_1, Q_2). Thus the minimal free resolution of $S/(Q_1, Q_2)$ has Betti diagram

	0	1	2
0	1	–	–
1	–	2	–
2	–	–	1

It follows that $S/(Q_1, Q_2)$ has the same Hilbert function as $S_X = S/I_X$. Since $I_X \supset (Q_1, Q_2)$ we have $I_X = (Q_1, Q_2)$.

In the remaining case, where precisely three of the points of X lie on a line, we have already seen that the ideal of X requires at least one cubic generator. Corollary 2.3 makes it easy to see from this that the Betti diagram of a minimal free resolution must be

	0	1	2
0	1	–	–
1	–	2	1
2	–	1	1

2. Points in the Plane and an Introduction to Castelnuovo–Mumford Regularity

2.1. Resolutions of points in the projective plane. This section gives a detailed description of the numerical invariants of a minimal free resolution of a finite set of points in the projective plane. To illustrate both the potential and the limitations of these invariants in capturing the geometry of the points we compute the Betti diagrams of all possible configurations of five points in the plane. In contrast to the example of four points worked out in the previous section, it is not possible to determine whether the points are in linearly general position from the Betti numbers alone. The presentation in this section is adapted from Chapter 3 of [Eisenbud $\geq$ 2004], to which we refer the reader who wishes to find proofs omitted here.

Let $X = \{p_1, \ldots, p_n\}$ be a set of distinct points in $\mathbb{P}^2$ and let I_X be the homogeneous ideal of X in $S = \mathbb{K}[x_0, x_1, x_2]$. Considering this situation has the virtue of simplifying the algebra to the point where one can describe a resolution of I_X quite explicitly while still retaining a lot of interesting geometry.

Fundamentally, the algebra is simple because the resolution of I_X is very short. In particular:

LEMMA 2.1. *If $I_X \subseteq \mathbb{K}[x_0, x_1, x_2]$ is the homogeneous ideal of a finite set of points in the plane, then a minimal resolution of I_X has length one.*

PROOF. Recall that if

$$0 \longrightarrow F_m \longrightarrow \cdots \longrightarrow F_1 \longrightarrow F_0 \longrightarrow S/I_X \longrightarrow 0$$

is a resolution of S/I_X then we get a resolution of I_X by simply deleting the term F_0 (which of course is just S). We will proceed by showing that S/I_X has a resolution of length two.

From the Auslander–Buchsbaum formula (see Theorem 3.1) we know that the length of a minimal resolution of S/I_X is:

$$\text{depth}\, S - \text{depth}\, S/I_X.$$

Since S is a polynomial ring in three variables, it has depth three. Our hypothesis is that S/I_X is the coordinate ring of a finite set points taken as a reduced subscheme of $\mathbb{P}^2$. The Krull dimension of S/I_X is one, and hence depth $S/I_X \leq 1$. Furthermore, since I_X is the ideal of all homogeneous forms in S that vanish on X, the irrelevant ideal is not associated. Therefore, we can find an element of S with positive degree that is a nonzerodivisor on S/I_X. We conclude that S/I_X has a free resolution of length two. $\square$

We see now that a resolution of I_X has the form

$$0 \longrightarrow \bigoplus_{i=1}^{t_1} S(-b_i) \xrightarrow{M} \bigoplus_{i=1}^{t_0} S(-a_i) \longrightarrow I_X \longrightarrow 0.$$

We can complete our description of the shape of the resolution via the following theorem:

THEOREM 2.2 (HILBERT–BURCH). *Suppose that an ideal I in a Noetherian ring R admits a free resolution of length one:*

$$0 \longrightarrow F \xrightarrow{M} G \longrightarrow I \longrightarrow 0.$$

If the rank of the free module F is t, then the rank of G is $t + 1$, and there exists a nonzerodvisor $a \in R$ such that I is $aI_t(M)$; in fact, regarding M as a matrix with respect to given bases of F and G, the generator of I that is the image of the i-th basis vector of G is $\pm a$ times the determinant of the submatrix of M formed by deleting the i-th row. Moreover, the depth of $I_t(M)$ is two.

Conversely, given a nonzerodivisor a of R and a $(t + 1) \times t$ matrix M with entries in R such that the depth of $I_t(M)$ is at least 2, the ideal $I = aI_t(M)$ admits a free resolution as above.

We will not prove the Hilbert–Burch Theorem here, or its corollary stated below; our main concern is with their consequences. (Proofs can be found in [Eisenbud ≥ 2004, Chapter 3]; alternatively, see [Eisenbud 1995, Theorem 20.15] for Hilbert–Burch and [Ciliberto et al. 1986] for the last statement of Corollary 2.3.)

As we saw in Section 1.1, the Hilbert function and the Hilbert polynomial of S/I_X are determined by the invariants of a minimal free resolution. So, for example, we expect to be able to compute the degree of X from the degrees of the entries of M. When X is a complete intersection this is already familiar to us from Bézout's theorem. In this case M is a 2×1 matrix whose entries generate I_X. Bézout's theorem says that the product of the degrees of the entries of M gives the degree of X.

The following corollary of the Hilbert–Burch Theorem generalizes Bézout's theorem and describes the relationships between the degrees of the generators of I_X and the degrees of the generators of the module of their syzygies. Since the map given by M has degree zero, the (i, j) entry of M has degree $b_j - a_i$. Let $e_i = b_i - a_i$ and $f_i = b_i - a_{i+1}$ denote the degrees of the entries on the two main diagonals of M. Schematically:

$$
\begin{pmatrix}
e_1 & & & & \\
f_1 & e_2 & & & \\
& f_2 & \cdot & & \\
& & & \cdot & \\
& & & & \cdot \\
& & & f_{t-1} & e_t \\
& & & & f_t
\end{pmatrix}
$$

COROLLARY 2.3. *Assume that* $a_1 \geq a_2 \geq \cdots \geq a_{t+1}$ *and* $b_1 \geq b_2 \geq \cdots \geq b_t$. *Then, for* $1 \leq i \leq t$, *we have*

$$
e_i \geq 1, \qquad f_i \geq 1, \qquad a_i = \sum_{j<i} e_j + \sum_{j \geq i} f_j, \qquad b_i = a_i + e_i.
$$

Moreover,

$$
n = \deg X = \sum_{i \leq j} e_i f_j. \tag{2–1}
$$

The last equality is due to Ciliberto, Geramita, and Orecchia [Ciliberto et al. 1986].

From Corollary 2.3 we can already bound the number of minimal generators of I_X given a little bit of information about the geometry of X.

COROLLARY 2.4 [Burch 1967]. *If X lies on a curve of degree d, then I_X requires at most $d+1$ generators.*

PROOF. Since X lies on a curve of degree d there is an element of I_X of degree d. Therefore, at least one of the a_i's must be at most d. By Corollary 2.3, each a_i is the sum of t integers that are all at least 1. Therefore, $t \leq a_i \leq d$, which implies that $t + 1$, the number of generators of I_X, is at most $d + 1$.   □

Using the information above one can show that for small values of n there are very few possibilities for the invariants of the resolution of I_X.

2.2. Resolutions of five points in the plane. We now show how to use these ideas to compute all possible Betti diagrams of X when X is a set of five distinct points in the projective plane. As before, let $S = \mathbb{K}[x_0, x_1, x_2]$, and let I_X be the saturated homogeneous ideal of X. In keeping with conventions, we give the Betti diagrams of the quotient S/I_X. From these computations it will be easy to determine the Hilbert function $H_X(d)$ as well.

First, we organize the possible configurations of the points into four categories based on their geometry:

(1) The five points are all collinear.
(2) Precisely four of the points are collinear.
(3) Some subset of three of the points lies on a line but no subset of four of the points lies on a line.
(4) The points are linearly general.

Case (1): Corollary 2.4 implies that I_X has at most two generators. Thus, $t = 1$, and the points are a complete intersection. By Bézout's Theorem, the generators of I_X have degrees $a_1 = 5$ and $a_2 = 1$. Furthermore, we see that I_X is resolved by a Koszul complex and hence $b_1 = 6$. The Betti diagram of the resolution of 5 collinear points is

	0	1	2
0	1	1	–
1	–	–	–
2	–	–	–
3	–	–	–
4	–	1	1

From Section 1.1 we see that the Hilbert function is given by

$$\binom{2+d}{2} - \binom{2+d-1}{2} - \binom{2+d-5}{2} + \binom{2+d-6}{2}.$$

Thus, $H_X(d)$ has values $1, 2, 3, 4, 5, 5, \ldots$ at $d = 1, 2, 3, 4, 5, \ldots$.

We claim that $t = 2$ in the remaining cases. Since the degree of X is prime, Bézout's theorem tells us that the points are a complete intersection if and only if they are collinear. Hence, $t \geq 2$. Since there is a 6-dimensional space of conics in three variables, any set of five points must lie on a conic. Thus Corollary 2.4 implies that I_X has at most three generators, so $t = 2$. We conclude that in Cases (2)–(4), the invariants of the resolution satisfy the relationships $a_1 \geq a_2 \geq a_3$, $b_1 \geq b_2$, and

$$a_1 = f_1 + f_2,$$
$$a_2 = e_1 + f_2.$$
$$a_3 = e_1 + e_2.$$

Case (2): If precisely four of the points are collinear, we see from the picture below that there are two conics containing the points: $L_1 \cup L_3$ and $L_2 \cup L_3$.

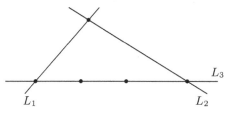

Since these conics are visibly different, their defining equations must be linearly independent. We conclude that I_X must have two minimal generators of degree two, and hence that

$$a_2 = e_1 + f_2 = 2,$$

$$a_3 = e_1 + e_2 = 2.$$

By Corollary 2.3, $e_1, e_2, f_2 \geq 1$, which implies that $e_1 = e_2 = f_2 = 1$. We also know that

$$5 = e_1 f_1 + e_1 f_2 + e_2 f_2 = f_1 + 1 + 1.$$

Hence, $f_1 = 3$, $a_1 = 4$, $b_1 = a_1 + e_1 = 5$, and $b_2 = a_2 + e_2 = 3$. In this case, the points have Betti diagram

	0	1	2
0	1	–	–
1	–	2	1
2	–	–	–
3	–	1	1

and Hilbert function

$$\binom{2+d}{2} - 2\binom{2+d-2}{2} - \binom{2+d-4}{2} + \binom{2+d-3}{2} + \binom{2+d-5}{2},$$

taking on the values $1, 3, 4, 5, 5, \ldots$ at $d = 0, 1, 2, 3, 4, \ldots$.

Case (3): We will show that the points lie on a unique reducible conic. By assumption there is a line L containing three of the points. Any conic containing all five must vanish at these three points and hence will vanish identically on L. Therefore, L must be a component of any conic that contains X. There are precisely two points not on L, and they determine a line L' uniquely. The union of L and L' is the unique conic containing these points.

If the five points lie on a unique conic, we can determine all of the remaining numerical invariants. We must have $a_3 = e_1 + e_2 = 2$, which implies that $e_1 = e_2 = 1$. We also know that $3 \leq a_2 = e_1 + f_2 = 1 + f_2$, which implies that $f_2 \geq 2$. Since

$$5 = e_1 f_1 + e_1 f_2 + e_2 f_2 = f_1 + f_2 + f_2,$$

we must have $f_1 = 1$ and $f_2 = 2$. Now the invariants a_1, a_2, b_1, and b_2 are completely determined: $a_1 = f_1 + f_2 = 3$, $a_2 = e_1 + f_2 = 3$, $b_1 = a_1 + e_1 = 4$ and $b_2 = a_2 + e_2 = 4$. The Betti diagram is

	0	1	2
0	1	–	–
1	–	1	–
2	–	2	2

We have the Hilbert function

$$\binom{2+d}{2} - \binom{2+d-2}{2} - 2\binom{2+d-3}{2} + 2\binom{2+d-4}{2},$$

which takes on values $1, 3, 5, 5, \ldots$ at $d = 0, 1, 2, 3, \ldots$.

Case (4): We claim that five points in linearly general position also lie on a unique conic. If the points lie on a reducible conic then it is the union of two lines, and one of the lines must contain at least three points. Therefore, if the points are linearly general, any conic containing them must be irreducible. By Bézout's theorem, five points cannot lie on two irreducible conics because the intersection of the conics contains only four points.

As we saw in Case (3), the Betti diagram of I_X was completely determined after we discovered that X lay on a unique conic. We conclude that the Betti numbers are not fine enough to distinguish between the geometric situations presented by Cases (3) and (4).

2.3. An introduction to Castelnuovo–Mumford regularity.

Let $S = \mathbb{K}[x_0, \ldots, x_r]$ and let M be a finitely generated graded S-module. One of the ways in which the Betti diagrams for the examples in Section 2.1 differ is in the number of rows. This apparently artificial invariant turns out to be fundamental. In this section we introduce it systematically via the notion of *Castelnuovo–Mumford regularity*. We follow along the lines of [Eisenbud $\geq$ 2004, Chapter 4].

DEFINITION 2.5. (i) If F is a finitely generated free module, we define $\operatorname{reg} F$, the *regularity of F*, as the maximum degree of a minimal generator of F:

$$\operatorname{reg} F = \max \{i \mid (F/(x_0, \ldots, x_r)F)_i \neq 0\}.$$

(The maximum over the empty set is $-\infty$.)

(ii) For an arbitrary finitely generated graded module M, we define the *regularity of M* as

$$\operatorname{reg} M = \max_i \{\operatorname{reg} F_i - i\},$$

where

$$0 \longrightarrow F_m \longrightarrow \cdots \longrightarrow F_1 \longrightarrow F_0 \longrightarrow M \longrightarrow 0$$

is a minimal free graded resolution of M. We say that M is *d-regular* if $d \geq \operatorname{reg} M$.

Notice that

$$\operatorname{reg} M \geq \operatorname{reg} F_0 - 0 = \operatorname{reg} F_0,$$

where $\operatorname{reg} F_0$ is the maximum degree of a generator of M. The regularity should be thought of as a stabilized version of this "generator degree" which takes into account the nonfreeness of M. One of the most fundamental results about the regularity is a reinterpretation in terms of cohomology. We begin with a special case where the cohomology has a very concrete meaning.

Suppose that we are given a finite set of points $X = \{p_1, \ldots, p_n\} \subset \mathbb{A}_{\mathbb{K}}^r$, where $\mathbb{K}$ is an infinite field. We claim that for any function $\phi : X \to \mathbb{K}$ there is a polynomial $f \in R = \mathbb{K}[x_1, \ldots, x_n]$ such that $f|_X = \phi$. As noted in Section 1.2, the set of all functions from X to $\mathbb{K}$ is spanned by characteristic functions $\phi_1, \ldots, \phi_n$, where

$$\phi_i(p_j) = \begin{cases} 1 & \text{if } i = j, \\ 0 & \text{if } i \neq j. \end{cases}$$

So it is enough to show that the characteristic functions can be given by polynomials. For each $i = 1, \ldots, n$, let L_i be a linear polynomial defining a line containing p_i but not any other point of X. Let $f_i = \prod_{j \neq i} L_i$. The restriction of f_i to X is the function ϕ_i up to a constant scalar.

DEFINITION 2.6. The *interpolation degree* of X is the least integer d such that for each $\phi : X \to \mathbb{K}$ there exists $f \in R$ with $\deg f \leq d$ such that $\phi = f|_X$.

When n is small, one can compute the interpolation degree of X easily from first principles:

EXAMPLE 2.7. Let X be a set of four points in $\mathbb{A}^2$ in linearly general position. None of the characteristic functions ϕ_i can be the restriction of a linear polynomial; each ϕ_i must vanish at three of the points, but no three are collinear. However, we can easily find quadratic polynomials whose restrictions give the ϕ_i. For instance, let L_{23} be a linear polynomial defining the line joining p_2 and p_3 and let L_{34} be a linear polynomial defining the line joining p_3 and p_4.

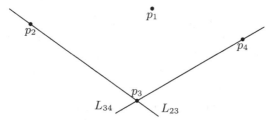

The product $L_{23}L_{34}|_X$ equals ϕ_1 up to a constant factor. By symmetry, we can repeat this procedure for each of the remaining ϕ_i. Thus, the interpolation degree of X is two.

It is clear that the interpolation degree of X depends on the cardinality of X. It will also depend on the geometry of the points. To see this, let's compute the interpolation degree of X when X consists of four points that lie on a line L. Any polynomial that restricts to one of the characteristic functions vanishes on three of the points and so intersects the line L at least three times. Therefore, the interpolation degree is at least three. We can easily find a polynomial f_i of degree three that restricts to each ϕ_i: For each $j \neq i$, let L_j be a linear polynomial that vanishes at p_j and no other points of X. Let f_i be the product of these three linear polynomials. Thus, if X consists of four collinear points, its interpolation degree is three.

To study what happens more generally we projectivize. We can view the affine r-space containing our n points as a standard affine open patch of $\mathbb{P}^r$ with coordinate ring $S = \mathbb{K}[x_0, \ldots, x_r]$, say, where x_0 is nonzero. A homogeneous polynomial does not have a well-defined value at a point of $\mathbb{P}^r$, so elements of S do not give functions on projective space. However, the notion of when a homogeneous polynomial vanishes at a point is well-defined. This observation shows that we can hope to find homogeneous polynomials that play the role that characteristic functions played for points in affine space and is the basis for the following definition.

DEFINITION 2.8. We say that X imposes *independent conditions* on forms of degree d if there exist $F_1, \ldots, F_n \in S_d$ such that $F_j(p_i)$ is nonzero if and only if $i = j$.

We may rephrase the condition in Definition 2.8 as follows: Suppose that F is a form of degree d, say $F = \sum a_\alpha x^\alpha$ with $a_\alpha \in \mathbb{K}$ for each $\alpha \in \mathbb{Z}_{\geq 0}^{n+1}$ such that $|\alpha| = d$. Fix a set of coordinates for each of the points $p_1, \ldots, p_n$ and substitute these values into F. Then $F(p_1) = 0, \ldots, F(p_n) = 0$ are n equations, linear in the a_α. These equations are the "conditions" that the points $p_1, \ldots, p_n$ impose.

We can translate the interpolation degree problem into the projective setting:

PROPOSITION 2.9. *The interpolation degree of X is the minimum degree d such that X imposes independent conditions on forms of degree d.*

PROOF. Let $\{f_1, \ldots, f_n\}$ be a set of polynomials in $\mathbb{K}[x_1, \ldots, x_r]$ of degree d whose restrictions to X are the characteristic functions of the points. Homogenizing the f_i with respect to x_0 gives us homogeneous forms satisfying the condition of Definition 2.8. So if the interpolation degree of X is d, then X imposes independent conditions of forms of degree d. Furthermore, if X imposes independent conditions on forms of degree d, then there exists a set of forms of degree d that are "homogeneous characteristic functions" and dehomogenizing by setting $x_0 = 1$ gives a set of polynomials of degree at most d whose restrictions to X are characteristic functions for the points in affine space.   □

To analyze the new problem in the projective setting we will use methods of coherent sheaf cohomology. (One could use local cohomology with respect to $(x_0, \ldots, x_r)$ equally well. See [Eisenbud $\geq$ 2004, Chapter 4] for this point of view.) Let $\mathscr{I}_X$ be the ideal sheaf of X and $\mathscr{O}_X$ its structure sheaf. These sheaves fit into the short exact sequence:

$$0 \longrightarrow \mathscr{I}_X \longrightarrow \mathscr{O}_{\mathbb{P}^r} \longrightarrow \mathscr{O}_X \longrightarrow 0.$$

The following proposition shows that the property that X imposes independent conditions on forms of degree d can be interpreted cohomologically.

PROPOSITION 2.10. *X imposes independent conditions on forms of degree d if and only if $H^1\mathscr{I}_X(d)$ vanishes.*

PROOF. We are interested in whether X imposes independent conditions on forms of degree d so we tensor (or "twist") the short exact sequence by $\mathscr{O}_{\mathbb{P}^r}(d)$. Exactness is clearly preserved on the level of stalks since the stalks of $\mathscr{O}_{\mathbb{P}^r}(d)$ are rank one free modules, and this suffices to show that exactness of the sequence of sheaves is also preserved. We write

$$0 \longrightarrow \mathscr{I}_X(d) \longrightarrow \mathscr{O}_{\mathbb{P}^r}(d) \longrightarrow \mathscr{O}_X(d) \longrightarrow 0.$$

The first three terms in the long exact sequence in cohomology have very concrete interpretations:

$$0 \longrightarrow H^0\mathscr{I}_X(d) \longrightarrow H^0\mathscr{O}_{\mathbb{P}^r}(d) \xrightarrow{\rho} H^0\mathscr{O}_X(d) \longrightarrow H^1\mathscr{I}_X(d) \longrightarrow \cdots$$

$$0 \longrightarrow (I_X)_d \longrightarrow S_d \longrightarrow \mathbb{K}^n \longrightarrow \cdots$$

with vertical isomorphisms $\cong$ on the first two terms.

Note that the map ρ, which is given by dehomogenizing with respect to x_0 and evaluating the resulting degree d polynomials at the points of X, is surjective if and only if X imposes independent conditions on forms of degree d. The equivalent cohomological condition is that $H^1\mathscr{I}_X(d) = 0$.   □

The next proposition is a first step in relating vanishings in cohomology with regularity.

PROPOSITION 2.11. *If $\operatorname{reg} I_X \leq d$ then $H^i\mathscr{I}_X(d-i) = 0$.*

PROOF. The point is to construct a short exact sequence of sheaves where $\mathscr{I}_X$ is either the middle or right-hand term and we can say a lot about the vanishing of the higher cohomology of the other two terms. We will get this short exact sequence (with $\mathscr{I}_X$ as the right-hand term) from a free resolution of $\mathscr{I}_X$. Let

$$0 \longrightarrow \bigoplus S(-a_{m,j}) \longrightarrow \cdots \bigoplus S(-a_{1,j}) \longrightarrow \bigoplus S(-a_{0,j}) \longrightarrow I_X \longrightarrow 0$$

be a minimal free resolution of I_X. Its sheafification is exact because localization is an exact functor. Splitting this complex into a series of short exact sequences,

and twisting by $d - i$ gives

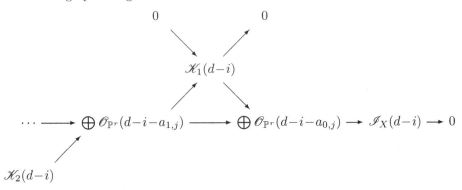

From the long exact sequence associated to each short exact sequence in the diagram we see that if $d - i \geq a_{i,j}$ for each i, j, then $H^i \mathscr{I}_X(d - i) = 0$ for each $i > 0$. □

These conditions on the vanishings of the higher cohomology of twists of a sheaf were first captured by Mumford by what we now call the *Castelnuovo–Mumford regularity* of a coherent sheaf. We give the definition for sheaves in terms of the regularity of the cohomology modules

$$H^i_* \mathscr{F} := \bigoplus_{d \geq 0} H^i \mathscr{F}(d),$$

which are finite-dimensional $\mathbb{K}$−vector spaces by a theorem of Serre. (See [Hartshorne 1977, Ch. III, Theorem 5.2] for a proof.)

DEFINITION 2.12. If $\mathscr{F}$ is a coherent sheaf on $\mathbb{P}^r$,

$$\operatorname{reg} \mathscr{F} = \max_{i > 0} \{ \operatorname{reg} H^i_* \mathscr{F} + i + 1 \}.$$

We say that $\mathscr{F}$ is d−*regular* if $d \geq \operatorname{reg} \mathscr{F}$.

One may also reformulate the definition of the regularity of a finitely generated graded module in a similar fashion.

THEOREM 2.13. *If H is an Artinian module, define*

$$\operatorname{reg} H = \max \{ i \mid H_i \neq 0 \}.$$

If M is an arbitrary finitely generated graded S-module define

$$\operatorname{reg} M = \max_{i \geq 0} \operatorname{reg} H^i_{(x_0, \ldots, x_r)} M + i.$$

For a proof, one may see [Eisenbud ≥ 2004].

The following theorem is Mumford's original definition of regularity.

THEOREM 2.14 [Mumford 1966, p. 99]. *If $\mathscr{F}$ is a coherent sheaf on $\mathbb{P}^r$ then $\mathscr{F}$ is d-regular if $H^i \mathscr{F}(d - i) = 0$ for all $i > 0$.*

From the proof of Proposition 2.11 it is clear that if an arbitrary homogeneous ideal $I \subseteq S$ is d-regular, then $\mathscr{I}$ is d-regular. In general, the relationship between the regularity of coherent sheaves and finitely generated modules is a bit technical. (One should expect this since each coherent sheaf on projective space corresponds to an equivalence class of finitely generated graded S-modules.) However, if we work with saturated homogeneous ideals of closed subsets of $\mathbb{P}^r$ the correspondence is quite nice:

THEOREM 2.15 [Bayer and Mumford 1993, Definition 3.2]. *If I_Z is the saturated homogeneous ideal of all elements of S that vanish on a Zariski-closed subset Z in $\mathbb{P}^r$ and $\mathscr{I}_Z$ is its sheafification, then* $\mathrm{reg}\,\mathscr{I}_Z = \mathrm{reg}\,I_Z$.

(See the technical appendix to Chapter 3 in [Bayer and Mumford 1993] for a proof.)

We return now to the problem of computing the interpolation degree of X. As a consequence of Proposition 2.10, the interpolation degree of X is the minimum d such that $H^1 \mathscr{I}_X(d) = 0$. We claim that $\mathscr{I}_X$ is $(d+1)$-regular if and only if $H^1 \mathscr{I}_X(d) = 0$. If $\mathscr{I}_X$ is $(d+1)$-regular, the vanishing is part of the definition. To see the opposite direction, look at the long exact sequence in cohomology associated to any positive twist of

$$0 \longrightarrow \mathscr{I}_X \longrightarrow \mathscr{O}_{\mathbb{P}^r} \longrightarrow \mathscr{O}_X \longrightarrow 0.$$

The higher cohomology of positive twists of $\mathscr{O}_{\mathbb{P}^r}$ always vanishes, and the higher cohomology of any twist of $\mathscr{O}_X$ vanishes because the support of $\mathscr{O}_X$ has dimension zero. Therefore, $H^i \mathscr{I}_X(k) = 0$ for any positive integer k and all $i \geq 2$.

We conclude that the interpolation degree of X equals d if and only if $\mathrm{reg}\,\mathscr{I}_X = d + 1$, if and only if $\mathrm{reg}\,I_X = d + 1$, if and only if $\mathrm{reg}\,S/I_X = d$. Thus, the interpolation degree of X is equal to the Castelnuovo–Mumford regularity of its homogeneous coordinate ring S/I_X. In Section 3 we will see many more ways in which regularity and geometry interact.

3. The Size of Free Resolutions

Throughout this section we set $S = \mathbb{K}[x_0, \ldots, x_r]$ and let M denote a finitely generated graded S-module. Let

$$\boldsymbol{F}: \quad 0 \longrightarrow F_m \xrightarrow{\phi_m} F_{m-1} \longrightarrow \cdots \longrightarrow F_1 \xrightarrow{\phi_1} F_0$$

be a minimal free resolution of M, and let $\beta_{i,j}$ be the graded Betti numbers— that is, $F_i = \bigoplus_j S(-j)^{\beta_{i,j}}$. In this section we will survey some results and conjectures related to the size of $\boldsymbol{F}$.

3.1. Projective dimension, the Auslander–Buchsbaum Theorem, and Cohen–Macaulay modules. The most obvious question is about the length m, usually called the *projective dimension* of M, written pd M. The Auslander–Buchsbaum Theorem gives a very useful characterization. Recall that a *regular sequence of length t on M* is a sequence of homogeneous elements $f_1, f_2, \ldots, f_t$ of positive degree in S such that f_{i+1} is a nonzerodivisor on $M/(f_1, \ldots, f_i)M$ for each $0 \le i < t$. (The definition usually includes the condition $(f_1, \ldots, f_t)M \neq M$, but this is superfluous because the f_i have positive degree and M is finitely generated.) The *depth of M* is the length of a maximal regular sequence on M (all such maximal regular sequences have the same length). For example, $x_0, \ldots, x_r$ is a maximal regular sequence on S and thus on any graded free S-module. In general, the depth of M is at most the (Krull) dimension of M; M is said to be a Cohen–Macaulay module when these numbers are equal, or equivalently when pd M = codim M.

THEOREM 3.1 (AUSLANDER–BUCHSBAUM). *If $S = \mathbb{K}[x_0, \ldots, x_r]$ and M is a finitely generated graded S-module, then the projective dimension of M is $r+1-t$, where t is the length of a maximal regular sequence on M.*

Despite this neat result there are many open problems related to the existence of modules with given projective dimension. Perhaps the most interesting concern Cohen–Macaulay modules:

PROBLEM 3.2. What is the minimal projective dimension of a module annihilated by a given homogeneous ideal I? From the Auslander–Buchsbaum Theorem this number is greater than or equal to codim I; is it in fact equal? That is, does every factor ring S/I have a Cohen–Macaulay module? If S/I does have a Cohen–Macaulay module, what is the smallest rank such a module can have?

If we drop the restriction that M should be finitely generated, then Hochster has proved that Cohen–Macaulay modules ("big Cohen–Macaulay modules") exist for all S/I, and the problem of existence of finitely generated ("small") Cohen–Macaulay modules was posed by him. The problem is open for most S/I of dimension ≥ 3. See [Hochster 1975] for further information.

One of the first author's favorite problems is a strengthening of this one. A module M is said to have *linear resolution* if its Betti diagram has just one row — that is, if all the generators of M are in some degree d, the first syzygies are generated in degree $d + 1$, and so on. For example the Koszul complex is a linear resolution of the residue class field $\mathbb{K}$ of S. Thus the following problem makes sense:

PROBLEM 3.3. What is the minimal projective dimension of a module *with linear resolution* annihilated by a given homogeneous ideal I? Is it in fact equal to codim I? That is, does every factor ring S/I have a Cohen–Macaulay module with linear resolution? If so, what is the smallest rank such a module can have?

Cohen–Macaulay modules with linear resolutions are often called *linear Cohen–Macaulay modules* or *Ulrich modules*. They appear, among other places, in the computation of resultants. See [Eisenbud et al. 2003b] for results in this direction and pointers to the literature. This last problem is open even when S/I is a Cohen–Macaulay ring — and the rank question is open even when I is generated by a single element. See [Brennan et al. 1987].

An interesting consequence of the Auslander–Buchsbaum theorem is that it allows us to compare projective dimensions of a module over different polynomial rings. A very special case of this argument gives us a nice interpretation of what it means to be a Cohen–Macaulay module or an Ulrich module. To explain it we need another notion:

Recall that a sequence of homogeneous elements $y_1, \ldots, y_d \in S$ is called a *system of parameters* on a graded module M of Krull dimension d if and only if $M/(y_1, \ldots, y_d)M$ has (Krull) dimension 0, that is, has finite length. (This happens if and only if $(y_1, \ldots, y_d) + \operatorname{ann} M$ contains a power of $(x_0, \ldots, x_r)$. For details see [Eisenbud 1995, Ch. 10], for example.) If $\mathbb{K}$ is an infinite field, M is a finitely generated graded module of dimension > 0, and y is a general linear form, then $\dim M/yM = \dim M - 1$. It follows that if M has Krull dimension d then any sufficiently general sequence of linear forms $y_1, \ldots, y_d$ is a system of parameters. Moreover, if M is a Cohen–Macaulay module then every system of parameters is a regular sequence on M.

COROLLARY 3.4. *Suppose that $\mathbb{K}$ is an infinite field, and that M is a finitely generated graded S-module of dimension d. Let $y_1, \ldots, y_d$ be general linear forms. The module M is a finitely generated module over the subring $T := \mathbb{K}[y_1, \ldots, y_d]$. It is a Cohen–Macaulay S-module if and only if it is free as a graded T-module. It is an Ulrich S-module if and only if, for some n, it is isomorphic to T^n as a graded T-module.*

PROOF. Because M is a graded S-module, it is also a graded T-module and M is zero in sufficiently negative degrees. It follows that M can be generated by any set of elements whose images generate $\overline{M} := M/(y_1, \ldots, y_d)M$. In particular, M is a finitely generated T module if $\overline{M}$ is a finite-dimensional vector space. Since $y_1, \ldots, y_d$ are general, the Krull dimension of $\overline{M}$ is 0. Since $\overline{M}$ is also a finitely generated S-module, it is finite-dimensional as a vector space, proving that M is a finitely generated T-module.

The module M is a Cohen–Macaulay S-module if and only if $y_1, \ldots y_d$ is an M-regular sequence, and this is the same as the condition that M be a Cohen–Macaulay T-module. Since T is regular and has the same dimension as M, the Auslander–Buchsbaum formula shows that M is a Cohen–Macaulay T-module if and only if it is a free graded T-module.

For the statement about Ulrich modules we need to use the characterization of Castelnuovo–Mumford regularity by local cohomology; see [Brodmann and Sharp 1998] or [Eisenbud $\geq$ 2004]. Since $y_1, \ldots, y_d$ is a system of parameters on M, the ideal $(y_1, \ldots, y_d) + \operatorname{ann} M$ has radical $(x_0, \ldots, x_r)$ and it follows that the local cohomology modules $H^i_{(x_0, \ldots, x_r)}(M)$ and $H^i_{(y_1, \ldots, y_d)}(M)$ are the same. Thus the regularity of M is the same as a T-module or as an S-module. We can rephrase the definition of the Ulrich property to say that M is Ulrich if and only if M is Cohen–Macaulay, $M_i = 0$ for $i < 0$ and M has regularity 0. Thus M is Ulrich as an S-module if and only if it is Ulrich as a T-module. Since M is a graded free T-module, we see that it is Ulrich if and only if, as a T-module, it is a direct sum of copies of T. $\hfill\square$

Since $S/\operatorname{ann} M$ acts on M as endomorphisms, we can say from Corollary 3.4 that there is an Ulrich module with annihilator I if and only if (for some n) the ring S/I admits a faithful representation as $n \times n$ matrices over a polynomial ring. Similarly, there is a Cohen–Macaulay module with annihilator I if S/I has a faithful representation as $\operatorname{End} F$ modules for some graded free module F.

3.2. Bounds on the regularity. The regularity of an arbitrary ideal $I \subset S$ can behave very wildly, but there is evidence to suggest that the regularity of ideals defining (nice) varieties is much lower. Here is a sampling of results and conjectures in this direction. See for example [Bayer and Mumford 1993] for the classic conjectures and [Chardin and D'Cruz 2003] and the papers cited there for a more detailed idea of current research.

Arbitrary ideals: Mayr–Meyer and Bayer–Stillman. Arguments going back to Hermann [1926] give a bound on the regularity of an ideal that depends only on the degrees of its generators and the number of variables — a bound that is extremely large.

THEOREM 3.5. *If I is generated by forms of degree d in a polynomial ring in $r + 1$ variables over a field of characteristic zero, then* $\operatorname{reg} I \leq (2d)^{2^{r-1}}$.

For recent progress in positive characteristic, see [Caviglia and Sbarra 2003].

An argument of Mayr and Meyer [1982], adapted to the case of ideals in a polynomial ring by Bayer and Stillman [1988], shows that the regularity can really be (roughly) as large as this bound allows: doubly exponential in the number of variables. These examples were improved slightly by Koh [1998] to give the following result.

THEOREM 3.6. *For each integer $n \geq 1$ there exists an ideal $I_n \subset \mathbb{K}[x_0, \ldots, x_r]$ with $r = 22n - 1$ that is generated by quadrics and has regularity*

$$\operatorname{reg} I_n \geq 2^{2^{n-1}}.$$

See [Swanson 2004] for a detailed study of the primary decomposition of the Bayer–Stillman ideals, which are highly nonreduced. See [Giaimo 2004] for a way of making reduced examples using these ideals as a starting point.

By contrast, for smooth or nearly smooth varieties, there are much better bounds, linear in each of r and d, due to Bertram, Ein and Lazarsfeld [Bertram et al. 1991] and Chardin and Ulrich [2002]: For example:

THEOREM 3.7. *If* $\mathbb{K}$ *has characteristic* 0 *and* $X \subset \mathbb{P}^r$ *is a smooth variety defined scheme-theoretically by equations of degree* $\leq d$, *then*

$$\operatorname{reg} I_X \leq 1 + (d-1)r.$$

More precisely, if X *has codimension* c *and* X *is defined scheme-theoretically by equations of degrees* $d_1 \geq d_2 \geq \cdots$, *then*

$$\operatorname{reg} I_X \leq d_1 + \cdots d_c - c + 1.$$

The hypotheses "smooth" and "characteristic 0" are used in the proof through the use of the Kawamata–Viehweg vanishing theorems; but there is no evidence that they are necessary to the statement, which might be true for any reduced algebraic set over an algebraically closed field.

The Bertram–Ein–Lazarsfeld bound is sharp for complete intersection varieties. But one might feel that, when dealing with the defining ideal of a variety that is far from a complete intersection, the degree of the variety is a more natural measure of complexity than the degrees of the equations. This point of view is borne out by a classic theorem proved by Castelnuovo in the smooth case and by Gruson, Lazarsfeld and Peskine [Gruson et al. 1983] in general: If $\mathbb{K}$ is algebraically closed and I is prime defining a projective curve X, then the regularity is linear in the degree of X. (Extending the ground field does not change the regularity, but may spoil primeness.)

THEOREM 3.8. *If* $\mathbb{K}$ *is algebraically closed, and* I *is the ideal of an irreducible curve* X *of degree* d *in* $\mathbb{P}^r$ *not contained in a hyperplane, then*

$$\operatorname{reg} I \leq d - r + 2.$$

Giaimo [2003] has proved a generalization of this bound when X is only assumed to be reduced, answering a conjecture of Eisenbud.

On the other hand, it is easy to see (or look up in [Eisenbud $\geq$ 2004, Ch. 4]) that if $X \subset \mathbb{P}^r$ is a scheme not contained in any hyperplane, and S/I_X is Cohen–Macaulay, then

$$\operatorname{reg} I_X \leq \deg X - \operatorname{codim} X + 1.$$

When X is an irreducible curve, this coincides with the Gruson–Lazarsfeld–Peskine Theorem. From this remark and some (scanty) further evidence, Eisenbud and Goto [1984] conjectured that the same bound holds for prime ideals:

CONJECTURE 3.9. *Let* $\mathbb{K}$ *be an algebraically closed field. If* $X \subset \mathbb{P}^r$ *is an irreducible variety not contained in a hyperplane, then*

$$\operatorname{reg} I_X \leq \deg X - \operatorname{codim} X + 1.$$

This is now known to hold for surfaces that are smooth [Bayer and Mumford 1993] and a little more generally [Brodmann 1999; Brodmann and Vogel 1993]; also for toric varieties of codimension two [Peeva and Sturmfels 1998] and a few other classes. Slightly weaker bounds, still linear in the degree, are known for smooth varieties up to dimension six [Kwak 1998; 2000]. Based on a similar analogy and and a little more evidence, Eisenbud has conjectured that the bound of Conjecture 3.9 holds if X is merely reduced and connected in codimension 1.

Both the connectedness and the reducedness hypothesis are necessary, as the following examples show:

EXAMPLE 3.10 (TWO SKEW LINES IN $\mathbb{P}^3$). Let

$$I = (s, t) \cap (u, v) = (s, t) \cdot (u, v) \subset S = \mathbb{K}[s, t, u, v]$$

be the ideal of the union X of two skew lines (that is, lines that do not meet) in $\mathbb{P}^3$. The degree of X is of course 2, and X is certainly not contained in a hyperplane. But the Betti diagram of the resolution of S/I is

	0	1	2	3
0	1	–	–	–
1	–	4	4	1

so reg $I = 2 > \deg X - \operatorname{codim} X + 1$.

EXAMPLE 3.11 (A MULTIPLE LINE IN $\mathbb{P}^3$). Let

$$I = (s, t)^2 + (p(u, v) \cdot s + q(u, v) \cdot t) \subset S = \mathbb{K}[s, t, u, v],$$

where $p(u, v)$ and $q(u, v)$ are relatively prime forms of degree $d \geq 1$. The ideal I has degree 2, independent of d, and no embedded components. The scheme X defined by I has degree 2; it is a double structure on the line $V(s, t)$, contained in the first infinitesimal neighborhood $V((s, t)^2)$ of the line in $\mathbb{P}^3$. It may be visualized as the thickening of the line along a "ribbon" that twists d times around the line. But the Betti diagram of the resolution of S/I is

	0	1	2	3
0	1	–	–	–
1	–	3	2	–
2	–	–	–	–
⋮	⋮	⋮	⋮	⋮
$d-1$	–	–	–	–
d	–	1	2	1

so reg $I = d + 1 > \deg X - \operatorname{codim} X + 1 = 1$.

This last example (and many more) shows that there is no bound on the regularity of a nonreduced scheme in terms of the degree of the scheme alone. But the problem for reduced schemes is much milder, and Bayer and Stillman [1988] have conjectured that the regularity of a reduced scheme over an algebraically closed field should be bounded by its degree (the sum of the degrees of its components). Perhaps the strongest current evidence for this assertion is the recent result of Derksen and Sidman [2002]:

THEOREM 3.12. *If $\mathbb{K}$ is an algebraically closed field and X is a union of d linear subspaces of $\mathbb{P}^r$, then* $\operatorname{reg} I_X \leq d$

3.3. Bounds on the ranks of the free modules. From the work of Hermann [1926], there are (very large) upper bounds known for the ranks of the free modules F_i in a minimal free resolution

$$\mathbf{F}: \quad 0 \longrightarrow F_m \xrightarrow{\phi_m} \ldots \to F_1 \xrightarrow{\phi_1} F_0$$

in terms of the ranks of the modules F_1 and F_0 and the degrees of their generators. However, recent work has focused on *lower* bounds. The only general result known is that of Evans and Griffith [1981; 1985]:

THEOREM 3.13. *If $F_m \neq 0$ then* $\operatorname{rank} \operatorname{im} \phi_i \geq i$ *for $i < m$; in particular,* $\operatorname{rank} F_i \geq 2i + 1$ *for $i < m - 1$, and* $\operatorname{rank} F_{m-1} \geq n$.

For an example, consider the Koszul complex resolving S/I, where I is generated by a regular sequence of length m. In this case $\operatorname{rank} F_{m-1} = m$, showing that the first statement of Theorem 3.13 is sharp for $i = m-1$. But in the Koszul complex case the "right" bound for the rank is a binomial coefficient. Based on many small examples, Horrocks (see [Hartshorne 1979, problem 24]), motivated by questions on low rank vector bundles, and independently Buchsbaum and Eisenbud [1977], conjectured that something like this should be true more generally:

CONJECTURE 3.14. *If M has codimension c, then the i-th map ϕ_i in the minimal free resolution of M has* $\operatorname{rank} \phi_i \geq \binom{c-1}{i-1}$, *so the i-th free module, has* $\operatorname{rank} F_i \geq \binom{c}{i}$. *In particular,* $\sum \operatorname{rank} F_i \geq 2^c$.

The last statement, slightly generalized, was made independently by the topologist Gunnar Carlsson [1982; 1983] in connection with the study of group actions on products of spheres.

The conjecture is known to hold for resolutions of monomial ideals [Charalambous 1991], for ideals in the linkage class of a complete intersection [Huneke and Ulrich 1987], and for small r (see [Charalambous and Evans 1992] for more information). The conjectured bound on the sum of the ranks holds for almost complete intersections by Dugger [2000] and for graded modules in certain cases by Avramov and Buchweitz [1993].

4. Linear Complexes and the Strands of Resolutions

As before we set $S = \mathbb{K}[x_0, \ldots, x_r]$. The free resolution of a finitely generated graded module can be built up as an iterated extension of *linear complexes*, its *linear strands*. These are complexes whose maps can be represented by matrices of linear forms. In this section we will explain the *Bernstein–Gelfand–Gelfand correspondence* (BGG) between linear free complexes and modules over a certain exterior algebra. We will develop an exterior algebra version of Fitting's Lemma, which connects annihilators of modules over a commutative ring with minors of matrices. Finally, we will use these tools to explain Green's proof of the Linear Syzygy Conjecture of Eisenbud, Koh, and Stillman.

4.1. Strands of resolutions. Let

$$\boldsymbol{F}: \quad 0 \longrightarrow F_m \xrightarrow{\phi_m} F_{m-1} \longrightarrow \cdots \longrightarrow F_1 \xrightarrow{\phi_1} F_0$$

be any complex of free graded modules, and write $F_i = \bigoplus_j S(-j)^{\beta_{i,j}}$. Although we do not assume that $\boldsymbol{F}$ is a resolution, we require it to be a *minimal complex* in the sense that F_i maps to a submodule not containing any minimal generator of F_{i-1}. By Nakayama's Lemma, this condition is equivalent to the condition that $\phi_i F_i \subset (x_0, \ldots, x_r) F_{i-1}$ for all $i > 0$.

Under these circumstances $\boldsymbol{F}$ has a natural filtration by subcomplexes as follows. Let $b_0 = \min_{i,j} \{j - i \mid \beta_{i,j} \neq 0\}$. For each i let L_i be the submodule of F_i generated by elements of degree $b_0 + i$. Because F_i is free and has no elements of degree $< b_0 + i$ the module L_i is free, $L_i \cong S(-b_0 - i)^{\beta_{i,b_0+i}}$. Further, since $\phi_i F_i$ does not contain any of the minimal degree elements of F_{i-1} we see that $\phi_i L_i \subset L_{i-1}$; that is, the modules L_i form a subcomplex of $\boldsymbol{F}$. This subcomplex $\boldsymbol{L}$ is a *linear free complex* in the sense that L_i is generated in degree 1 more than L_{i-1}, so that the differential $\psi_i = \phi_i|_{L_i} : L_i \to L_{i-1}$ can be represented by a matrix of linear forms.

We will denote this complex $\mathscr{L}_{b_0}$, and call it the first strand of $\boldsymbol{F}$. Factoring out $\boldsymbol{L}_{b_0}$ we get a new minimal free complex, so we can repeat the process to get a filtration of $\boldsymbol{F}$ by these strands.

We can see the numerical characteristics of the strands of $\boldsymbol{F}$: it follows at once from the definition that the Betti diagrams of the linear strands of $\boldsymbol{F}$ are the rows of the Betti diagram of $\boldsymbol{F}$! This is perhaps the best reason of all for writing the Betti diagram in the form we have given.

Now suppose that $\boldsymbol{F}$ is actually the minimal free resolution of I_X for some projective scheme X. It turns out in many interesting cases that the lengths of the individual strands of $\boldsymbol{F}$ carry much deeper geometric information than does the length of $\boldsymbol{F}$ itself. A first example of this can be seen in the case of the four points, treated in Section 1. The Auslander–Buchsbaum Theorem shows that the resolution of S_X has length exactly r for any finite set of points $X \subset \mathbb{P}^r$. But for four points not all contained in a line, the first linear strand of the minimal resolution of the ideal I_X had length 1 if and only if some three of the points were collinear (else it had length 0). In fact, the line itself was visible in the first strand: in case it had length 1, it had the form

$$0 \to S(-3) \xrightarrow{\phi} S(-2)^2 \longrightarrow 0,$$

and the entries of a matrix representing ϕ were exactly the generators of the ideal of the line.

A much deeper example is represented by the following conjecture of Mark Green. After the genus, perhaps the most important invariant of a smooth algebraic curve is its *Clifford index*. For a smooth curve of genus $g \geq 3$ this may be defined as the minimum, over all degree d maps α (where $d \leq g - 1$) from X to a projective space $\mathbb{P}^r$, with image not contained in a hyperplane, as $\deg \alpha - 2r$. The number Cliff X is always nonnegative (Clifford's Theorem). The smaller the Clifford index is, the more special X is. For example, the Clifford index Cliff X is 0 if and only if X is a double cover of $\mathbb{P}^1$. If it is not 0, it is 1 if and only if X is either a smooth plane curve or a triple cover of the line. For "most curves" the Clifford index is simply $d - 2$, where d is the smallest degree of a nonconstant map from X to $\mathbb{P}^1$; see [Eisenbud et al. 1989].

A curve X of genus $g \geq 3$ that is not hyperelliptic has a distinguished embedding in the projective space $X \subset \mathbb{P}^{g-1}$ called the *canonical embedding*, obtained by taking the complete linear series of canonical divisors. Any invariant derived from the canonical embedding of a curve is thus an invariant of the abstract (nonembedded) curve. It turns out that the Hilbert functions of all canonically embedded curves of genus g are the same. This is true also of the projective dimension and the regularity of the ideals of such curves. But the graded Betti numbers seem to reflect quite a lot of the geometry of the curve. In particular, Green conjectured that the length of the first linear strand of the resolution of I_X gives precisely the Clifford index:

CONJECTURE 4.1. Suppose that $\mathbb{K}$ has characteristic 0, and let X be a smooth curve of genus g, embedded in $\mathbb{P}^{g-1}$ by the complete canonical series. The length of the first linear strand of the minimal free resolution of I_X is $g - 3 -$ Cliff X.

The conjecture has been verified by Schreyer [1989] for all curves of genus $g \leq 8$. It was recently proved for a generic curve of each Clifford index by Teixidor [2002] and Voisin [2002a; 2002b] (this may not prove the whole conjecture, because the family of curves of given genus and Clifford index may not be irreducible).

For a version of the conjecture involving high-degree embeddings of X instead of canonical embeddings, see [Green and Lazarsfeld 1988]. See also [Eisenbud 1992; Schreyer 1991] for more information.

4.2. How long is a linear strand? With these motivations, we now ask for bounds on the length of the first linear strand of the minimal free resolution of an arbitrary graded module M. One of the few general results in this direction is due to Green. To prepare for it, we give two examples of minimal free resolutions with rather long linear strands — in fact they will have the maximal length allowed by Green's Theorem:

EXAMPLE 4.2 (THE KOSZUL COMPLEX). The first example is already familiar: the Koszul complex of the linear forms $x_0, \ldots, x_s$ is equal to its first linear strand.

Notice that it is a resolution of a module with just one generator, and that the length of the resolution (or its linear strand) is the dimension of the vector space $\langle x_0, \ldots, x_s \rangle$ of linear forms that annihilate that generator.

EXAMPLE 4.3 (THE CANONICAL MODULE OF THE RATIONAL NORMAL CURVE). Let X be the *rational normal curve of degree d*, the image in $\mathbb{P}^d$ of the map

$$\mathbb{P}^1 \to \mathbb{P}^d : \qquad (s, t) \mapsto (s^d, s^{d-1}t, \ldots, st^{d-1}, t^d).$$

It is not hard to show (see [Harris 1992, Example 1.16] or [Eisenbud 1995, Exercise A2.10]) that the ideal I_X is generated by the 2×2 minors of the matrix

$$\begin{pmatrix} x_0 & x_1 & \cdots & x_{d-1} \\ x_1 & x_2 & \cdots & x_d \end{pmatrix}$$

and has minimal free resolution $\boldsymbol{F}$ with Betti diagram

	0	1	$\cdots$	$d-2$	$d-1$
0	1	$-$	$\cdots$	$-$	$-$
1	$-$	$\binom{d}{2}$	$\cdots$	$(d-2)d$	$d-1$

Let ω be the module of twisted global sections of the canonical sheaf $\omega_{\mathbb{P}^1} = \mathscr{O}_{\mathbb{P}^1}(-2 \text{ points})$:

$$\mathrm{H}^0 \mathscr{O}_{\mathbb{P}^1}(-2 + d \text{ points}) \oplus \mathrm{H}^0 \mathscr{O}_{\mathbb{P}^1}(-2 + 2d \text{ points}) \oplus \cdots$$

By duality, ω can be expressed as $\mathrm{Ext}_S^{d-1}(S_X, S(-d-1))$, which is the cokernel of the dual of the last map in the resolution of S_X, twisted by $-d-1$. From the length of the resolution and the Auslander–Buchsbaum Theorem we see that S_X is Cohen–Macaulay, and it follows that the twisted dual of the resolution, $\boldsymbol{F}^*(-d-1)$, is the resolution of ω, which has Betti diagram

	0	1	$\cdots$	$d-2$	$d-1$
1	$d-1$	$(d-2)d$	$\cdots$	$\binom{d}{2}$	$-$
2	$-$	$-$	$\cdots$	$-$	1

In particular, we see that ω has $d-1$ generators, and the length of the first linear strand of its resolution is $d-2$. Since ω is the module of twisted global sections of a line bundle on X, it is a torsion-free S_X-module. In particular, since the ideal of X does not contain any linear forms, no element of ω is annihilated by any linear form.

These examples hint at two factors that might influence the length of the first linear strand of the resolution of M: the generators of M that are annihilated by linear forms; and the sheer number of generators of M. We can pack both these numbers into one invariant. For convenience we will normalize M by shifting the grading until $M_i = 0$ for $i < 0$ and $M_0 \neq 0$. Let $W = S_1$ be the space of linear forms in S, and let $\mathbb{P} = \mathbb{P}(M_0^*)$ be the projective space of 1-dimensional subspaces of M_0 (we use the convention that $\mathbb{P}(V)$ is the projective space of 1-dimensional quotients of V). Let $A(M) \subset W \times \mathbb{P}$ be the set

$$A(M) := \{(x, \langle m \rangle) \in W \times \mathbb{P} \mid xm = 0\}.$$

The set $A(M)$ contains $0 \times \mathbb{P}$, so it has dimension $\geq \dim_{\mathbb{K}} M_0 - 1$. In Example 4.3 this is all it contains. In Example 4.2 however it also contains $\langle x_0, \ldots, x_s \rangle \times \langle 1 \rangle$, a variety of dimension $s+1$, so it has dimension $s+1$. In both cases its dimension is the same as the length of the first linear strand of the resolution. The following result was one of those conjectured by Eisenbud, Koh and Stillman (the "Linear Syzygy Conjecture") based on a result of Green's covering the torsion-free case, and then proved in general by Green [1999]:

THEOREM 4.4 (LINEAR SYZYGY CONJECTURE). *Let M be a graded S-module, and suppose for convenience that $M_i = 0$ for $i < 0$ while $M_0 \neq 0$. The length of the first linear strand of the minimal free resolution of M is at most $\dim A(M)$.*

Put differently, the only way that the length of the first linear strand can be $> \dim_{\mathbb{K}} M_0$ is if there are "many" nontrivial pairs $(x, m) \in W \times M_0$ such that $xm = 0$.

The statement of Theorem 4.4 is only one of several conjectures in the paper of Eisenbud, Koh, and Stillman. For example, they also conjecture that if the resolution of M has first linear strand of length $> \dim_{\mathbb{K}} M_0$, and M is minimal in a suitable sense, then every element of M_0 must be annihilated by some linear form. See [Eisenbud et al. 1988] for this and other stronger forms.

Though it explains the length of the first linear strand of the resolutions of the residue field $\mathbb{K}$ or its first syzygy module $(x_0, \ldots, x_r)$, Theorem 4.4 is far from sharp in general. A typical case where one would like to do better is the following: the second syzygy module of $\mathbb{K}$ has no torsion and $\binom{r+1}{2}$ generators, so the theorem bounds the length of its first linear strand by $\binom{r+1}{2} - 1$. However, its first linear strand only has length $r - 1$. We have no theory — not even a conjecture — capable of predicting this.

We will sketch the proof after developing some basic theory connecting the question with questions about modules over exterior algebras.

4.3. Linear free complexes and exterior modules.

The Bernstein–Gelfand–Gelfand correspondence is usually thought of as a rather abstract isomorphism between some derived categories. However, it has at its root a very simple observation about linear free complexes: A linear free complex over S is "the same thing" as a module over the exterior algebra on the dual of S_1.

To simplify the notation, we will (continue to) write W for the vector space of linear forms S_1 of S, and we write $V := W^*$ for its vector space dual. We set $E = \bigwedge V$, the exterior algebra on V. Since W consists of elements of degree 1, we regard elements of V as having degree -1, and this gives a grading on E with $\bigwedge^i V$ in degree $-i$. We will use the element $\sum_i x_i \otimes e_i \in W \otimes V$, where $\{x_i\}$ and

$\{e_i\}$ are dual bases of W and V. This element does not depend on the choice of bases — it is the image of $1 \in \mathbb{K}$ under the dual of the natural contraction map

$$V \otimes W \to \mathbb{K} : \qquad v \otimes w \mapsto v(w).$$

Although E is not commutative in the usual sense, it is *strictly commutative* in the sense that

$$ef = (-1)^{(\deg e)(\deg f)} fe$$

for any two homogeneous elements $e, f \in E$. In nearly all respects, E behaves just like a finite-dimensional commutative graded ring. The following simple idea connects graded modules over E with linear free complexes over S:

PROPOSITION 4.5. *Let $\{x_i\}$ and $\{e_i\}$ be dual bases of W and V, and let $P = \bigoplus P_i$ be a graded E-module. The maps*

$$\phi_i : S \otimes P_i \to S \otimes P_{i-1}$$

$$1 \otimes p \mapsto \sum_j x_j \otimes e_j p$$

make

$$\mathbf{L}(P): \qquad \cdots \longrightarrow S \otimes P_i \xrightarrow{\phi_i} S \otimes P_{i-1} \longrightarrow \cdots,$$

into a linear complex of free S-modules. Every complex of free S modules $\mathbf{L} : \cdots \to F_i \to F_{i-1} \to \cdots$, where F_i is a sum of copies of $S(-i)$ has the form $\mathbf{L}(P)$ for a unique graded E-module P.

PROOF. Given an E-module P, we have

$$\phi_{i-1}\phi_i(p) = \phi_{i-1}\left(\sum_j x_j \otimes e_j p\right) = \sum_k \left(\sum_j x_j x_k \otimes e_j e_k p\right).$$

The terms $x_j^2 \otimes e_j^2 p$ are zero because $e_j^2 = 0$. Each other term occurs twice, with opposite signs, because of the skew-commutativity of E, so $\phi_{i-1}\phi_i = 0$ and $\mathbf{L}(P)$ is a linear free complex as claimed.

Conversely, given a linear free complex $\mathbf{L}$, we set $P_i = F_i/(x_0, \ldots, x_r)F_i$. Because $\mathbf{L}$ is linear, differentials of $\mathbf{L}$ provide maps $\psi_i : P_i \to W \otimes P_{i-1}$. Suppose $p \in P_i$. If $\psi_i(p) = \sum x_j \otimes p_j$ then we define $\mu_i : E \otimes P_i \to E \otimes P_{i-1}$ by $1 \otimes p \mapsto \sum e_j \otimes p_j$. Using the fact that the differentials of $\mathbf{L}$ compose to 0, it is easy to check that these "multiplication" maps make P into a graded E-module, and that the two operations are inverse to one another. $\quad\square$

EXAMPLE 4.6. The Koszul complex

$$0 \to \textstyle\bigwedge^{r+1} S^{r+1} \to \cdots \to \textstyle\bigwedge^1 S^{r+1} \to S \to 0$$

is a linear free complex over S. To make the maps natural, we should think of S^{r+1} as $S \otimes W$. Applying the recipe in Proposition 4.5 we see that $P_i = \bigwedge^i W$. The module structure on P is that given by contraction, $e \otimes x \mapsto e \neg x$. This module P is canonically isomorphic to $\mathrm{Hom}(E, \mathbb{K})$, which is a left E-module via the right-module structure of E. It is also noncanonically isomorphic to E; to define the isomorphism $E \to P$ we must choose a nonzero element of $\bigwedge^{r+1} W$, an *orientation*, to be the image of $1 \in E$. See [Eisenbud 1995, Ch. 17], for details.

The BGG point of view on linear complexes is well-adapted to studying the linear strand of a resolution, as one sees from the following result. For any finitely generated left module P over E we write $\hat{P}$ for the module $\mathrm{Hom}(P, \mathbb{K})$; it is naturally a right module, but we make it back into a left module via the involution $\iota : E \to E$ sending a homogeneous element $a \in E$ to $\iota(a) = (-1)^{\deg a} a$.

PROPOSITION 4.7. *Let $\boldsymbol{L} = \boldsymbol{L}(P)$ be a finite linear free complex over S corresponding to the graded E-module P.*

1. *$\boldsymbol{L}$ is a subcomplex of the first linear strand of a minimal free resolution if and only if $\hat{P}$ is generated in degree 0.*
2. *$\boldsymbol{L}$ is the first linear strand of a minimal free resolution if and only if $\hat{P}$ has a linear presentation matrix.*

...

∞. *$\boldsymbol{L}$ is a free resolution if and only if $\hat{P}$ has a linear free resolution.*

Here the infinitely many parts of the proposition correspond to the infinitely many degrees in which $\boldsymbol{L}$ could have homology. For the proof, which depends on the Koszul homology formula $\mathrm{H}_i(\boldsymbol{L})_{i+d} = \mathrm{Tor}_d(\mathbb{K}, \hat{P})_{-d-i}$, see [Eisenbud et al. 2003a].

4.4. The exterior Fitting Lemma and the proof of the Linear Syzygy Conjecture.
The strategy of Green's proof of the Linear Syzygy Conjecture is now easy to describe. We first reformulate the statement slightly. The algebraic set $A(M)$ consists of $\mathbb{P}$ and the set

$$A' = \big\{ (x, \langle y \rangle) \mid 0 \neq x \in W,\ y \in M_0 \text{ and } xm = 0 \big\}.$$

Supposing that the first linear strand $\boldsymbol{L}$ of the resolution of M has length k greater than $\dim A'$, we must show that $k < \dim M_0$. We can write $\boldsymbol{L} = \boldsymbol{L}(P)$ for some graded E-module P, and we must show that $P_m = 0$, where $m = \dim_{\mathbb{K}} M_0$.

From Proposition 4.7 we know that $\hat{P}$ is generated in degree 0. Thus to show $P_m = 0$, it is necessary and sufficient that we show that $\hat{P}$ is annihilated by the m-th power of the maximal ideal E_+ of E. In fact, we also know that $\hat{P}$ is linearly presented. Thus we need to use the linear relations on $\hat{P}$ to produce enough elements of the annihilator of P to generate $(E_+)^m$.

If E were a commutative ring, this would be exactly the sort of thing where we would need to apply the classical Fitting Lemma (see for example [Eisenbud 1995, Ch. 20]), which derives information about the annihilator of a module from a free presentation. We will explain a version of the Fitting Lemma that can be used in our exterior situation.

First we review the classical version. We write $I_m(\phi)$ for the ideal of $m \times m$ minors of a matrix ϕ

LEMMA 4.8 (FITTING'S LEMMA). *If*

$$M = \operatorname{coker}(S^n(-1) \xrightarrow{\phi} S^m)$$

is a free presentation, then $I_m(\phi) \subset \operatorname{ann} M$. *In the generic case, when* ϕ *is represented by a matrix of indeterminates, the annihilator is equal to* $I_m(\phi)$.

To get an idea of the analogue over the exterior algebra, consider first the special case of a module with presentation

$$P = \operatorname{coker}(E(1) \xrightarrow{\begin{pmatrix} e_1 \\ \vdots \\ e_t \end{pmatrix}} E^t),$$

where the e_i are elements of $V = E_{-1}$. If we write $p_1, \ldots, p_t$ for the elements of P that are images of the basis vectors of E^t, then the defining relation is $\sum e_i p_i = 0$. We claim that $\prod_i e_i$ annihilates P. Indeed,

$$\left(\prod_i e_i\right) p_j = \left(\prod_{i \neq j} e_i\right) e_j p_j = \left(\prod_{i \neq j} e_i\right) \sum_{i \neq j} -e_i p_i = 0$$

since $e_i^2 = 0$. Once can show that, if the e_i are linearly independent, then $\prod_i e_i$ actually generates $\operatorname{ann} P$. Thus the product is the analogue of the "Fitting ideal" in this case.

In general, if

$$P = \operatorname{coker}(E(1)^s \xrightarrow{\phi} E^t),$$

then the product of the elements of every column of the matrix ϕ annihilates P for the same reason. The same is true of the *generalized columns* of ϕ — that is, the linear combinations with $\mathbb{K}$ coefficients of the columns. In the generic case these products generate the annihilator. Unfortunately it is not clear from this description which — or even how many — generalized columns are required to generate this ideal.

To get a more usable description, recall that the *permanent* $\operatorname{perm} \phi$ of a $t \times t$ matrix ϕ is the sum over permutations σ of the products $\phi_{1,\sigma(1)} \cdots \phi_{t,\sigma(t)}$ (the "determinant without signs"). At least in characteristic zero, the product $\prod e_i$ in our first example is $t!$ times the permanent of the $t \times t$ matrix obtained by repeating the same column t times. More generally, if we make a $t \times t$ matrix ϕ using a_1 copies of a column ϕ^1, a_2 copies of a second column ϕ^2, and in general a_u copies of ϕ^u, so that $\sum a_i = t$, we find that, in the exterior algebra over the integers, the permanent is divisible by $a_1! a_2! \cdots a_u!$. We will write

$$(\phi_1^{(a_1)}, \ldots, \phi_u^{(a_u)}) = \frac{1}{a_1! a_2! \cdots a_u!} \operatorname{perm} \phi$$

for this expression, and we call it a $t \times t$ divided permanent of the matrix ϕ. It is easy to see that the divided permanents are in the linear span of the products of the generalized columns of ϕ. This leads us to the desired analogue of the Fitting Lemma:

THEOREM 4.9. *Let P be a module over E with linear presentation matrix $\phi : E^s(1) \to E^t$. The divided permanents $(\phi_1^{(a_1)}, \ldots, \phi_u^{(a_u)})$ are elements of the annihilator of P. If the st entries of the matrix ϕ are linearly independent in V, then these elements generate the annihilator.*

The ideal generated by the divided permanents can be described without recourse to the bases above as the image of a certain map $D_t(E^s(1)) \otimes \bigwedge^t E^t \to E$ defined from ϕ by multilinear algebra, where $D_t(F) = (\mathrm{Sym}_t F^*)^*$ is the t-th divided power. This formula first appears in [Green 1999] For the fact that the annihilator is generated by the divided permanents, and a generalization to matrices with entries of any degree, see [Eisenbud and Weyman 2003].

References

[Avramov and Buchweitz 1993] L. L. Avramov and R.-O. Buchweitz, "Lower bounds for Betti numbers", *Compositio Math.* **86**:2 (1993), 147–158.

[Bayer and Mumford 1993] D. Bayer and D. Mumford, "What can be computed in algebraic geometry?", pp. 1–48 in *Computational algebraic geometry and commutative algebra* (Cortona, 1991), edited by D. Eisenbud and L. Robbiano, Sympos. Math. **34**, Cambridge Univ. Press, Cambridge, 1993.

[Bayer and Stillman 1988] D. Bayer and M. Stillman, "On the complexity of computing syzygies", *J. Symbolic Comput.* **6**:2-3 (1988), 135–147.

[Bertram et al. 1991] A. Bertram, L. Ein, and R. Lazarsfeld, "Vanishing theorems, a theorem of Severi, and the equations defining projective varieties", *J. Amer. Math. Soc.* **4**:3 (1991), 587–602.

[Brennan et al. 1987] J. P. Brennan, J. Herzog, and B. Ulrich, "Maximally generated Cohen–Macaulay modules", *Math. Scand.* **61**:2 (1987), 181–203.

[Brodmann 1999] M. Brodmann, "Cohomology of certain projective surfaces with low sectional genus and degree", pp. 173–200 in *Commutative algebra, algebraic geometry, and computational methods* (Hanoi, 1996), edited by D. Eisenbud, Springer, Singapore, 1999.

[Brodmann and Sharp 1998] M. P. Brodmann and R. Y. Sharp, *Local cohomology: an algebraic introduction with geometric applications*, Cambridge Studies in Advanced Math. **60**, Cambridge University Press, Cambridge, 1998.

[Brodmann and Vogel 1993] M. Brodmann and W. Vogel, "Bounds for the cohomology and the Castelnuovo regularity of certain surfaces", *Nagoya Math. J.* **131** (1993), 109–126.

[Buchsbaum and Eisenbud 1977] D. A. Buchsbaum and D. Eisenbud, "Algebra structures for finite free resolutions, and some structure theorems for ideals of codimension 3", *Amer. J. Math.* **99**:3 (1977), 447–485.

[Burch 1967] L. Burch, "A note on ideals of homological dimension one in local domains", *Proc. Cambridge Philos. Soc.* **63** (1967), 661–662.

[Carlsson 1982] G. Carlsson, "On the rank of abelian groups acting freely on $(S^n)^k$", *Invent. Math.* **69**:3 (1982), 393–400.

[Carlsson 1983] G. Carlsson, "On the homology of finite free $(\mathbf{Z}/2)^n$-complexes", *Invent. Math.* **74**:1 (1983), 139–147.

[Caviglia and Sbarra 2003] G. Caviglia and E. Sbarra, "Characteristic-free bounds for the Castelnuovo-Mumford regularity", 2003. Available at arXiv:math.AC/0310122.

[Charalambous 1991] H. Charalambous, "Betti numbers of multigraded modules", *J. Algebra* **137**:2 (1991), 491–500.

[Charalambous and Evans 1992] H. Charalambous and E. G. Evans, Jr., "Problems on Betti numbers of finite length modules", pp. 25–33 in *Free resolutions in commutative algebra and algebraic geometry* (Sundance, UT, 1990), edited by edited by David Eisenbud and C. Huneke, Res. Notes Math. **2**, Jones and Bartlett, Boston, MA, 1992.

[Chardin and D'Cruz 2003] M. Chardin and C. D'Cruz, "Castelnuovo–Mumford regularity: examples of curves and surfaces", *J. Algebra* **270**:1 (2003), 347–360.

[Chardin and Ulrich 2002] M. Chardin and B. Ulrich, "Liaison and Castelnuovo–Mumford regularity", *Amer. J. Math.* **124**:6 (2002), 1103–1124.

[Ciliberto et al. 1986] C. Ciliberto, A. V. Geramita, and F. Orecchia, "Remarks on a theorem of Hilbert–Burch", pp. 1–25 (Exp. E) in *The curves seminar at Queen's* (Kingston, Ont., 1985–1986), vol. 4, edited by A. V. Geramita, Queen's Papers in Pure and Appl. Math. **76**, Queen's Univ., Kingston, ON, 1986.

[Derksen and Sidman 2002] H. Derksen and J. Sidman, "A sharp bound for the Castelnuovo–Mumford regularity of subspace arrangements", *Adv. Math.* **172**:2 (2002), 151–157.

[Dugger 2000] D. Dugger, "Betti numbers of almost complete intersections", *Illinois J. Math.* **44**:3 (2000), 531–541.

[Eisenbud 1992] D. Eisenbud, "Green's conjecture: an orientation for algebraists", pp. 51–78 in *Free resolutions in commutative algebra and algebraic geometry* (Sundance, UT, 1990), edited by edited by David Eisenbud and C. Huneke, Res. Notes Math. **2**, Jones and Bartlett, Boston, MA, 1992.

[Eisenbud 1995] D. Eisenbud, *Commutative algebra, with a view toward algebraic geometry*, Graduate Texts in Mathematics **150**, Springer, New York, 1995.

[Eisenbud $\geq$ 2004] D. Eisenbud, *The geometry of syzygies*, Graduate Texts in Math., Springer, New York. Available at http://www.msri.org/~de. To appear.

[Eisenbud and Goto 1984] D. Eisenbud and S. Goto, "Linear free resolutions and minimal multiplicity", *J. Algebra* **88**:1 (1984), 89–133.

[Eisenbud and Weyman 2003] D. Eisenbud and J. Weyman, "Fitting's lemma for $\mathbf{Z}/2$-graded modules", *Trans. Amer. Math. Soc.* **355**:11 (2003), 4451–4473 (electronic).

[Eisenbud et al. 1988] D. Eisenbud, J. Koh, and M. Stillman, "Determinantal equations for curves of high degree", *Amer. J. Math.* **110**:3 (1988), 513–539.

[Eisenbud et al. 1989] D. Eisenbud, H. Lange, G. Martens, and F.-O. Schreyer, "The Clifford dimension of a projective curve", *Compositio Math.* **72**:2 (1989), 173–204.

[Eisenbud et al. 2003a] D. Eisenbud, G. Fløystad, and F.-O. Schreyer, "Sheaf cohomology and free resolutions over exterior algebras", *Trans. Amer. Math. Soc.* **355**:11 (2003), 4397–4426 (electronic).

[Eisenbud et al. 2003b] D. Eisenbud, F.-O. Schreyer, and J. Weyman, "Resultants and Chow forms via exterior syzygies", *J. Amer. Math. Soc.* **16**:3 (2003), 537–579 (electronic).

[Evans and Griffith 1981] E. G. Evans and P. Griffith, "The syzygy problem", *Ann. of Math.* (2) **114**:2 (1981), 323–333.

[Evans and Griffith 1985] E. G. Evans and P. Griffith, *Syzygies*, London Mathematical Society Lecture Note Series **106**, Cambridge University Press, Cambridge, 1985.

[Giaimo 2003] D. M. Giaimo, "On the Castelnuovo–Mumford regularity of connected curves", 2003. Available at arXiv:math.AG/0309051.

[Giaimo 2004] D. M. Giaimo, *On the Castelnuovo–Mumford regularity of curves and reduced schemes*, Ph.D. thesis, University of California, Berkeley, CA, 2004.

[Goto and Watanabe 1978a] S. Goto and K. Watanabe, "On graded rings. I", *J. Math. Soc. Japan* **30**:2 (1978), 179–213.

[Goto and Watanabe 1978b] S. Goto and K. Watanabe, "On graded rings. II. ($\mathbf{Z}^n$-graded rings)", *Tokyo J. Math.* **1**:2 (1978), 237–261.

[Green 1999] M. L. Green, "The Eisenbud–Koh–Stillman conjecture on linear syzygies", *Invent. Math.* **136**:2 (1999), 411–418.

[Green and Lazarsfeld 1988] M. Green and R. Lazarsfeld, "Some results on the syzygies of finite sets and algebraic curves", *Compositio Math.* **67**:3 (1988), 301–314.

[Gruson et al. 1983] L. Gruson, R. Lazarsfeld, and C. Peskine, "On a theorem of Castelnuovo, and the equations defining space curves", *Invent. Math.* **72**:3 (1983), 491–506.

[Harris 1992] J. Harris, *Algebraic geometry: A first course*, Graduate Texts in Mathematics **133**, Springer, New York, 1992.

[Hartshorne 1977] R. Hartshorne, *Algebraic geometry*, Graduate Texts in Mathematics **52**, Springer, New York, 1977.

[Hartshorne 1979] R. Hartshorne, "Algebraic vector bundles on projective spaces: a problem list", *Topology* **18**:2 (1979), 117–128.

[Hermann 1926] G. Hermann, "Die Frage der endlich vielen Schritte in der Theorie der Polynomideale", *Math. Ann.* **95** (1926), 736– 788.

[Hilbert 1890] D. Hilbert, "Ueber die Theorie der algebraischen Formen", *Math. Ann.* **36** (1890), 473–534.

[Hochster 1975] M. Hochster, *Topics in the homological theory of modules over commutative rings*, CBMS Regional Conference Series in Mathematics **24**, American Math. Society, Providence, 1975.

[Huneke and Ulrich 1987] C. Huneke and B. Ulrich, "The structure of linkage", *Ann. of Math.* (2) **126**:2 (1987), 277–334.

[Koh 1998] J. Koh, "Ideals generated by quadrics exhibiting double exponential degrees", *J. Algebra* **200**:1 (1998), 225–245.

[Kwak 1998] S. Kwak, "Castelnuovo regularity for smooth subvarieties of dimensions 3 and 4", *J. Algebraic Geom.* **7**:1 (1998), 195–206.

[Kwak 2000] S. Kwak, "Generic projections, the equations defining projective varieties and Castelnuovo regularity", *Math. Z.* **234**:3 (2000), 413–434.

[Lang 2002] S. Lang, *Algebra*, third ed., Graduate Texts in Mathematics **211**, Springer, New York, 2002.

[Mayr and Meyer 1982] E. W. Mayr and A. R. Meyer, "The complexity of the word problems for commutative semigroups and polynomial ideals", *Adv. in Math.* **46**:3 (1982), 305–329.

[Mumford 1966] D. Mumford, *Lectures on curves on an algebraic surface*, Annals of Math. Studies **59**, Princeton University Press, Princeton, 1966.

[Peeva and Sturmfels 1998] I. Peeva and B. Sturmfels, "Syzygies of codimension 2 lattice ideals", *Math. Z.* **229**:1 (1998), 163–194.

[Rotman 1979] J. J. Rotman, *An introduction to homological algebra*, Pure and Applied Mathematics **85**, Academic Press, New York, 1979.

[Schreyer 1989] F.-O. Schreyer, "Green's conjecture for general p-gonal curves of large genus", pp. 254–260 in *Algebraic curves and projective geometry* (Trento, 1988), edited by E. Ballico and C. Ciliberto, Lecture Notes in Math. **1389**, Springer, Berlin, 1989.

[Schreyer 1991] F.-O. Schreyer, "A standard basis approach to syzygies of canonical curves", *J. Reine Angew. Math.* **421** (1991), 83–123.

[Swanson 2004] I. Swanson, "On the embedded primes of the Mayr–Meyer ideals", *J. Algebra* **275**:1 (2004), 143–190.

[Teixidor 2002] M. Teixidor I Bigas, "Green's conjecture for the generic r-gonal curve of genus $g \geq 3r - 7$", *Duke Math. J.* **111**:2 (2002), 195–222.

[Voisin 2002a] C. Voisin, "Green's generic syzygy conjecture for curves of even genus lying on a K3 surface", *J. Eur. Math. Soc.* **4**:4 (2002), 363–404.

[Voisin 2002b] C. Voisin, "Green's canonical syzygy conjecture for generic curves of odd genus", 2002. Available at math.AG/0301359.

DAVID EISENBUD
MATHEMATICAL SCIENCES RESEARCH INSTITUTE
and
DEPARTMENT OF MATHEMATICS
UNIVERSITY OF CALIFORNIA
970 EVANS HALL
BERKELEY, CA 94720-3840
UNITED STATES
de@msri.org

JESSICA SIDMAN
DEPARTMENT OF MATHEMATICS AND STATISTICS
MOUNT HOLYOKE COLLEGE
SOUTH HADLEY, MA 01075
UNITED STATES
jsidman@mtholyoke.edu

Trends in Commutative Algebra
MSRI Publications
Volume **51**, 2004

Commutative Algebra of n Points in the Plane

MARK HAIMAN

WITH AN APPENDIX BY EZRA MILLER

ABSTRACT. We study questions arising from the geometry of configurations of n points in the affine plane $\mathbb{C}^2$. We first examine the ideal of the locus where some two of the points coincide, and then study the rings of invariants and coinvariants for the action of the symmetric group S_n permuting the points among themselves. We also discuss the ideal of relations among the slopes of the lines that connect the n points pairwise, which is the subject of beautiful and surprising results by Jeremy Martin.

CONTENTS

Introduction

These lectures address commutative algebra questions arising from the geometry of configurations of n points in the affine plane $\mathbb{C}^2$. In the first lecture, we study the ideal of the locus where some two of the points coincide. We are led naturally to consider the action of the symmetric group S_n permuting the points among themselves. This provides the topic for the second lecture, in which we study the rings of invariants and coinvariants for this action. As you can see, we have chosen to study questions that involve rather simple and naive geometric considerations. For those who have not encountered this subject before, it may come as a surprise that the theorems which give the answers are quite remarkable, and seem to be hard.

One reason for the subtlety of the theorems is that lurking in the background is the more subtle geometry of the Hilbert scheme of points in the plane. The special properties of this algebraic variety play a role in the proofs of the theorems. The involvement of the Hilbert scheme in the proofs means that at present the theorems apply only to points in the plane, even though we could equally well raise the same questions for points in $\mathbb{C}^d$, and conjecturally we expect them to have similar answers.

In the third lecture, we change perspective slightly, by introducing the $\binom{n}{2}$ lines connecting the points in pairs, and asking for the ideal of relations among the slopes of these lines when the points are in general position (that is, no two points coincide). We present a synopsis of the beautiful and surprising results on this problem found by my former student, Jeremy Martin.

Lecture 1: A Subspace Arrangement

We consider ordered n-tuples of points in the plane, denoted by

$$P_1, \ldots, P_n \in \mathbb{C}^2.$$

We work over $\mathbb{C}$ to keep things simple and geometrically concrete, although some of the commutative algebra results remain true over more general ground rings. Assigning the points coordinates

$$x_1, y_1, \ldots, x_n, y_n,$$

we identify the space E of all n-tuples $(P_1, \ldots, P_n)$ with $\mathbb{C}^{2n}$. The coordinate ring of E is then the polynomial ring

$$\mathbb{C}[E] = \mathbb{C}[\boldsymbol{x}, \boldsymbol{y}] = \mathbb{C}[x_1, y_1, \ldots, x_n, y_n]$$

in $2n$ variables. Let V_{ij} be the locus where $P_i = P_j$, that is, the codimension-2 subspace of E defined by the equations $x_i = x_j$ and $y_i = y_j$. The locus

$$V = \bigcup_{i<j} V_{ij}$$

where some two points coincide is a *subspace arrangement* of $\binom{n}{2}$ codimension-2 subspaces in E. Evidently, V is the zero locus of the radical ideal

$$I = I(V) = \bigcap_{i<j} (x_i - x_j, \ y_i - y_j).$$

The central theme of today's lecture is: *What does the ideal I look like?*

As a warm-up, we consider the much easier case of n points on a *line*. Then we only have coordinates $x_1, \ldots, x_n$, and the analog of I is the ideal

$$J = \bigcap_{i<j} (x_i - x_j) \subseteq \mathbb{C}[\boldsymbol{x}].$$

This ideal has some easily checked properties.

(1) J is the principal ideal $(\Delta(\boldsymbol{x}))$ generated by the Vandermonde determinant

$$\Delta(\boldsymbol{x}) = \prod_{i<j}(x_i - x_j) = \det \begin{bmatrix} 1 & x_1 & \cdots & x_1^{n-1} \\ 1 & x_2 & \cdots & x_2^{n-1} \\ \vdots & \vdots & & \vdots \\ 1 & x_n & \cdots & x_n^{n-1} \end{bmatrix}.$$

(2) J is (trivially) a free $\mathbb{C}[\boldsymbol{x}]$ module with generator $\Delta(\boldsymbol{x})$.

(3) $J^m = J^{(m)} \stackrel{\text{def}}{=} \bigcap_{i<j}(x_i - x_j)^m$, that is, the powers of J are equal to its symbolic powers. This is clear, since both ideals are equal to $(\Delta(\boldsymbol{x})^m)$.

(4) The Rees algebra $\mathbb{C}[\boldsymbol{x}][tJ]$ is Gorenstein. In fact, it's just a polynomial ring in $n+1$ variables.

All this follows from the fact that J is the ideal of a hyperplane arrangement. In general, one cannot say much about the ideal of an arrangement of subspaces of codimension 2 or more. However, our ideal I is rather special, so let's try to compare its properties with those listed above for J.

Beginning with property (1), we can observe that I has certain obvious elements. The symmetric group S_n acts on E, permuting the points P_i. In coordinates, this is the *diagonal action*

$$\sigma x_i = x_{\sigma(i)}, \quad \sigma y_i = y_{\sigma(i)} \quad \text{for } \sigma \in S_n.$$

We denote the sign character of S_n by

$$\varepsilon(\sigma) = \begin{cases} 1 & \text{if } \sigma \text{ is even,} \\ -1 & \text{if } \sigma \text{ is odd.} \end{cases}$$

Let

$$\mathbb{C}[\boldsymbol{x}, \boldsymbol{y}]^\varepsilon = \{f \in \mathbb{C}[\boldsymbol{x}, \boldsymbol{y}] : \sigma f = \varepsilon(\sigma)f \text{ for all } \sigma \in S_n\}$$

be the space of *alternating polynomials*. Any alternating polynomial f satisfies

$$f(x_1, y_1, \ldots, x_i, y_i, \ldots, x_j, y_j, \ldots, x_n, y_n)$$
$$= -f(x_1, y_1, \ldots, x_j, y_j, \ldots, x_i, y_i, \ldots, x_n, y_n),$$

which immediately implies that f vanishes on every V_{ij}, that is, f belongs to I.

There is a natural vector space basis for $\mathbb{C}[\boldsymbol{x}, \boldsymbol{y}]^\varepsilon$. Namely, let $\boldsymbol{x}^\alpha \boldsymbol{y}^\beta = x_1^{\alpha_1} y_1^{\beta_1} \ldots x_n^{\alpha_n} y_n^{\beta_n}$ be a monomial, and put

$$A(\boldsymbol{x}^\alpha \boldsymbol{y}^\beta) = \sum_{\sigma \in S_n} \varepsilon(\sigma)\sigma(\boldsymbol{x}^\alpha \boldsymbol{y}^\beta).$$

If the exponent pairs (α_i, β_i) are not all distinct, then $A(\boldsymbol{x}^\alpha \boldsymbol{y}^\beta) = 0$. If they are all distinct, set $D = \{(\alpha_1, \beta_1), \ldots, (\alpha_n, \beta_n)\} \subseteq \mathbb{N} \times \mathbb{N}$. Then $A(\boldsymbol{x}^\alpha \boldsymbol{y}^\beta)$ is given by a bivariate analog of the Vandermonde determinant

$$A(\boldsymbol{x}^\alpha \boldsymbol{y}^\beta) = \Delta_D = \det \begin{bmatrix} x_1^{\alpha_1} y_1^{\beta_1} & \cdots & x_1^{\alpha_n} y_1^{\beta_n} \\ \vdots & & \vdots \\ x_n^{\alpha_1} y_n^{\beta_1} & \cdots & x_n^{\alpha_n} y_n^{\beta_n} \end{bmatrix},$$

which only depends on D, up to sign. It is easy to see that the set of all such polynomials

$$\{\Delta_D : D \subseteq \mathbb{N} \times \mathbb{N}, \ |D| = n\}$$

is a vector space basis of $\mathbb{C}[\boldsymbol{x}, \boldsymbol{y}]^\varepsilon$. In particular, the ideal they generate is the same as the ideal generated by all alternating polynomials. We have just seen that this ideal is contained in I.

THEOREM 1.1. *We have $I = (\Delta_D : D \subseteq \mathbb{N} \times \mathbb{N}, \ |D| = n)$.*

As far as I know, this is not an easy theorem. We will say something about its proof later on. Before that, we discuss briefly the question of finding a minimal set of generators for I, and take up the analogs of the other properties (2)–(4) that we had for J.

Note that I is a homogeneous ideal — in fact it is doubly homogeneous, with respect to the double grading given by degrees in the $\boldsymbol{x}$ and $\boldsymbol{y}$ variables separately. It follows that a set of homogeneous generators for I, for example a subset of the Δ_D's, is minimal if and only if its image is a vector space basis of

$$I/(\boldsymbol{x}, \boldsymbol{y})I.$$

It turns out that we know exactly what the size of such a minimal generating set must be, although no one has yet succeeded in finding an explicit choice of minimal generators.

THEOREM 1.2. *The dimension of $I/(\boldsymbol{x}, \boldsymbol{y})I$ is equal to the Catalan number*

$$C_n = \frac{1}{n+1}\binom{2n}{n}.$$

Indeed, quite a bit more can be said. The space $M = I/(\boldsymbol{x}, \boldsymbol{y})I$ is doubly graded, say $M = \bigoplus_{r,s} M_{r,s}$. Define a "$q,t$-analog" of the Catalan number by

$$C_n(q,t) = \sum_{r,s} t^r q^s \dim M_{r,s}.$$

According to Theorem 1.2 we then have $C_n(1,1) = C_n$. From geometric considerations involving the Hilbert scheme we have a formula for $C_n(q,t)$ [Haiman 1998; 2002], and Theorem 1.2 is proved by specializing the formula to $q = t = 1$. The formula gives $C_n(q,t)$ as a complicated rational function of q, t that on its face is not even obviously a polynomial. However, Garsia and Haglund [Garsia and Haglund 2001; 2002] discovered a simple combinatorial interpretation of the formula, as follows. Let $\mathscr{D}$ be the set of integer sequences

$$\lambda_1 \geq \lambda_2 \geq \cdots \geq \lambda_{n-1} \geq 0$$

satisfying

$$\lambda_i \leq n - i \quad \text{for all } i.$$

In other words, $\mathcal{D}$ is the set of partitions whose Young diagram fits inside that of the partition $(n-1, n-2, \ldots, 1)$. It is well-known that the number of these is the Catalan number C_n. For each $\lambda \in \mathcal{D}$, define

$$a(\lambda) = \sum_i (n-i-\lambda),$$

$$b(\lambda) = \left| \{ i < j : \lambda_i - \lambda_j + i - j \in \{0,1\} \} \right|.$$

Garsia and Haglund showed that

$$C_n(q,t) = \sum_{\lambda \in \mathcal{D}} q^{a(\lambda)} t^{b(\lambda)}.$$

PROBLEM 1.3. Find a rule associating to each $\lambda \in \mathcal{D}$ an n-element subset $D(\lambda) \subseteq \mathbb{N} \times \mathbb{N}$ in such a way that $\deg_y \Delta_{D(\lambda)} = a(\lambda)$, $\deg_x \Delta_{D(\lambda)} = b(\lambda)$, and the set $\{ \Delta_{D(\lambda)} : \lambda \in \mathcal{D} \}$ generates I.

A solution to this problem would give a new and in some sense improved proof of the Garsia–Haglund result. One can proceed similarly for the powers of I, defining

$$M^{(m)} = I^m / (x, y) I^m$$

and

$$C_n^{(m)}(q,t) = \sum_{r,s} t^r q^s \dim M_{r,s}^{(m)}.$$

Again there is a formula for $C_n^{(m)}(q,t)$ from geometry. There is also a *conjectured* combinatorial interpretation, as follows. Let $\mathcal{D}^{(m)}$ be the set of integer sequences

$$\lambda_1 \geq \lambda_2 \geq \cdots \geq \lambda_{n-1} \geq 0$$

satisfying

$$\lambda_i \leq m(n-i) \quad \text{for all } i.$$

In other words, we now allow partitions whose Young diagram fits inside that of $m \cdot (n-1, n-2, \ldots, 1)$. For each $\lambda \in \mathcal{D}^{(m)}$, define

$$a^{(m)}(\lambda) = \sum_i (m(n-i) - \lambda),$$

$$b^{(m)}(\lambda) = \left| \{ i < j : \lambda_i - \lambda_j + m(i-j) \in \{0, 1, \ldots, m\} \} \right|.$$

CONJECTURE 1.4. *We have* $C_n^{(m)}(q,t) = \sum_{\lambda \in \mathcal{D}^{(m)}} q^{a^{(m)}(\lambda)} t^{b^{(m)}(\lambda)}$.

PROBLEM 1.5. Find generators for I^m indexed by elements $\lambda \in \mathcal{D}^{(m)}$, with y-degree equal to $a^{(m)}(\lambda)$ and x-degree equal to $b^{(m)}(\lambda)$.

It is known that $C_n^{(m)}(q,1) = \sum_{\lambda \in \mathcal{D}^{(m)}} q^{a^{(m)}(\lambda)}$, and hence in particular that $\dim I^m/(x,y)I^m = C_n(1,1) = |\mathcal{D}^{(m)}|$. The generating set given by a solution to Problem 1.5 would therefore be minimal, so Conjecture 1.4 would follow automatically.

Now we ask whether I has an analog of property (2) for J. It certainly cannot be that I is a free $\mathbb{C}[\boldsymbol{x}, \boldsymbol{y}]$-module, for then $\mathbb{C}[\boldsymbol{x}, \boldsymbol{y}]/I$ would have depth $2n-1$, whereas it has dimension $2n-2$. What we have instead is that I is a free module with respect to either set of variables alone.

THEOREM 1.6. *The ideal I is a free $\mathbb{C}[\boldsymbol{y}]$-module.*

This theorem is best possible, modulo one detail. The ideal I has an extra degree of freedom: it is invariant with respect to $\boldsymbol{x}$-translations mapping each x_i to $x_i + a$. This invariance holds for $I/(\boldsymbol{y})I$ as well, and implies that $I/(\boldsymbol{y})I$ is a free $\mathbb{C}[x_1]$-module (say). Hence Theorem 1.6 actually implies that I is a free $\mathbb{C}[\boldsymbol{y}, x_1]$-module, and in particular has depth at least $n+1$. On the other hand, it is easy to see that $\Delta(\boldsymbol{y})$ represents a nonzero element of $I/(\boldsymbol{y})I$ annihilated by $(x_1 - x_2, \ldots, x_{n-1} - x_n)$. This implies that depth $I/(\boldsymbol{y})I \leq 1$ and hence depth $I = n+1$.

Next we turn to property (3), the coincidence of powers with symbolic powers.

THEOREM 1.7. *We have $I^m = I^{(m)} \stackrel{\text{def}}{=} \bigcap_{i<j}(x_i - x_j, y_i - y_j)^{(m)}$ for all m.*

In fact, Theorems 1.1, 1.6, and 1.7 are all plainly corollaries to the following two statements.

THEOREM 1.8. *For all m, the m-th power of the ideal $(\Delta_D : D \subseteq \mathbb{N} \times \mathbb{N}, |D| = n)$ is a free $\mathbb{C}[\boldsymbol{y}]$-module.*

COROLLARY 1.9. *For all m, we have $I^{(m)} = (\Delta_D : D \subseteq \mathbb{N} \times \mathbb{N}, |D| = n)^m$.*

On the maxim that every mathematics lecture should contain one proof, we sketch how Theorem 1.8 implies Corollary 1.9. Abbreviating $(\Delta_D : D \subseteq \mathbb{N} \times \mathbb{N}, |D| = n)$ to (Δ_D), we clearly have

$$(\Delta_D)^m \subseteq I^{(m)}.$$

Localizing at any point $\mathbf{P} \in E$ with not all P_i equal, one shows that both $(\Delta_D)_{\mathbf{P}}$ and $I_{\mathbf{P}}^{(m)}$ factor locally into products of the corresponding ideals in subsets of the variables. On the open set U where some $P_i \neq P_j$ we can therefore assume locally that $(\Delta_D)_{\mathbf{P}}^m = I_{\mathbf{P}}^{(m)}$, by induction on n.

Now Theorem 1.8 implies that $\mathbb{C}[\boldsymbol{x}, \boldsymbol{y}]/(\Delta_D)^m$ has depth $\geq n-1$ as a $\mathbb{C}[\boldsymbol{y}]$-module. In particular, $(\Delta_D)^m$ cannot have an associated prime supported in $V(y_1 - y_2, \ldots, y_{n-1} - y_n)$, if $n \geq 3$. In other words, if $f \in \mathbb{C}[\boldsymbol{x}, \boldsymbol{y}]$ belongs to the localization $(\Delta_D)_Q^m$ for all $Q \in (\operatorname{Spec} \mathbb{C}[\boldsymbol{y}]) \setminus V(y_1 - y_2, \ldots, y_{n-1} - y_n)$, then $f \in (\Delta_D)^m$. By induction this holds for all $f \in I^{(m)}$. The induction step assumes $n \geq 3$. The base cases $n = 1, 2$ are trivial.   □

Finally, we discuss property (4). Take the Rees algebra $R = \mathbb{C}[\boldsymbol{x}, \boldsymbol{y}][t(\Delta_D)]$, and put $X = \operatorname{Proj} R$, that is, the blowup of E at the ideal (Δ_D). Here, as above, (Δ_D) is shorthand for the ideal generated by all the alternating polynomials Δ_D.

In view of Theorem 1.1, we can also identify X with the blowup of E along V, but it is preferable for geometric reasons not to take this as the definition.

The symmetric group S_n acts equivariantly on both X and E, giving a diagram

$$\begin{array}{ccc} X & \longrightarrow & E \\ \downarrow & & \downarrow \\ X/S_n & \longrightarrow & E/S_n. \end{array}$$

Now it develops that X/S_n is nothing else but the *Hilbert scheme* $\mathrm{Hilb}^n(\mathbb{C}^2)$ parametrizing 0-dimensional subschemes of length n in $\mathbb{C}^2$, or equivalently, ideals $\mathscr{I} \subseteq \mathbb{C}[x, y]$ such that $\dim_\mathbb{C} \mathbb{C}[x, y]/\mathscr{I} = n$. This is in fact not difficult to show, using explicit local coordinates on $\mathrm{Hilb}^n(\mathbb{C}^2)$ and the definition of X.

By a classical theorem of Fogarty [1968], $\mathrm{Hilb}^n(\mathbb{C}^2)$ is non-singular and irreducible — see the Appendix for another proof using explicit local coordinates. It is also known that the locus in $\mathrm{Hilb}^n(\mathbb{C}^2)$ where the y-coordinates vanish, that is, the locus describing subschemes of $\mathbb{C}^2$ supported on the x-axis, has codimension n. From this it follows easily that $\dim R/(\boldsymbol{y}) = n+1$.

We come now to the most important theorem from the geometric point of view.

THEOREM 1.10. *The blowup scheme X is arithmetically Gorenstein, that is, R is a Gorenstein ring.*

Let us pause to understand how this result is related to Theorem 1.8. The dimension count above shows that $(\boldsymbol{y})$ is a complete intersection ideal in R. Hence, if we assume Theorem 1.10 holds, then R is a free $\mathbb{C}[\boldsymbol{y}]$-module, which is merely a restatement of Theorem 1.8. So Theorem 1.8 is a simple corollary to Theorem 1.10.

Unfortunately for this logic, the only proof of Theorem 1.10 known at present *uses* Theorem 1.8. Specifically, although the main argument of the proof given in [Haiman 2001] is an induction based on elementary geometry of the Hilbert schemes, there is a key technical step that depends on Theorem 1.8. So for now we cannot elegantly deduce Theorem 1.8 from Theorem 1.10, as above, but must prove Theorem 1.8 directly.

PROBLEM 1.11. Find an "intrinsic" proof of Theorem 1.10 that does not rely on Theorem 1.8.

In this connection we may note that there are classical theorems in commutative algebra for showing that Rees algebras are Cohen–Macaulay or Gorenstein. In particular, as W. Vasconcelos pointed out to me, since our ideal has codimension 2 it is enough to show that the Rees algebra R is Cohen–Macaulay, and it is then automatically Gorenstein (this consequence also follows from the geometry). Unfortunately, as far as I am aware, the theorems one might use to show that R is Cohen–Macaulay tend to require hypotheses on the blowup ideal, such as strong

Cohen–Macaulayness, or small analytic spread, that fail drastically for our ideal
I. It is natural to inquire whether advances in singularity theory might even
make it possible to show that our Rees algebra R has singularities better than
Cohen–Macaulay. Could one hope to prove, for instance, that R is of F-rational
type?

I'll conclude with some remarks concerning the existing proof and possible generalizations of Theorem 1.8, which for the moment remains the linchpin among
the results. To prove Theorem 1.8, we first show that $(\Delta_D)^m$ is a direct summand
as a graded $\mathbb{C}[x, y]$-module of the coordinate ring $\mathbb{C}[W]$ of an auxiliary subspace
arrangement $W \subseteq E \times \mathbb{C}^{2mn}$, called a "polygraph." Then we show that $\mathbb{C}[W]$
is a free $\mathbb{C}[y]$-module by explicitly constructing a basis. This requires a horri-
bly complicated and not very illuminating induction. The basis construction is
secretly modeled on a combinatorial interpretation of a formula from geometry
for the Hilbert series of $\mathbb{C}[W]$. In the end, however, both the formula and the
combinatorics are suppressed from the proof, as they must be, since one can only
prove such formulas by assuming the theorem *a priori*.

I think that some of the complexity of the existing proof may eventually be
removed. I also think that most of the phenomena concerning the ideal I should
persist if we take points in $\mathbb{C}^d$ for general d, instead of $\mathbb{C}^2$. If so, we will need
proofs that do not refer to the Hilbert scheme, secretly or otherwise. Here are
some specific problems motivated by my thoughts along these lines.

PROBLEM 1.12. Is it possible to dispense with the polygraph and construct a
free $\mathbb{C}[y]$-module basis of $(\Delta_D)^m$ directly? It would already be interesting to
accomplish this for $d = 2$. In this case, the geometry does provide a formula for
the Hilbert series, but an obstacle to using it is that we don't have a combinatorial
interpretation, and therefore no clue how to index the basis elements.

PROBLEM 1.13. Our subspace arrangement V can be written as $\mathbb{C}^2 \otimes V'$, where
V' is the hyperplane arrangement $V' = \bigcup_{i<j} V(x_i - x_j)$ in $\mathbb{C}^n$. Here, for any
subspace arrangement $A = \bigcup_k A_k \subseteq \mathbb{C}^n$, we denote by $\mathbb{C}^d \otimes A$ the arrangement
of subspaces $\mathbb{C}^d \otimes A_k \subseteq \mathbb{C}^d \otimes \mathbb{C}^n = \mathbb{C}^{dn}$.

(a) Is it true more generally that for all d, the ideal of $\mathbb{C}^d \otimes V'$ is a free $\mathbb{C}[x]$-
    module, where x is one of the d sets of n coordinates on $\mathbb{C}^{dn}$?
(b) The hyperplane arrangement V' is the Coxeter arrangement of type A_{n-1}.
    What if we consider instead the Coxeter arrangements of other types?
(c) Are there general criteria for a hyperplane arrangement $A \subseteq \mathbb{C}^n$ to have the
    property that the ideal I_d of $\mathbb{C}^d \otimes A$ is a free module over the coordinate ring
    of $\mathbb{C}^n$, for all d?
(d) Exercise: show that a hyperplane arrangement with the property in (c) must
    be *free* in the sense used in the theory of hyperplane arrangements [Orlik and
    Terao 1992]. Freeness as a hyperplane arrangement is not sufficient for (c),
    however.

Lecture 2: A Ring of Invariants

As in Lecture 1, let $E = \mathbb{C}^{2n}$ be the space of n-tuples $(P_1, \ldots, P_n)$ of points in the plane. The action of the symmetric group S_n on E has already made an appearance in our study of the ideal of the locus where points coincide. In this lecture we will discuss some other features of this action. We will begin with a review of some general theory of invariants and coinvariants of linear representations of finite groups, then turn to particulars of the representation of S_n on E.

For the moment, we consider an arbitrary finite group G, acting linearly on a finite-dimensional vector space $V = k^n$. Our only assumption will be that $\operatorname{char} k$ does not divide $|G|$. Then all finite-dimensional representations of G are completely reducible, that is, they are direct sums of irreducible representations. In particular, each homogeneous component of the ring $k[V]$ of polynomial functions on V is completely reducible. Of special interest is the subring of *invariants* $k[V]^G$. It follows from complete reducibility that $k[V]^G$ is a direct summand of $k[V]$ as a G-module, and also as a $k[V]^G$-module. The projection of $k[V]$ on its summand $k[V]^G$ is given explicitly by the *Reynolds operator*

$$Rf = \frac{1}{|G|} \sum_{g \in G} g \cdot f,$$

which will be important in what follows.

A second ring associated with the action of G on V is the ring of *coinvariants*, defined as

$$R_G = k[V]/I_G,$$

where $I_G = k[V] \cdot (k[V]^G_+)$ is the ideal generated by all homogeneous invariants of positive degree. Geometrically, these rings have the following interpretation (at least when k is algebraically closed). The space of G-orbits V/G has a natural structure of algebraic variety, with regular functions given by the G-invariant functions on V. Thus its coordinate ring is the ring of invariants:

$$k[V]^G = k[V/G].$$

The homogeneous maximal ideal $k[V]^G_+$ in $k[V]^G$ is the ideal of the origin $0 \in V/G$ (the G-orbit consisting only of the origin in V). Then the scheme-theoretic fiber $\pi^{-1}(0)$ of the natural projection

$$\pi: V \to V/G$$

has coordinate ring equal to the ring of coinvariants,

$$R_G = k[\pi^{-1}(0)].$$

The two constructions are related by a famous lemma of Hilbert.

LEMMA 2.1 (HILBERT). *Homogeneous invariants $f_1, \ldots, f_r$ of positive degree generate $k[V]^G$ as a k-algebra if and only if they generate I_G as an ideal.*

PROOF. If $k[V]^G = k[f_1, \ldots, f_r]$, then every homogeneous invariant of positive degree is a polynomial without constant term in the f_i's. This shows that $I_G \subseteq (f_1, \ldots, f_r)$, and the reverse inclusion is trivial.

For the converse, suppose to the contrary that $I_G = (f_1, \ldots, f_r)$ but $k[V]^G \neq k[f_1, \ldots, f_r]$. Let h be a homogeneous invariant of minimal degree, say d, not contained in $k[f_1, \ldots, f_r]$. Certainly $d > 0$, so $h \in I_G$, and we can write

$$h = \sum_i a_i f_i,$$

where we can assume without loss of generality that a_i is homogeneous of degree $d - \deg f_i$. Applying the Reynolds operator to both sides gives

$$h = \sum_i (\boldsymbol{R} a_i) f_i.$$

But each $\boldsymbol{R} a_i$ is a homogeneous invariant of degree $< d$, hence belongs to $k[f_1, \ldots, f_r]$. This contradicts the assumption $h \notin k[f_1, \ldots, f_r]$. $\square$

It is natural to ask for a bound on the degrees of a minimal set of homogeneous generators for $k[V]^G$, or equivalently for I_G. To give precise bounds for particular G and V is in general a difficult problem. One has the following global bound, which was proved by Noether in characteristic 0.

THEOREM 2.2. *The ring of invariants $k[V]^G$ is generated by homogeneous elements of degree at most $|G|$.*

Let us pause to discuss a more modern proof of this theorem, based on a beautiful lemma of Harm Derksen. To state the lemma we need some additional notation. Let $x_1, \ldots, x_n$ be a basis of coordinates on V, so $k[V] = k[\boldsymbol{x}]$. We introduce a second copy of V, with coordinates $y_1, \ldots, y_n$. Then the coordinate ring $k[V \times V]$ is identified with the polynomial ring $k[\boldsymbol{x}, \boldsymbol{y}]$. For each $g \in G$, let

$$J_g = (x_i - g y_i : 1 \leq i \leq n) \subseteq k[\boldsymbol{x}, \boldsymbol{y}] \tag{2-1}$$

be the ideal of the subspace $W_g = \{(v, gv) : v \in V\} \subseteq V \times V$.

LEMMA 2.3 [Derksen 1999]. *Let $J = \bigcap_{g \in G} J_g$, with J_g as above. Then $k[\boldsymbol{x}] \cap (J + (\boldsymbol{y})) = I_G$.*

PROOF. If $f(\boldsymbol{x})$ is a homogeneous invariant of positive degree, then $f(\boldsymbol{y}) \in (\boldsymbol{y})$, and $f(\boldsymbol{x}) - f(\boldsymbol{y}) \in J$, since $f(\boldsymbol{x}) - f(\boldsymbol{y})$ vanishes on setting $\boldsymbol{y} = g\boldsymbol{x}$ for any $g \in G$. This shows $I_G \subseteq k[\boldsymbol{x}] \cap (J + (\boldsymbol{y}))$.

For the reverse inclusion, suppose $f(\boldsymbol{x}) \in J + (\boldsymbol{y})$, so

$$f(\boldsymbol{x}) = \sum_i a_i(\boldsymbol{x}) b_i(\boldsymbol{y}) + p(\boldsymbol{x}, \boldsymbol{y}), \tag{2-2}$$

where $p(\boldsymbol{x}, \boldsymbol{y}) \in J$ and we can assume $b_i(\boldsymbol{y})$ homogeneous of positive degree. Let $\boldsymbol{R}_{\boldsymbol{y}}$ be the Reynolds operator for the action of G on the $\boldsymbol{y}$ variables only. The ideal J is invariant for this action, so $R_y J \subseteq J$. Hence, applying R_y to both sides in (2–2) yields

$$f(x) = \sum_i a_i(x) R_y b_i(y) + q(x, y)$$

with $q(x, y) \in J$. In particular, $q(x, x) = 0$. Substituting $y \mapsto x$ on both sides now exhibits f as an element of I_G. $\qquad\square$

We remark that J is the ideal of the subspace arrangement $W = \bigcup_g W_g$, which we will call *Derksen's arrangement*. It is the arrangement in $V \times V$ whose projection on the first factor V has finite fiber over each point v, identified set-theoretically with the orbit Gv (by projecting on the second factor). Derksen's Lemma says that the scheme-theoretic 0-fiber of the projection $W \to V$ is isomorphic to the scheme-theoretic 0-fiber of $\pi \colon V \to V/G$, that is, to $\operatorname{Spec} R_G$.

Derksen's lemma has the following easy analog for the product ideal.

LEMMA 2.4. *Let $d = |G|$ and let $J' = \prod_g J_g$, with J_g as in (2–1). Then $k[x] \cap (J' + (y)) = (x)^d$.*

PROOF. Note that $k[x] \cap (J' + (y))$ is the set of polynomials $\{f(x, 0) : f(x, y) \in J'\}$ (this holds with any ideal in the role of J'). Since J' is generated by products of d linear forms, this shows $k[x] \cap (J' + (y)) \subseteq (x)^d$. For the reverse inclusion, fix any monomial x^α of degree d, and write it as a product of individual variables

$$x^\alpha = x_{i_1} x_{i_2} \dots x_{i_d}.$$

Let $g_1, \dots, g_d$ be an enumeration of all the elements of G, and consider the polynomial

$$f(x, y) = \prod_j (x_{i_j} - g_j y_{i_j}).$$

The j-th factor belongs to J_{g_j}, so $f(x, y) \in J'$, and clearly $f(x, 0) = x^\alpha$. $\qquad\square$

Now $J' \subseteq J$, so Lemmas 2.3 and 2.4 imply $(x)^d \subseteq I_G$. Hence I_G is generated by its homogeneous elements of degree at most d, proving Theorem 2.2. In fact, we have proved something stronger.

COROLLARY 2.5. *The ring of coinvariants R_G is zero in degrees $\geq |G|$.*

The degree bound in Theorem 2.2 is tight only when G is a cyclic group. For arbitrary G and V, rather little is known about how to describe $k[V]^G$ and R_G more fully. Of the two, the ring of invariants is better understood. In particular, we have the Eagon–Hochster theorem:

THEOREM 2.6 [Hochster and Eagon 1971]. *The ring of invariants $k[V]^G$ is Cohen–Macaulay.*

My hope in this lecture is to persuade you that $k[V]^G$ and R_G can have surprisingly rich structure for naturally occurring group representations, and that the problem of describing them is deserving of further study. We now turn to the particular case $G = S_n$, and fix $k = \mathbb{C}$. As we did in Lecture 1, let's warm up in the easier situation of n points on a line. This means we consider the representation of S_n on $V = \mathbb{C}^n$, permuting the coordinates $x_1, \ldots, x_n$. We make several observations.

(I) The ring of invariants $\mathbb{C}[\boldsymbol{x}]^{S_n}$ is the polynomial ring $\mathbb{C}[e_1, \ldots, e_n]$ freely generated by the elementary symmetric functions $e_j = e_j(\boldsymbol{x})$. This is the *fundamental theorem of symmetric functions*. Its Hilbert series is

$$\frac{1}{(1-q)(1-q^2)\cdots(1-q^n)},$$

which can also be written as

$$h_n(1, q, q^2, \ldots),   \qquad (2\text{--}3)$$

where $h_n(z_1, z_2, \ldots)$ denotes the complete homogeneous symmetric function of degree n in infinitely many variables.

(II) By Lemma 2.1, $I_{S_n}(\boldsymbol{x}) = (e_1, \ldots, e_n)$. In particular it is a complete intersection ideal. Hence $R_{S_n}(\boldsymbol{x})$ is an Artinian local complete intersection ring. It can be described quite precisely. For example, since $\deg e_j = j$, the Hilbert series of $R_{S_n}(\boldsymbol{x})$ is given by the q-analog of $n!$, namely,

$$[n]_q! = \frac{(1-q)(1-q^2)\cdots(1-q^n)}{(1-q)^n} = [n]_q[n-1]_q\cdots[1]_q,$$

where $[k]_q = 1 + q + \cdots + q^{k-1}$. Hence

$$\dim_{\mathbb{C}} R_{S_n}(\boldsymbol{x}) = n!.$$

(III) Since $\mathbb{C}[\boldsymbol{x}]$ is a graded Cohen–Macaulay ring, and $e_1, \ldots, e_n$ is a homogeneous system of parameters, it follows that $\mathbb{C}[\boldsymbol{x}]$ is a free $\mathbb{C}[\boldsymbol{x}]^{S_n}$-module, with basis given by any $n!$ homogeneous elements forming a vector space basis of $R_{S_n}(\boldsymbol{x})$. It is easy using standard techniques to determine the character of the polynomial ring $\mathbb{C}[\boldsymbol{x}]$ as a graded S_n representation, and from this to determine the corresponding graded character of $R_{S_n}(\boldsymbol{x})$. The answer can be expressed as follows. The irreducible representations V_λ of S_n are indexed by partitions λ of the integer n. For each λ, define

$$f_\lambda(q) = (1-q)(1-q^2)\cdots(1-q^n)s_\lambda(1, q, q^2, \ldots),$$

where $s_\lambda(z_1, z_2, \ldots)$ is the *Schur symmetric function* indexed by λ in infinitely many variables. Then $f_\lambda(q)$ is a polynomial with positive integer coefficients, and $f_\lambda(1)$ is the number of standard Young tableau of shape λ, which is also equal to $\dim V_\lambda$. Let $m(V_\lambda, R_{S_n}(\boldsymbol{x})_d)$ denote the multiplicity of V_λ in a decomposition of the degree d homogeneous component $R_{S_n}(\boldsymbol{x})_d$ as a direct sum of irreducible representations of S_n. Then these multiplicities are given by

$$\sum_d m(V_\lambda, R_{S_n}(\boldsymbol{x})_d)q^d = f_\lambda(q).$$

This is a very precise answer, as $f_\lambda(q)$ has an explicit combinatorial description, and it is possible to produce a correspondingly explicit decomposition of $R_{S_n}(\boldsymbol{x})$ into irreducibles with generators indexed by suitable combinatorial data. It would take us too far afield to go into this here, but see [Allen 1993], for example, for more details. We only note that ignoring the grading gives

$$m(V_\lambda, R_{S_n}(\boldsymbol{x})) = f_\lambda(1) = \dim V_\lambda,$$

so $R_{S_n}(\boldsymbol{x})$ is a graded version of the regular representation of S_n (the representation of S_n by left multiplication on its group algebra $\mathbb{C}S_n$).

(IV) Derksen's arrangement W is a complete intersection in $\mathbb{C}^n \times \mathbb{C}^n$, defined by the ideal $(e_i(\boldsymbol{x}) - e_i(\boldsymbol{y}) : 1 \leq i \leq n)$. In particular, its coordinate ring $\mathbb{C}[W]$ is Cohen–Macaulay, and since $(\boldsymbol{y})$ is obviously a system of parameters, $\mathbb{C}[W]$ is a free $\mathbb{C}[\boldsymbol{y}]$-module.

These special properties of the invariants and coinvariants of S_n on $\mathbb{C}^n$ are consequences of the fact that S_n acts on $\mathbb{C}^n$ as a group generated by *complex reflections*: linear transformations that fix a hyperplane pointwise. In the case of S_n, the reflections are the transpositions (i, j), which fix every vector on the hyperplane $x_i = x_j$. By general results of Steinberg, Chevalley, Shepard and Todd, every complex reflection group G has $k[V]^G$ a polynomial ring, I_G a complete intersection ideal, and R_G isomorphic to a graded version of the regular representation of G. Moreover, each of these properties holds *only* for complex reflection groups, and there is a complete classification of such groups [Chevalley 1955; Shephard and Todd 1954; Steinberg 1960; 1964].

Finally we come to the situation that we set out to study in the first place, namely, the action of S_n on $E = \mathbb{C}^{2n}$. Note that this is *not* an action generated by complex reflections. In fact, every element of S_n acts on E with determinant 1, while a nontrivial complex reflection has determinant $\neq 1$. The determinant 1 property does have a useful consequence, however, owing to the following refinement of the Eagon–Hochster theorem.

THEOREM 2.7 [Watanabe 1974]. *The canonical module of $k[V]^G$ is the module of covariants $k[V]^\varepsilon$, where ε denotes the determinant character $\varepsilon(g) = \det_V(g)$. In particular if G acts on V by endomorphisms with determinant 1, then $k[V]^G$ is Gorenstein.*

There is an old theorem of Weyl giving a (minimal) generating set for the ring of invariants $\mathbb{C}[E]^{S_n}$.

THEOREM 2.8 [Weyl 1939]. *The ring of invariants* $\mathbb{C}[\boldsymbol{x}, \boldsymbol{y}]^{S_n}$ *is generated by the polarized power sums*

$$p_{r,s} = \sum_{i=1}^{n} x_i^r y_i^s, \quad 1 \leq r+s \leq n.$$

The analogous theorem holds in d sets of variables. Note that the actual degree bound on the generators in this case, namely n, is very much smaller than the order of the group!

It turns out to be almost as easy to determine the Hilbert series of $\mathbb{C}[E]^{S_n} = \mathbb{C}[\boldsymbol{x}, \boldsymbol{y}]^{S_n}$ as it is for $\mathbb{C}[\boldsymbol{x}]^{S_n}$. In fact, we can compute its Hilbert series as a *doubly* graded ring, by degree in the $\boldsymbol{x}$ and $\boldsymbol{y}$ variables separately. It is given by the following analog of (2–3).

$$\sum_{r,s} \dim(\mathbb{C}[\boldsymbol{x}, \boldsymbol{y}]^{S_n})_{r,s} q^r t^s = h_n(1, q, q^2, \ldots, t, qt, q^2t, \ldots, t^2, qt^2, q^2t^2, \ldots).$$

There is a also similar formula for the Hilbert series of the ring of invariants $\mathbb{C}[\boldsymbol{x}, \boldsymbol{y}, \ldots, \boldsymbol{z}]^{S_n}$ in d sets of variables, as an $\mathbb{N}^d$-graded ring. So we have good analogs of observation (I) for the invariants of n points in the plane or more generally in $\mathbb{C}^d$.

The interesting surprises appear when we turn to analogs of observations (II) and (III), on the ring of coinvariants. We now drop the modifier $\boldsymbol{x}$ from the notation and write simply R_{S_n} for the ring of coinvariants $\mathbb{C}[E]/I_{S_n}$.

Around 1991, Garsia and I were led to investigate R_{S_n} because of its connection with a problem on Macdonald polynomials. For small values of n, we used a computer to determine its dimension and S_n character in each (double) degree. Immediately we noticed some amazing coincidences between our data and well-known combinatorial numbers. We publicized our early findings informally, leading various other people, especially Ira Gessel and Richard Stanley, to discover still more such coincidences. Eventually I published a compilation of these discoveries, all of which were then just conjectures, in [Haiman 1994].

Later, Procesi pointed out to us the fact that the Hilbert scheme $\mathrm{Hilb}^n(\mathbb{C}^2)$ provides a nice resolution of singularities of E/S_n, as discussed in Lecture 1, and observed how this should be useful in attacking the conjectures. Assuming the validity of some geometric hypotheses that would make Procesi's method work, I was soon able to find a formula for the doubly graded character of R_{S_n} in terms of Macdonald polynomials. Garsia and I then proved that the earlier combinatorial conjectures would all follow from the master formula. Recently I succeeded in proving the needed geometric hypotheses, which by this time were the only missing pieces remaining [Haiman 2001; 2002].

There is not room here to discuss in full the geometry of the Hilbert scheme and the combinatorial theory of Macdonald polynomials. I will only summarize some of the facts about R_{S_n} that have been established using these methods.

THEOREM 2.9. *The coinvariant ring R_{S_n} for S_n acting on $\mathbb{C}^{2n}$ has length*

$$\dim_{\mathbb{C}} R_{S_n} = (n+1)^{n-1}.$$

Ignoring the grading, the representation of S_n on R_{S_n} is isomorphic to the sign representation tensored by the obvious permutation representation of S_n on the finite Abelian group $Q/(n+1)Q$, where $Q = \mathbb{Z}^n/\mathbb{Z} \cdot (1,1,\dots,1)$. Retaining the grading by $\boldsymbol{x}$ degree only, one has the Hilbert series

$$\sum_d \dim(R_{S_n})_{(d,-)} q^d = F_n(q),$$

where $F_n(q)$ is the generating function enumerating rooted forests on the vertex set $\{1,\dots,n\}$ by number of inversions, or equivalently, enumerating parking functions on n cars by weight (see [Haiman 1994] for definitions and details).

Here we should mention the connection between R_{S_n} and the ideal I studied in the previous lecture, given by the following proposition, which is easy to prove.

PROPOSITION 2.10. *Homogeneous S_n-alternating polynomials $f_1,\dots,f_r \in \mathbb{C}[E]$ minimally generate the ideal I in Theorem 1.1 if and only if their images modulo I_{S_n} form a basis of the space of S_n-alternating elements of R_{S_n}.*

In particular, Theorem 1.2 is really a statement about the character of R_{S_n}. Like Theorem 2.9, it follows from the master formula for the character of R_{S_n} given by the geometry of the Hilbert scheme.

I think it should be possible to obtain at least some of the above results on R_{S_n}, and maybe some new ones, or analogous ones for other groups, without invoking Hilbert scheme and Macdonald polynomial machinery. In particular, it seems to me that there is room for purely algebraic approaches. One encouraging sign is recent work by Iain Gordon [2003], where he obtains an extension of the $(n+1)^{n-1}$ theorem, in a slightly weakened form, to any Weyl group. This is especially notable in that for the Weyl groups of type G_2, F_4, and D_n, it is known that there is no suitable geometric analog of the Hilbert scheme.

To close, let me suggest some open problems that might repay further study.

PROBLEM 2.11. Can one determine the dimension and Hilbert series of R_{S_n} inductively by fitting it into an exact complex with other terms built out of the coinvariant rings R_{S_k} for $k < n$? A specific conjecture along these lines in [Haiman 1994] remains open.

PROBLEM 2.12. Describe the minimal free resolution of $\mathbb{C}[\boldsymbol{x},\boldsymbol{y}]^{S_n}$ with respect to the minimal generators given by Theorem 2.8. One could also consider this problem in d sets of variables, although $d = 2$ may be nicer, since the ring of invariants is Gorenstein. I don't think a good description is known even for the first syzygies.

PROBLEM 2.13. Let W be the Derksen arrangement for S_n acting on E, say with coordinates x, y, x', y' on $E \times E$. Is $\mathbb{C}[W]$ a free $\mathbb{C}[y]$-module? What about the same problem for fiber powers $W \times_E W \times_E \cdots \times_E W$? An affirmative answer would be equivalent to sheaf cohomology vanishing properties for certain vector bundles on the Hilbert scheme. Are there similar results in d sets of variables, with E replaced by $\mathbb{C}^{dn}$? Are there similar results for other Weyl groups G, with E the direct sum of two (or more) copies of the defining representation?

Lecture 3: A Remarkable Gröbner Basis

This lecture will be an overview of some results by Jeremy Martin. I'll give less detailed notes here than for the previous two lectures, referring you to [Martin 2003a; 2003b] for the full story. Martin's results concern the situation where we introduce not only the points $P_1, \ldots, P_n \in \mathbb{C}^2$ but also lines L_{ij} connecting them in pairs. That is, L_{ij} is a line passing through P_i and P_j. When P_i and P_j are distinct, of course, L_{ij} is determined. When they coincide, the line L_{ij} can pass through them with any slope, introducing an extra degree of freedom.

The locus of all configurations of points and lines as above is the *picture space* $\mathscr{X}(K_n)$. One thinks of these configurations as plane "pictures" of the complete graph K_n on n vertices, with edges represented by lines. To specify a picture, we need to give the coordinates $x_1, y_1, \ldots, x_n, y_n$ of the n points, together with the slopes m_{ij} of the $\binom{n}{2}$ lines. In principle, the slopes m_{ij} lie on a projective line $\mathbb{P}^1$. However, we will be interested only in local questions, so we will consider the affine open set in $\mathscr{X}(K_n)$ where $m_{ij} \neq \infty$. It is the locus cut out (set-theoretically, at least) by the equations

$$y_j - y_i = m_{ij}(x_j - x_i) \quad \text{for all } i, j.$$

Now $\mathscr{X}(K_n)$ is in general not irreducible. For example, $\mathscr{X}(K_4)$ has two irreducible components, each of dimension 8: the generic component — the closure of the locus where all the points are distinct, and the lines are determined — and another component where all four points coincide, and the six lines have arbitrary slopes. Martin has given a complete combinatorial description of the component structure of $\mathscr{X}(G)$ for any graph G, which we won't discuss in this lecture. Instead we will concentrate on his results describing the generic component $\mathscr{V}(K_n)$ of $\mathscr{X}(K_n)$, which we call the *graph variety*. Note that $\mathscr{V}(K_n)$ is, essentially by definition, the simultaneous blowup of $\mathbb{C}^{2n}$ along the coincidence subspaces $V_{ij} = V(x_i - x_j, y_i - y_j)$ discussed in Lecture 1. This is, however, quite a different thing from the blowup along the union of these subspaces, which is the variety X from Lecture 1.

PROPOSITION 2.1. *The graph variety $\mathscr{V}(K_n)$ is cut out set-theoretically in $\mathscr{X}(K_n)$ by the equations in the variables m_{ij} giving the algebraic relations among the slopes that hold when the points P_i are in general position (no two coincide).*

In view of this proposition, the key issue is to understand the ideal of relations among the slope variables m_{ij}. Although the problem of describing all relations among the slopes of the $\binom{n}{2}$ lines connecting n points in general position in the plane is very classical in nature, there seems to have been almost no earlier work on it. In more geometric terms, the projection of the graph variety $\mathscr{V}(K_n)$ on the slope coordinates is a variety $\mathscr{S}(K_n)$, called the *slope variety*, whose ideal $I(\mathscr{S}_n)$ is the ideal of all algebraic dependencies among the rational functions $(y_j - y_i)/(x_j - x_i)$. We want to describe this ideal.

The first result tells us which subsets of the variables m_{ij} are minimally algebraically dependent — that is, are circuits of the algebraic dependence matroid of the quantities $(y_j - y_i)/(x_j - x_i)$.

THEOREM 2.2. *The variables m_{ij} corresponding to a set of edges $E \subseteq E(K_n)$ are minimally algebraically dependent if and only if*

(1) $|E| = 2\,|V(E)| - 2$, *and*
(2) $|F| \leq 2\,|V(F)| - 3$ *for all nonempty $F \subsetneq E$,*

where $V(E)$ denotes the set of all endpoints of the edges in E.

This result is particularly interesting because there is another well-known algebraic dependence matroid whose characterization (due to Laman) is exactly the same: that is the *rigidity* matroid of algebraic dependencies among the squared-lengths $(x_i - x_j)^2 + (y_i - y_j)^2$ of the line segments connecting the points (for points with real coordinates).

The next result, which is a key one, is an explicit description of the polynomial giving the algebraic dependence among the slopes in a rigidity-circuit. First one shows that every rigidity circuit is the edge-disjoint union of two spanning trees on a common set of vertices. Conversely, every minimal such union is a rigidity circuit.

Now consider any two disjoint spanning trees S and T on the same vertex set, and fix an arbitrary orientation of the edges of each tree. For each edge $f \in S$, there are unique coefficients $c_{ef} \in \{0, \pm 1\}$ such that

$$f - \sum_{e \in T} c_{ef} e \tag{2-1}$$

is a directed cycle. Let us abbreviate $x_e = x_j - x_i$, $y_e = y_j - y_i$ for a directed edge $e = (i, j)$. Then for a cycle as in (2–1), we have

$$y_f = \sum_{e \in T} c_{ef} y_e, \qquad x_f = \sum_{e \in T} c_{ef} x_e.$$

Now since $y_f = m_f x_f$ and $y_e = m_e x_e$, we have an identity between two expressions for y_f

$$\sum_{e \in T} c_{ef} m_e x_e = m_f \sum_{e \in T} c_{ef} x_e,$$

or

$$\sum_{e \in T} c_{ef}(m_e - m_f)x_e = 0.$$

This of course is not yet an equation among the variables m_{ij}. However, if S and T are trees on $d+1$ vertices, then we have d such equations, one for each f, which we can regard as linear equations in the d "unknowns" x_e. When the points are in general position, they obviously have a nonzero solution, since the x_e's do not vanish. Hence the $d \times d$ matrix

$$M_{ST} = [c_{ef}(m_e - m_f)]_{f \in S, e \in T}$$

must be singular. Its determinant

$$D_{S \cup T}(m)$$

is a polynomial of degree d in the slope variables m_e for e in our rigidity circuit $S \cup T$, and this polynomial belongs to $I(\mathscr{S}_n)$.

THEOREM 2.3. *The determinants $D_{S \cup T}$ enjoy the following properties:*

(1) *Up to sign, $D_{S \cup T}$ depends only on the union $S \cup T$, and not on the decomposition into trees S, T.*
(2) *Every term of $D_{S \cup T}$ is a square-free monomial $\pm \prod_{e \in S'} m_e$, where S' is a spanning tree in $S \cup T$ whose complement is also a spanning tree.*
(3) *$D_{S \cup T}$ is irreducible if and only if $S \cup T$ is a rigidity circuit, and in that case it generates the principal ideal of algebraic dependencies among the slope variables m_e for $e \in S \cup T$.*

One particularly simple class of rigidity circuits consists of the *wheels*. A wheel is a graph consisting of a cycle (the *rim*) and one additional vertex (the *hub*) with edges to all the rim vertices (the *spokes*). With this terminology established, we can state Martin's main theorem.

THEOREM 2.4. *The polynomials D_W for W a wheel generate $I(\mathscr{S}_n)$. In fact, they form a Gröbner basis for this ideal, with respect to the graded lexicographic term order on the obvious lexicographic ordering of the variables m_{ij}. Moreover, the initial ideal $\mathrm{in}(I(\mathscr{S}_n))$, and hence also $I(\mathscr{S}_n)$ itself, is Cohen–Macaulay, of dimension $2n - 3$ and degree*

$$M_{2n-4} = (2n-5)(2n-7)\cdots 3 \cdot 1.$$

Let us say just a few words about the proof of this theorem, which involves a beautiful interplay of commutative algebra and combinatorics. By Theorem 2.3, the initial term of D_W is a square-free "tree monomial" $m_T = \prod_{e \in T} m_e$, for some tree. Martin proves first that for wheels, the initial terms belong, not to arbitrary trees, but to trees which are *paths*, of the following special form.

DEFINITION 2.5. A *Martin path* in the graph K_n on vertices $\{1, \ldots, n\}$ is a path $Q = (x, v, \ldots, w, y)$ such that (1) x and y are the two largest vertices of Q, and (2) assuming without loss of generality that $x < y$, then $v < w$.

Now the initial ideal $\mathrm{in}(D_W)$ of the ideal generated by wheel polynomials D_W is the square-free monomial ideal generated by monomials M_Q for Q a Martin path. Hence

$$R_\Delta = \mathbb{C}[\mathbf{m}]/\mathrm{in}(D_W)$$

is the *Stanley–Reisner ring* of the simplicial complex Δ on the edge set of K_n, whose faces are those subgraphs $H \subseteq K_n$ that contain no Martin path. Martin proves next that this simplicial complex has the most optimal properties one could desire.

PROPOSITION 2.6. *Every maximal subgraph of K_n containing no Martin path — that is, every facet of the simplicial complex Δ — has $2n - 3$ edges. The number of these facets is M_{2n-4}. Moreover the complex Δ is shellable.*

Shellability is a combinatorial property of a simplicial complex which implies in particular that it is *Cohen–Macaulay*, that is, the link of each face has only one nonzero reduced homology group. By a theorem of Hochster (see [Stanley 1996]), the latter property is equivalent to the Stanley–Reisner ring being Cohen–Macaulay. So Proposition 2.6 shows that the ideal

$$J = \mathrm{in}(D_W : \text{all wheels } W)$$

is Cohen–Macaulay, of dimension $2n - 3$ and degree M_{2n-4}.

Finally, Martin uses a geometric argument to give a lower bound on the degree of the slope variety $\mathscr{S}_n$.

PROPOSITION 2.7. *The slope variety $\mathscr{S}_n$ has dimension $2n - 3$ and degree at least M_{2n-4}.*

Let us see where the above results leave us. We have two ideals, $J = \mathrm{in}(D_W)$, and $I = \mathrm{in}\, I(\mathscr{S}_n)$, and from the facts established so far we have:

(i) $J \subseteq I$,
(ii) J is unmixed (since it is Cohen–Macaulay),
(iii) $\dim J = \dim I$,
(iv) $\deg J \leq \deg I$.

Together, these imply $J = I$, and Theorem 2.4 follows.

To close, I'll mention a striking combinatorial fact, which Martin left as a conjecture at the end of his thesis, but has since proved. The number M_{2n-4} is the number of *matchings* on $2n-4$ vertices, that is, graphs in which every vertex is the endpoint of exactly one edge. The Hilbert series of the slope variety may be written

$$\frac{h_n(q)}{(1-q)^{2n-3}},$$

where $h_n(q)$ is a polynomial with positive integer coefficients (because the ring is Cohen–Macaulay) and $h_n(1) = M_{2n-4}$. Hence

$$h_n(q) = a_0 + a_1 q + a_2 q_2 + \cdots$$

is a q-*analog* of the number of matchings M_{2n-4}. It turns out that it coincides with a combinatorial q-analog studied long ago by Kreweras and Poupard [1978].

THEOREM 2.8.   *The coefficient a_l in the polynomial $h_n(q)$ is the number of matchings on the integers $\{1, \ldots, 2n-4\}$ with l long edges, where an edge i, j is long if $|i - j| \neq 1$.*

Appendix: Hilbert Schemes of Points in the Plane

BY EZRA MILLER

Consider the polynomial ring $\mathbb{C}[x, y]$ in two variables over the complex numbers. As a set, the *Hilbert scheme* $H_n = \mathrm{Hilb}^n(\mathbb{C}^2)$ of n points in the plane consists of those ideals $I \subseteq \mathbb{C}[x, y]$ such that the quotient $\mathbb{C}[x, y]/I$ has dimension n as a vector space over $\mathbb{C}$. This appendix provides some background on how this set can be considered naturally as a smooth algebraic variety of dimension $2n$. The goal is to orient the reader rather than to give a complete introduction. Therefore some details are omitted from the exposition to make the intuition more clear (and short). The material here, which is based loosely on the introductory parts of [Haiman 1998], reflects what was presented at the help session for Haiman's lectures; in particular, the Questions were all asked by participants at the help session.

To begin, let's get a feeling for what an ideal I of colength n can look like. If $P_1, \ldots, P_n \in \mathbb{C}^2$ are distinct (reduced) points, for example, then the ideal of functions vanishing on these n points has colength n. This is because the ring of functions on n points has a vector space basis $\{f_1, \ldots, f_n\}$ in which $f_i(P_j) = 0$ unless $i = j$, and $f_i(P_i) = 1$. Ideals of the form $I(P_1, \ldots, P_n)$ are called *generic* colength n ideals.

At the opposite end of the spectrum, I could be an ideal whose (reduced) zero set consists of only one point $P \in \mathbb{C}^2$. In this case, $\mathbb{C}[x, y]/I$ is a local ring with lots of nilpotent elements. In geometric terms, this means that P carries a nonreduced scheme structure. Such a nonreduced scheme structure on P is far from unique; in other words, there are many length n local rings $\mathbb{C}[x, y]/I$ supported at P. In fact, they come in an $(n-1)$-dimensional family.

Among the ideals supported at single points, the *monomial ideals* are the most special. These ideals have the form $I = \langle x^{a_1} y^{b_1}, \ldots, x^{a_m} y^{b_m} \rangle$ for some nonnegative integers $a_1, b_1, \ldots, a_m, b_m$, and are supported at $(0, 0) \in \mathbb{C}^2$. Note that if $x^h y^k$ is a monomial outside of I and $x^{h'} y^{k'}$ is a monomial dividing $x^h y^k$ (so $h' \leq h$ and $k' \leq k$), then $x^{h'} y^{k'}$ also lies outside of I. This makes it convenient to draw the monomials outside of I as the boxes "under a staircase".

EXAMPLE A.1. For the ideal $I = \langle x^2, xy, y^3 \rangle$ of colength $n = 4$, the diagram of boxes under the staircase is L-shaped:

$$\begin{array}{|c|c|}
\hline
y^2 & \\
\hline
y & \\
\hline
1 & x \\
\hline
\end{array}$$

Note that the monomial x^2 would be the first box after the bottom row, while xy would nestle in the nook of the 'L', and y^3 would lie atop the first column. Thus the minimal generators of I specify where to draw the staircase.

If the diagram of monomials outside I has λ_i boxes in row i under the staircase, then $\sum_i \lambda_i = n$ is by definition a *partition* λ of n, and we write $I = I_\lambda$.

EXAMPLE A.2. In Example A.1, there are 2 boxes in row 0, and 1 box in each of rows 1 and 2, yielding the partition $2 + 1 + 1 = 4$ of $n = 4$. Thus the ideal is $I = I_{2+1+1}$.

In full generality, the quotient $\mathbb{C}[x, y]/I$ is a product of local rings with maximal ideals corresponding to a finite set $P_1, \ldots, P_r$ of distinct points in $\mathbb{C}^2$, with the lengths $\ell_1, \ldots, \ell_r$ of these local rings satisfying satisfying $\ell_1 + \cdots + \ell_r = n$ (do not confuse this partition of n with the partitions obtained from monomial ideals, where $r = 1$). When $r = n$ it must be that $\ell_i = 1$ for all i, so the ideal I is generic.

QUESTION 1. Is there some transformation of the plane so that every colength n ideal has a basis of monomials?

ANSWER 1. This question can be interpreted in two different ways, because the word "basis" has multiple meanings. Thinking of "basis" as "generating set", the question asks if given I, there is a coordinate system for $\mathbb{C}^2$ in which I is a monomial ideal. The answer is no, in general; for instance, if $\mathbb{C}[x, y]/I$ is not a local ring, then I can't be a monomial ideal in any coordinates. The second meaning of "basis" is "$\mathbb{C}$-vector space basis". Even though I itself may not be expressible in some coordinates as a monomial ideal, the quotient $\mathbb{C}[x, y]/I$ always has a $\mathbb{C}$-vector space basis of (images of) monomials. This observation will be crucial later on.

If all colength n ideals were generic, then the set H_n would be easy to describe, as follows. Every unordered list of n distinct points in $\mathbb{C}^2$ corresponds to a set of $n!$ points in $(\mathbb{C}^2)^n$, or alternatively to a single point in the quotient $S^n\mathbb{C}^2 := (\mathbb{C}^2)^n/S_n$ by the symmetric group. Of course, not every point of $S^n\mathbb{C}^2$ corresponds to an unordered list of *distinct* points; for that, one needs to remove the diagonals

$$\{(P_1, \ldots, P_n) \in (\mathbb{C}^2)^n \mid P_i = P_j\} \tag{A-1}$$

of $(\mathbb{C}^2)^n$ before quotienting by S_n. Since S_n acts freely on the complement $((\mathbb{C}^2)^n)^\circ$ of the diagonals (A-1), the complement $(S^n\mathbb{C}^2)^\circ$ of the diagonals in the quotient $S^n \mathbb{C}^2$ is smooth. Therefore, whatever variety structure we end up using, H_n will contain an open smooth subvariety $(S^n \mathbb{C}^2)^\circ$ of dimension $2n$ parametrizing generic ideals.

The variety structure on H_n arises by identifying it as an algebraic subvariety of a more familiar variety: a grassmannian. Consider the vector subspace V_d inside of $\mathbb{C}[x, y]$ spanned by the $\binom{d+2}{2}$ monomials of degree at most d.

LEMMA A.3. *Fix $d \geq n$. Given any colength n ideal I, the image of V_d spans the quotient $\mathbb{C}[x, y]/I$ as a vector space.*

PROOF. The n monomials outside any initial ideal of I span the quotient $\mathbb{C}[x, y]/I$, and these monomials must lie inside V_d.   □

The intersection $I \cap V_d$ is a vector subspace of codimension n. Thus H_n is (as a set, at least) contained in the grassmannian $\mathrm{Gr}^n(V_d)$ of codimension n subspaces of V_d.

DEFINITION A.4. Given a partition λ of n, write $U_\lambda \subset H_n$ for the set of ideals I such that the monomials outside I_λ map to a vector space basis for $\mathbb{C}[x, y]/I$.

The set of codimension n subspaces $W \subset V_d$ for which the monomials outside I_λ span V_d/W constitutes a standard open affine subvariety of $\mathrm{Gr}^n(V_d)$, defined by the nonvanishing of the corresponding Plücker coordinate. This means that W has a unique basis consisting of vectors of the form

$$x^r y^s - \sum_{hk \in \lambda} c_{hk}^{rs} x^h y^k \quad \text{for } 0 \leq r + s \leq d. \tag{A-2}$$

Here, we write $hk \in \lambda$ to mean $x^h y^k \notin I_\lambda$, so the box labeled (h, k) lies under the staircase for I_λ. The affine open inside $\mathrm{Gr}^n(V_d)$ is actually a cell — namely, the variety whose coordinate ring is the polynomial ring in the coefficients c_{hk}^{rs} from (A–2).

The intersection of each ideal $I \in U_\lambda$ with V_d is a codimension n subspace of V_d spanned by vectors of the form (A–2), by definition of U_λ. Of course, if $W \subset V_d$ is to be expressible as the intersection of V_d with some ideal I, the coefficients c_{hk}^{rs} can't be chosen completely at will. Indeed, the fact that I is an ideal imposes relations on the coefficients that say "multiplication by x takes $x^r y^s$ to $x^{r+1} y^s$ and preserves I, and similarly for multiplication by y."

Explicitly, if $x^{r+1} y^s \in V_d$, then multiplying (A–2) by x yields another polynomial $x^{r+1} y^s - \sum_{hk \in \lambda} c_{hk}^{rs} x^{h+1} y^k$ inside $I \cap V_d$. Some of the terms $x^{h+1} y^k$ no longer lie outside I_λ, so we have to expand them again using (A–2) to get

$$x^{r+1} y^s - \left(\sum_{h+1, k \in \lambda} c_{hk}^{rs} x^{h+1} y^k + \sum_{h+1, k \notin \lambda} c_{hk}^{rs} \sum_{h'k' \in \lambda} c_{h'k'}^{h+1,k} x^{h'} y^{k'} \right) \in I. \tag{A-3}$$

Equating the coefficients on $x^h y^k$ in (A–3) to those in

$$x^{r+1} y^s - \sum_{hk \in \lambda} c_{hk}^{r+1,s} x^h y^k$$

from (A–2) yields relations in the polynomial ring $\mathbb{C}[\{c_{hk}^{rs}\}]$. These relations, taken along with their counterparts that result by switching the roles of x and y, cut out U_λ. Though we have yet to see that these relations generate a radical ideal, we can at least conclude that U_λ is an algebraic subset of an open cell in the grassmannian.

THEOREM A.5. *Fix $d \geq n+1$. The affine varieties U_λ cover the subset $H_n \subset \mathrm{Gr}^n(V_d)$, thereby endowing H_n with the structure of quasiprojective algebraic variety.*

PROOF. The sets U_λ cover H_n by Lemma A.3, and each set U_λ is locally closed in $\mathrm{Gr}^n(V_d)$ by the discussion above. □

In summary: H_n is a quasiprojective variety because it is locally obtained by the intersection of a Zariski open condition (certain monomials span mod I) and a Zariski closed condition ($W \subset V_d$ is closed under multiplication by x and y).

Theorem A.5 does not claim that the variety structure is independent of d, although it is true (and important), and can be deduced using smoothness of H_n (Theorem A.14) along with the fact that projection $V_{d+1} \to V_d$ maps H_n to itself by sending $I \cap V_{d+1} \mapsto I \cap V_d$. Had we allowed $d = n$, however, where Proposition A.12 can fail, the variety structure might be different. In any case, fix $d \geq n+1$ in the forthcoming discussion.

Having endowed H_n with an algebraic variety structure, let us explore its properties.

LEMMA A.6. *Every point $I \in H_n$ is connected to a monomial ideal by a rational curve.*

PROOF. Choosing a term order and taking a Gröbner basis of I yields a family of ideals parametrized by the coordinate variable t on the affine line. When $t = 1$ we get I back, and when $t = 0$ we get the initial ideal of I, which is a monomial ideal. □

This proof is stated somewhat vaguely, but can be made quite precise using the notion of flat family and the fact that Gröbner degenerations are flat families over the affine line [Eisenbud 1995, Proposition 15.17]. Here is an example, for more concrete intuition.

EXAMPLE A.7. Suppose $I = \langle x^2,\, xy + \sqrt{2}x,\, y^3 - 2y \rangle$, and consider the ideal

$$I_t = \langle x^2,\, xy + \sqrt{2}tx,\, y^3 - 2ty \rangle \subset \mathbb{C}[x, y][t].$$

This new ideal should be thought of as a family of ideals in $\mathbb{C}[x, y]$, parametrized by the coordinate t. The ideal at $\alpha \in \mathbb{C}$ is obtained by setting $t = \alpha$ in the generators for I_t. Every one of these ideals has colength 4, because they all have the ideal $\langle x^2, xy, y^3 \rangle$ from Example A.1 as an initial ideal. It follows that this family of ideals (or better yet, the family $\mathbb{C}[x, y][t]/I_t$ of quotients) is flat over $\mathbb{C}[t]$.

Lemma A.6 allows us to conclude the following:

PROPOSITION A.8. *The Hilbert scheme H_n is connected.*

QUESTION 2. Lemma A.6 only says that every ideal connects to some monomial ideal. How do you know that you can get from one monomial ideal to another?

ANSWER 2. They're all connected to generic ideals:

LEMMA A.9. *For every partition λ of n, the point $I_\lambda \in H_n$ lies in the closure of the generic locus $(S^n\mathbb{C}^2)^\circ$.*

PROOF. Consider the set of exponent vectors (h, k) on monomials $x^h y^k$ outside I as a subset of $\mathbb{Z}^2 \subset \mathbb{C}^2$. These exponent vectors constitute a collection of n points in $\mathbb{C}^2$. The colength n ideal of these points is called the *distraction* I_λ' of I_λ. If $I_\lambda = \langle x^{a_1} y^{b_1}, \ldots, x^{a_m} y^{b_m} \rangle$, then $I_\lambda' = \langle f_1, \ldots, f_m \rangle$, where

$$f_i = x(x-1)(x-2) \cdots (x - a_i + 2)(x - a_i + 1) y(y - 1) \cdots (y - b_i + 1).$$

Indeed, this ideal has colength n because every term of f_i divides its leading term $x^{a_i} y^{b_i}$, forcing I_λ to be the unique initial ideal of $\langle f_1, \ldots, f_m \rangle$; and each polynomial f_i clearly vanishes on the exponent set of I_λ, so each f_i lies in I_λ'. $\square$

EXAMPLE A.10. The distraction of $I_{2+1+1} = \langle x^2, xy, y^3 \rangle$ is the ideal

$$I_{2+1+1}' = \langle x(x-1), \, xy, \, y(y-1)(y-2) \rangle.$$

The zero set of every generator of the distraction is a union of lines, namely integer translates of one of the two coordinate axes in $\mathbb{C}^2$. The zero set of our ideal I_{2+1+1}' is

The groups of lines on the right hand side are the zero sets of $x(x-1)$, xy, and $y(y-1)(y-2)$, respectively.

REMARK A.11. Proposition A.8 holds for Hilbert schemes of n points in $\mathbb{C}^m$ even when m is arbitrary, with the same proof. Hartshorne's connectedness theorem [Hartshorne 1966] says that it holds for certain more general Hilbert schemes, under the $\mathbb{Z}$-grading. However, the result does not extend to Hilbert schemes under arbitrary gradings [Haiman and Sturmfels 2002; Santos 2002].

PROPOSITION A.12. *For each λ, the local ring of $H_n \subset \mathrm{Gr}^n(V_d)$ at I_λ has embedding dimension at most $2n$; that is, the maximal ideal $\mathfrak{m}_{I_\lambda}$ satisfies*

$$\dim_{\mathbb{C}}(\mathfrak{m}_{I_\lambda} / \mathfrak{m}_{I_\lambda}^2) \le 2n.$$

PROOF. Identify each variable c_{hk}^{rs} with an arrow pointing from the box $hk \in \lambda$ to the box $rs \notin \lambda$ (see Example A.13). Allow arrows starting in boxes with $h < 0$ or $k < 0$, but set them equal to zero. The arrows lie inside — and in fact generate — the maximal ideal $\mathfrak{m}_{I_\lambda}$ at the point $I_\lambda \in H_n$. As each term in the double sum in (A–3) has two c's in it, the double sum lies inside $\mathfrak{m}_{I_\lambda}^2$. Moving both the tail and head of any given arrow one box to the right therefore does not change the arrow's residue class modulo $\mathfrak{m}_{I_\lambda}^2$, as long as the tail of the original arrow does not end up past the last box in a row of λ, and the head of the arrow does not end up on a monomial of degree strictly larger than d. Switching the roles of x and y, we conclude that an arrow's residue class mod $\mathfrak{m}_{I_\lambda}^2$ is unchanged by moving vertically or horizontally, as long as the tail stays under the staircase, while the head stays above it (but still inside the set of monomials of degree at most d). This analysis includes the case where the tail of the arrow crosses either axis, in which case the arrow is zero.

Using the fact that $d \geq n+1$ in Theorem A.5 to pass the head through corners $(h+1, k+1)$ for $(h, k) \in \lambda$, every arrow can be moved horizontally and vertically until either

(i) the tail crosses an axis; or
(ii) there is a box $hk \in \lambda$ such that the tail lies just inside row k of λ while the head lies just above column h outside λ; or
(iii) there is a box $hk \in \lambda$ such that the tail lies just under the top of column h in λ while the head lies in the first box to the right outside row k of λ.

Arrows of the first sort do not contribute at all to $\mathfrak{m}_{I_\lambda}/\mathfrak{m}_{I_\lambda}^2$. On the other hand, there are exactly n northwest-pointing arrows of the second sort, and exactly n southeast-pointing arrows of the third sort. Therefore $\mathfrak{m}_{I_\lambda}/\mathfrak{m}_{I_\lambda}^2$ has dimension at most $2n$.   □

EXAMPLE A.13. All three figures below depict the same partition λ: $8+8+5+3+3+3+3+2 = 35$. In the left figure, the middle of the five arrows represents $c_{31}^{54} \in \mathfrak{m}_{I_\lambda}$. As in the proof of Proposition A.12, all of the arrows in the left figure are equal modulo $\mathfrak{m}_{I_\lambda}^2$. Since the bottom one is manifestly zero as in item (i) from the proof of Proposition A.12, all of the arrows in the left figure represent zero in $\mathfrak{m}_{I_\lambda}/\mathfrak{m}_{I_\lambda}^2$.

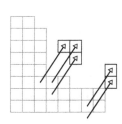
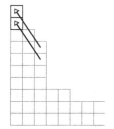
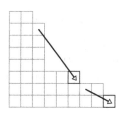

The two arrows in the middle figure are equal, and the bottom one c_{25}^{08} provides an example of a regular parameter in $\mathfrak{m}_{I_\lambda}$ as in (ii). Finally, the two arrows in the rightmost figure represent unequal regular parameters as in (iii).

Now we finally have enough prerequisites to prove the main result.

THEOREM A.14. *The Hilbert scheme H_n is a smooth and irreducible subvariety of dimension $2n$ inside $\operatorname{Gr}^n(V_d)$ for $d \geq n+1$.*

PROOF. Since the intersection of two irreducible components would be contained in the singular locus of H_n, it is enough by Proposition A.8 to prove smoothness. Lemma A.9 implies that the dimension of the local ring of H_n at any monomial ideal I_λ is at least $2n$, because the generic locus has dimension $2n$. On the other hand, Proposition A.12 shows that the maximal ideal of that local ring can be generated by $2n$ polynomials. Therefore H_n is regular in a neighborhood of any point I_λ.

The two-dimensional torus acting on $\mathbb{C}^2$ by scaling the coordinates has an induced action on H_n. Under this action, Lemma A.6 and its proof say that every orbit on H_n contains a monomial ideal (= torus-fixed point) in its closure. By general principles, the singular locus of H_n must be torus-fixed (though not necessarily pointwise, of course) and closed. Since every torus orbit on H_n contains a smooth point of H_n in its closure, the singular locus must be empty. □

The proof of Theorem A.14 used the fact that Gröbner degenerations are accomplished by taking limits of one-parameter torus actions on H_n. In plain language, this means simply that if appropriate powers of t are used in the equations defining the family I_t, the variable t can be thought of as a coordinate on $\mathbb{C}^*$ for nonzero values of t.

REMARK A.15. Theorem A.14 fails for Hilbert schemes $\operatorname{Hilb}^n(\mathbb{C}^m)$ of points in spaces of dimension $m \geq 3$, as proved by Iarrobino [Iarrobino 1972]. If it were irreducible, then $\operatorname{Hilb}^n(\mathbb{C}^m)$ would have dimension mn, the dimension of the open subset of configurations of n distinct points. But Iarrobino constructed a dimension e family of ideals of colength n in the polynomial ring, where e is proportional to $n^{(2-2/m)}$. It follows that $\operatorname{Hilb}^n(\mathbb{C}^m)$ is in fact reducible for $m \geq 3$ and n sufficiently large. On the other hand, $\operatorname{Hilb}^n(\mathbb{C}^m)$ is connected by reasoning as in the case $n = 2$ (Lemma A.6 and Lemma A.9).

QUESTION 3. Is the open set $U_\lambda \subset H_n$ the locus of colength n ideals having I_λ as an initial ideal?

ANSWER 3. When λ is the partition $1 + \cdots + 1 = n$, then yes. Otherwise, no, since the set of such ideals has dimension strictly less than $2n$. However, the locus in H_n of ideals having initial ideal I_λ is cell — that is, isomorphic to $\mathbb{C}^m$ for some m. Lemma A.6 can be interpreted as saying that H_n is the disjoint union of these cells. This is the *Białynicki-Birula decomposition* of H_n [Białynicki-Birula 1976; Ellingsrud and Strømme 1987]. It exists essentially because H_n has an action of the torus $(\mathbb{C}^*)^2$ with isolated fixed points. Knowledge of the Białynicki-Birula decomposition allows one to compute the cohomology ring of H_n, which was the purpose of [Ellingsrud and Strømme 1987].

References

[Allen 1993] E. E. Allen, "A conjecture of Procesi and a new basis for the decomposition of the graded left regular representation of S_n", *Adv. Math.* **100**:2 (1993), 262–292.

[Białynicki-Birula 1976] A. Białynicki-Birula, "Some properties of the decompositions of algebraic varieties determined by actions of a torus", *Bull. Acad. Polon. Sci. Sér. Sci. Math. Astronom. Phys.* **24**:9 (1976), 667–674.

[Chevalley 1955] C. Chevalley, "Invariants of finite groups generated by reflections", *Amer. J. Math.* **77** (1955), 778–782.

[Derksen 1999] H. Derksen, "Computation of invariants for reductive groups", *Adv. Math.* **141**:2 (1999), 366–384.

[Eisenbud 1995] D. Eisenbud, *Commutative algebra, with a view toward algebraic geometry*, Graduate Texts in Mathematics **150**, Springer, New York, 1995.

[Ellingsrud and Strømme 1987] G. Ellingsrud and S. A. Strømme, "On the homology of the Hilbert scheme of points in the plane", *Invent. Math.* **87**:2 (1987), 343–352.

[Fogarty 1968] J. Fogarty, "Algebraic families on an algebraic surface", *Amer. J. Math* **90** (1968), 511–521.

[Garsia and Haglund 2001] A. M. Garsia and J. Haglund, "A positivity result in the theory of Macdonald polynomials", *Proc. Natl. Acad. Sci. USA* **98**:8 (2001), 4313–4316.

[Garsia and Haglund 2002] A. M. Garsia and J. Haglund, "A proof of the q, t-Catalan positivity conjecture", *Discrete Math.* **256**:3 (2002), 677–717.

[Gordon 2003] I. Gordon, "On the quotient ring by diagonal invariants", *Invent. Math.* **153**:3 (2003), 503–518.

[Haiman 1994] M. D. Haiman, "Conjectures on the quotient ring by diagonal invariants", *J. Algebraic Combin.* **3**:1 (1994), 17–76.

[Haiman 1998] M. Haiman, "t, q-Catalan numbers and the Hilbert scheme", *Discrete Math.* **193**:1-3 (1998), 201–224.

[Haiman 2001] M. Haiman, "Hilbert schemes, polygraphs and the Macdonald positivity conjecture", *J. Amer. Math. Soc.* **14**:4 (2001), 941–1006.

[Haiman 2002] M. Haiman, "Vanishing theorems and character formulas for the Hilbert scheme of points in the plane", *Invent. Math.* **149**:2 (2002), 371–407.

[Haiman and Sturmfels 2002] M. Haiman and B. Sturmfels, "Multigraded Hilbert schemes", 2002. Available at arXiv:math.AG/0201271.

[Hartshorne 1966] R. Hartshorne, "Connectedness of the Hilbert scheme", *Inst. Hautes Études Sci. Publ. Math.* **29** (1966), 5–48.

[Hochster and Eagon 1971] M. Hochster and J. A. Eagon, "Cohen–Macaulay rings, invariant theory, and the generic perfection of determinantal loci", *Amer. J. Math.* **93** (1971), 1020–1058.

[Iarrobino 1972] A. Iarrobino, "Reducibility of the families of 0-dimensional schemes on a variety", *Invent. Math.* **15** (1972), 72–77.

[Kreweras and Poupard 1978] G. Kreweras and Y. Poupard, "Sur les partitions en paires d'un ensemble fini totalement ordonné", *Publ. Inst. Statist. Univ. Paris* **23**:1-2 (1978), 57–74.

[Martin 2003a] J. L. Martin, "Geometry of graph varieties", *Trans. Amer. Math. Soc.* **355**:10 (2003), 4151–4169.

[Martin 2003b] J. L. Martin, "The slopes determined by n points in the plane", 2003. Available at arXiv:math.AG/0302106.

[Orlik and Terao 1992] P. Orlik and H. Terao, *Arrangements of hyperplanes*, Grundlehren Math. Wiss. **300**, Springer, Berlin, 1992.

[Santos 2002] F. Santos, "Non-connected toric Hilbert schemes", 2002. Available at arXiv:math.CO/0204044.

[Shephard and Todd 1954] G. C. Shephard and J. A. Todd, "Finite unitary reflection groups", *Canadian J. Math.* **6** (1954), 274–304.

[Stanley 1996] R. P. Stanley, *Combinatorics and commutative algebra*, Second ed., Progress in Mathematics **41**, Birkhäuser, Boston, MA, 1996.

[Steinberg 1960] R. Steinberg, "Invariants of finite reflection groups", *Canad. J. Math.* **12** (1960), 616–618.

[Steinberg 1964] R. Steinberg, "Differential equations invariant under finite reflection groups", *Trans. Amer. Math. Soc.* **112** (1964), 392–400.

[Watanabe 1974] K. Watanabe, "Certain invariant subrings are Gorenstein, I and II", *Osaka J. Math.* **11** (1974), 1–8, 379–388.

[Weyl 1939] H. Weyl, *The classical groups: their invariants and representations*, Princeton University Press, Princeton, 1939.

MARK HAIMAN
DEPARTMENT OF MATHEMATICS
UNIVERSITY OF CALIFORNIA
970 EVANS HALL
BERKELEY, CA 94720-3840
UNITED STATES
mhaiman@math.berkeley.edu

EZRA MILLER
SCHOOL OF MATHEMATICS
UNIVERSITY OF MINNESOTA
127 VINCENT HALL
206 CHURCH STREET, SE
MINNEAPOLIS, MN 55455
UNITED STATES
ezra@math.umn.edu

Trends in Commutative Algebra
MSRI Publications
Volume **51**, 2004

Tight Closure Theory and Characteristic p Methods

MELVIN HOCHSTER

WITH AN APPENDIX BY GRAHAM J. LEUSCHKE

ABSTRACT. We give an introductory overview of the theory of tight closure, which has recently played a primary role among characteristic-p methods. We shall see that such methods can be used even when the ring contains a field of characteristic 0.

CONTENTS

Introduction

The theory of tight closure has recently played a primary role among commutative algebraic methods in characteristic p. We shall see that such methods can be used even when the ring contains a field of characteristic 0.

Hochster was supported in part by a grant from the National Science Foundation. Leuschke was partially supported by an NSF Postdoctoral Fellowship. He also gratefully acknowledges the support of MSRI in attending the workshop where this material was presented.

Unless otherwise specified, the rings that we consider here will be Noetherian rings R containing a field. Frequently, we restrict, for simplicity, to the case of domains finitely generated over a field K. The theory of tight closure exists in much greater generality. For the development of the larger theory and its applications, and for discussion of related topics such as the existence of big Cohen–Macaulay algebras, we refer the reader to the joint works by Hochster and Huneke listed in the bibliography, to [Hochster 1994a; 1994b; 1996], to the expository accounts [Bruns 1996; Huneke 1996; 1998], and to the appendix to this paper by Graham Leuschke.

Here, in reverse order, are several of the most important reasons for studying tight closure theory, which gives a closure operation on ideals and on submodules. We focus mostly on the case of ideals here, although there is some discussion of modules. We shall elaborate on the themes brought forth in the list below in the sequel.

11. Tight closure can be used to shorten difficult proofs of seemingly unrelated results. The results turn out to be related after all. Often, the new results are stronger than the original results.

10. Tight closure provides algebraic proofs of several results that can otherwise be proved only in equal characteristic 0, and whose original proofs depended on analytic techniques.

9. In particular, tight closure can be used to prove the Briançon–Skoda theorem on integral closures of ideals in regular rings.

8. Likewise, tight closure can be used to prove that rings of invariants of linearly reductive algebraic groups acting on regular rings are Cohen–Macaulay.

7. Tight closure can be used to prove several of the local homological conjectures.

6. Tight closure can be used to "control" certain cohomology modules: in particular, one finds that the Jacobian ideal kills them.

5. Tight closure implies several vanishing theorems that are very difficult from any other point of view.

4. Tight closure controls the behavior of ideals when they are expanded to a module-finite extension ring and then contracted back to the original ring.

3. Tight closure controls the behavior of certain colon ideals involving systems of parameters.

2. Tight closure provides a method of compensating for the failure of ambient rings to be regular.

1. If a ring is already regular, the tight closure is very small: it coincides with the ideal (or submodule). This gives an extraordinarily useful test for when an element is in an ideal in regular rings.

One way of thinking about many closure operations is to view them as arising from necessary conditions for an element to be in an ideal. If the condition fails, the element is not in the ideal. If the condition is not both necessary and sufficient, then when it holds, the element might be in the ideal, but it may only be in some larger ideal, which we think of as a kind of closure.

Tight closure in positive characteristic can be thought of as arising from such a necessary but not sufficient condition for ideal membership. One of the reasons that it is so useful for proving theorems is that in some rings, the condition is both necessary and sufficient. In particular, that is true in regular rings. In consequence, many theorems can be proved about regular rings that are rather surprising. They have the following nature: one can see that in a regular ring a certain element is "almost" in an ideal. Tight closure permits one to show that the element actually is in the ideal. This technique works like magic on several major results that seemed very difficult before tight closure came along.

One has to go to some considerable trouble to get a similar theory working in rings that contain the rationals, but this has been done, and the theory works extremely well for "nice" Noetherian rings like the ones that come up in algebraic and analytic geometry.

It is still a mystery how to construct a similar theory for rings that do not contain a field. This is not a matter of thinking about anything pathological. Many conjectures could be resolved if one had a good theory for domains finitely generated as algebras over the integers.

Before proceeding to talk about tight closure, we give some examples of necessary and/or sufficient conditions for membership in an ideal. The necessary conditions lead to a kind of closure.

(1) A necessary condition for $r \in R$ to be in the ideal I is that the image of r be in IK for every homomorphism of R to a field K. This is not sufficient: the elements that satisfy the condition are precisely the elements with a power in I, the *radical* of I.

(2) A necessary condition for $r \in R$ to be in the ideal I is that the image of r be in IV for every homomorphism of R to a valuation ring V. This is not sufficient: the elements that satisfy the condition are precisely the elements in $\bar{I}$, the *integral closure* of I. If R is Noetherian, one gets the same integral closure if one only considers Noetherian discrete valuation rings V. There are many alternative definitions of integral closure.

(3) If R has positive prime characteristic p let S_e denote R viewed as an R-algebra via the e-th iteration F^e of the Frobenius endomorphism F (thus, $S_e = R$, but the structural homomorphism $R \to S_e = R$ sends r to r^{p^e}). A necessary condition that $r \in I$ is that for some integer e, $r^{p^e} \in IS_e$. Note that when S_e is identified with R, IS_e becomes the ideal generated by all elements i^{p^e} for $i \in I$. This ideal is denoted $I^{[p^e]}$. This condition is not sufficient for membership in I. The corresponding closure operation is the Frobenius closure I^F of I: it consists of all elements $r \in R$ such that $r^{p^e} \in I^{[p^e]}$ for some nonnegative integer e. (Once this holds for once choice of e, it holds for all larger choices.) For example, in

$$K[x, y, z] = K[X, Y, Z]/(X^3 + Y^3 + Z^3),$$

if K has characteristic 2 (quite explicitly, if $K = \mathbb{Z}_2 = \mathbb{Z}/2\mathbb{Z}$) then with $I = (x, y)$, we have $z^2 \in I^F - I$. In fact, $(z^2)^2 \in I^{[2]} = (x^2, y^2)$ here, since $z^4 = z^3 z = -(x^3 + y^3)z \in (x^2, y^2)$.

Finally, here is a test for ideal membership that is sufficient but *not* necessary. It was used in the first proof of the Briançon–Skoda theorem, and so we mention it, although easier proofs by analytic methods are available now.

(4) Skoda's analytic criterion. Let Ω be a pseudoconvex open set in $\mathbb{C}^n$ and ϕ a plurisubharmonic function[1] on Ω. Let f and $g_1, \ldots, g_k$ be holomorphic functions on Ω. Let $\gamma = (|g_1|^2 + \cdots + |g_k|^2)^{1/2}$. Let X be the set of common zeros of the g_j. Let $d = \max\{n, k-1\}$. Let λ denote Lebesgue measure on $\mathbb{C}^n$. Skoda's criterion asserts that *if either*

$$\int_{\Omega - X} \frac{|f|^2}{\gamma^{2\alpha d+2}} e^{-\phi} \, d\lambda < +\infty,$$

for some real $\alpha > 1$, or

$$\int_{\Omega - X} \frac{|f|^2}{\gamma^{2d}} \left(1 + \Delta \log(\gamma)\right) e^{-\phi} \, d\lambda < +\infty,$$

then there exist $h_1, \ldots, h_k$ holomorphic on Ω such that $f = \sum_{j=1}^{k} h_j g_j$. Hilbert's Nullstellensatz states that if f vanishes at the common zeros of the g_j then $f \in \operatorname{Rad} I$ where $I = (g_1, \ldots, g_k)$. The finiteness of any of the integrals above conveys the stronger information that, in some sense, f is "small" whenever all the g_j are "small" (or the integrand will be too "large" for the integral to converge), and we get the stronger conclusion that $g \in I$.

1. Reasons for Thinking About Tight Closure

We give here five results valid in any characteristic (i.e., over any field) that can be proved using tight closure theory. The tight closure proofs are remarkably simple, at least in the main cases. The terminology used in the following closely related theorems is discussed briefly after their statements.

THEOREM 1.1 (HOCHSTER AND ROBERTS). *Let S be a regular ring that is an algebra over the field K, and let G be a linearly reductive algebraic group over K acting on S. Then the ring of invariants $R = S^G$ is a Cohen–Macaulay ring.*

---

[1]We won't explain these terms from complex analysis here: the definitions are not so critical for us, because in the application to the Briançon–Skoda theorem, which we discuss later, we work in the ring of germs of holomorphic functions at, say, the origin in complex n-space, $\mathbb{C}^n$ ($\cong$ convergent power series $\mathbb{C}\{z_1, \ldots, z_n\}$) — we can pass to a smaller, pseudoconvex neighborhood; likewise, ϕ becomes unimportant.

THEOREM 1.1° (HOCHSTER AND HUNEKE). *If R is a direct summand (as an R-module) of a regular ring S containing a field, then R is Cohen–Macaulay.*

Theorem 1.1° implies Theorem 1.1. Both apply to many examples from classical invariant theory. Recall that an algebraic group (i.e., a Zariski closed subgroup of $GL(n, K)$) is called *linearly reductive* if every representation is completely reducible. In characteristic 0, these are the same as the reductive groups and include finite groups, products of $GL(1, K)$ (algebraic tori), and semi-simple groups. Over $\mathbb{C}$ such a group is the complexification of compact real Lie group. A key point is that when a linearly reductive algebraic group acts on a K-algebra S, if S^G is the ring of invariants or fixed ring $\{s \in S : g(s) = s$ for all $g \in G\}$ there is a canonical retraction map map $S \to S^G$, called the *Reynolds operator*, that is S^G-linear. Thus, $R = S^G$ is a direct summand of S as an R-module.

In particular, if S is a polynomial ring over a field K and G is a linearly reductive linear algebraic group acting on S_1, the vector space of 1-forms of S, and, hence, all of S (the action should be an appropriate one, i.e., determined by a K-morphism of G into the automorphisms of the vector space S_1), then the fixed ring S^G is a Cohen–Macaulay ring R. What is a Cohen–Macaulay ring? The issue is local: for a local ring the condition means that some (equivalently, every) system of parameters is a regular sequence. In the graded case the Cohen–Macaulay condition has the following pleasant interpretation: when R is represented as a finitely generated module over a graded polynomial subring A, R is free over A. This is a very restrictive and useful condition on R, especially in higher dimension. The Cohen–Macaulay condition is very important in intersection theory. Notice that since moduli spaces are frequently constructed as quotients of smooth varieties by actions of reductive groups, Theorem 1.1 implies the Cohen–Macaulay property for many moduli spaces.

Theorem 1.1 was first proved by a complicated reduction to characteristic $p > 0$ [Hochster and Roberts 1974]. Boutot [1987] gave a shorter proof for affine algebras in characteristic 0 using resolution of singularities and the Grauert–Riemenschneider vanishing theorem. The tight closure proof of Theorem 1.1° is the simplest in many ways.

THEOREM 1.2 (BRIANÇON–SKODA THEOREM). *Let R be a regular ring and I an ideal of R generated by n elements. Then $\overline{I^n} \subseteq I$.*

We gave one characterization of what $u \in \bar{J}$ means earlier. It turns out to be equivalent to require that there be an equation

$$u^h + j_1 u^{h-1} + \cdots + j_h = 0$$

such that every $j_t \in J^t$, $1 \leq t \leq h$. We shall give a third characterization later. Theorem 1.2 was first proved by analytic techniques; compare (4) on page 184. See [Skoda and Briançon 1974; Skoda 1972]: in the latter paper the analytic criteria needed were proved. The first algebraic proofs were given in [Lipman and Teissier 1981] (for a very important special case) and [Lipman and Sathaye 1981]. There are several instances in which tight closure can be used to prove results that were first proved either by analytic techniques or by results like the Kodaira vanishing theorem and related characteristic 0 vanishing theorems in algebraic geometry. See [Huneke and Smith 1997] for a discussion of the connection with the Kodaira vanishing theorem.

THEOREM 1.3 (EIN–LAZARSFELD–SMITH COMPARISON THEOREM). *Let P be a prime ideal of codimension h in a regular ring. Then $P^{(hn)} \subseteq P^n$ for every integer n.*

This was most unexpected. The original proof, valid in characteristic 0, ultimately depends on resolution of singularities and deep vanishing theorems, as well as a theory of asymptotic multiplier ideals. See [Ein et al. 2001]. The tight closure proof in [Hochster and Huneke 2002] permits one to extend the results to characteristic p as well as recovering the characteristic 0 result. There are other connections between tight closure theory and the theory of multiplier ideals: see [Smith 2000; Hara 2001; Hara and Yoshida 2003].

THEOREM 1.4 (HOCHSTER AND HUNEKE). *Let R be a reduced equidimensional finitely generated K-algebra, where K is algebraically closed. Let $f_1, \ldots, f_h$ be elements of R that generate an ideal I of codimension (also called height) h mod every minimal prime of R. Let J be the Jacobian ideal of R over K. Then J annihilates the Koszul cohomology $H^i(f_1, \ldots, f_h; R)$ for all $i < h$, and hence the local cohomology $H^i_I(R)$ for $i < h$.*

This result is a consequence of phantom homology theory, test element theory for tight closure, and the Lipman–Sathaye Jacobian theorem [Lipman and Sathaye 1981], all of which we will describe eventually. If

$$R \cong K[x_1, \ldots, x_n]/(f_1, \ldots, f_m)$$

has codimension r in $\mathbb{A}^n_K$, then J is the ideal of R generated by the images of the size r minors of the Jacobian matrix $(\partial f_j / \partial x_j)$, and defines the non-smooth (over K) locus in Spec R. The ideal $J \subseteq R$ turns out to be independent of which presentation of R one chooses.

Here is a more geometrically flavored corollary.

COROLLARY 1.5. *Let R be a finitely generated graded domain of dimension $n+1$ over an algebraically closed field K, so that $X = \mathrm{Proj}(R)$ is a projective variety of dimension n over an algebraically closed field K. Let g denote a homogeneous element of the Jacobian ideal $J \subseteq R$ of degree d (so that g gives a global section of $\mathcal{O}_X(d)$). Then for $1 \le j \le n-1$, the map $H^j(X, \mathcal{O}_X(t)) \to H^j(X, \mathcal{O}_X(t+d))$ induced by multiplication by g is 0.*

The reason this follows from Theorem 1.4 is that for $j \ge 1$, if we let $M = \bigoplus_{t \in \mathbb{Z}} H^j(X, \mathcal{O}_X(t))$, then M is isomorphic (as an R-module) with $H^{j+1}_m(R)$

which may be viewed as an R-module. We may replace m by the ideal generated by a homogeneous system of parameters, since the two have the same radical. Then Theorem 1.4 implies that the Jacobian ideal of R kills M for $1 \le j \le n-1$. See also Corollary 8.3.

2. The Definition of Tight Closure in Positive Characteristic

One of our guidelines towards a heuristic feeling for when an element u of a Noetherian ring R should be viewed as "almost" in an ideal $I \subseteq R$ will be this: if R has a module-finite extension S such that $u \in IS$ then u is "almost" in I.

Notice that if R is a normal domain (i.e., integrally closed in its field of fractions) containing the rational numbers and S is a module-finite extension, then $IS \cap R = I$, so that for normal rings containing $\mathbb{Q}$ we are not allowing any new elements into the ideal. One can see this as follows. By first killing a minimal prime ideal of S disjoint from $R - \{0\}$ we may assume that S is a domain. Let $L \to L'$ be the corresponding finite algebraic extension of fraction fields, and suppose it has degree d. Let $\operatorname{tr}_{L'/L}$ denote field trace. Then $\dfrac{1}{d}\operatorname{tr}_{L'/L} : S \to R$ gives an R-linear retraction when R is normal. This implies that $IS \cap R = I$ for every ideal I of R. (We only need the invertibility of the single integer d in R for this argument.)

The situation for normal domains of positive characteristic is very different, where it is an open question whether the elements that are "almost" in an ideal in this sense may coincide with the tight closure in good cases. Our definition of tight closure may seem unrelated to the notion above at first, but there is a close connection.

For simplicity we start with the case of ideals in Noetherian domains of characteristic $p > 0$. Recall that in characteristic p the Frobenius endomorphism $F = F_R$ on R maps r to r^p, and is a ring endomorphism. When R is reduced, we denote by R^{1/p^e} the ring obtained by adjoining p^e-th roots for all elements of R: it is isomorphic to R, using the e-th iterate of its Frobenius endomorphism with the image restricted to R. Recall that in a ring of positive characteristic p, when $q = p^e$, we denote by $I^{[q]}$ the ideal of R generated by all q-th powers of elements of I. It is easy to see that this ideal is generated by q-th powers of generators of I. Notice that it is *much* smaller, typically, than the ordinary power I^q. I^q is generated by *all monomials* of degree q in the generators of I, not just q-th powers of generators.

DEFINITION 2.1. Let R be a Noetherian domain of characteristic $p > 0$, let I be an ideal of R, and let u be an ideal of R. We say that $u \in R$ is in the *tight closure* I^* of I in R if there exists an element $c \in R - \{0\}$ such that for all sufficiently large $q = p^e$, we have $cu^q \in I^{[q]}$.

It is equivalent in the definition above to say "for all q" instead of "for all sufficiently large q". We discuss why this condition should be thought of as placing u "almost" in I in some sense. Let $I = (f_1, \ldots, f_h)R$. Note that for every large $q = p^e$ one has

$$cu^q = r_{1q}f_1^q + \cdots r_{hq}f_h^q$$

and if we take q-th roots we have

$$c^{1/q}u = r_{1q}^{1/q}f_1 + \cdots r_{hq}^{1/q}f_h,$$

an equation that holds in the ring $S_q = R[c^{1/q}, r_{iq}^{1/q} : 1 \leq i \leq h]$. S is a module-finite extension of R. But this is not quite saying that u is in IS: rather, it says that $c^{1/q}u$ is in IS. But for very large q, for heuristic purposes, one may think of $c^{1/q}$ as being close to 1: after all, the exponent is approaching 0. Thus, u is multiplied into IS_q in a sequence of module-finite extensions by elements that are getting closer and closer to being a unit, in a vague heuristic sense. This may provide some motivation for the idea that elements that are in the tight closure of an ideal are "almost" in the ideal.

It is ironic that tight closure is an extremely useful technique for proving theorems about regular rings, because it turns out that in regular rings the tight closure of any ideal I is simply I itself. In some sense, the reason that tight closure is so useful in regular rings is that it gives a criterion for being in an ideal that, on the face of it, is considerably weaker than being in the ideal. We shall return to this point later.

We may extend the definition to Noetherian rings R of positive prime characteristic p that are not necessarily integral domains in one of two equivalent ways:

(1) Define u to be in I^* if the image of u in R/P is in the tight closure of $I(R/P)$ in R/P for every minimal prime P of R.
(2) Define u to be in I^* if there is an element $c \in R$ and not in any minimal prime of R such that $cu^q \in I^{[q]}$ for all $q = p^e \gg 0$.

3. Basic Properties of Tight Closure and the Briançon–Skoda Theorem

The following facts about tight closure in a Noetherian ring R of positive prime characteristic p are reasonably easy to verify from the definition.

(a) For any ideal I of R, $(I^*)^* = I^*$.
(b) For any ideals $I \subseteq J$ of R, $I^* \subseteq J^*$.

We shall soon need the following characterization of integral closure of ideals in Noetherian domains.

FACT 3.0. *Let R be a Noetherian domain and let J be an ideal. Then $u \in R$ is in $\bar{J}$ if and only if for some $c \in R - \{0\}$ and every integer positive integer n, $cu^n \in I^n$. It suffices if $cu^n \in I^n$ for infinitely many values of n.*

Comparing this with the definition of tight closure and using the fact that $I^{[q]} \subseteq I^q$ for all $q = p^e$, we immediately get

(c) For any ideal I of R, $I^* \subseteq \bar{I}$. In particular, I^* is contained in the radical of I.

Less obvious is the following theorem that we will prove later.

THEOREM 3.1. *If R is regular, every ideal of R is tightly closed.*

Assuming this fact for a moment, we can prove the following result:

THEOREM 3.2 (TIGHT CLOSURE FORM OF THE BRIANÇON–SKODA THEOREM IN CHARACTERISTIC p). *Let $I = (f_1, \ldots, f_n)R$ be an ideal of a regular ring R of characteristic $p > 0$. Then $\bar{I^n} \subseteq I$. When R is not necessarily regular, it is still true that $\bar{I^n} \subseteq I^*$.*

PROOF. Assuming Theorem 3.1 for the moment, we need only check the final assertion. It suffices to work modulo each minimal prime of R in turn, so we may assume that R is a domain. Then $u \in \bar{I^n}$ implies that for some nonzero c, $cu^m \in (I^n)^m$ for all m. Restricting $m = q = p^e$ we find that $cu^m \in I^{nq} \subseteq I^{[q]}$ for all q, since a monomial in n elements of degree nq must have a factor in which one of the elements is raised to the q-th power. □

Why is every ideal in a regular ring tightly closed? We first need the following:

FACT 3.3. *If R is regular of positive characteristic p, the Frobenius endomorphism is flat.*

PROOF. The issue is local on R. In the local case it suffices to prove it for the completion $\hat{R}$ because $R \to \hat{R}$ is faithfully flat. We have therefore reduced to considering the case $R \cong K[[x_1, \ldots, x_n]]$. The Frobenius map is then isomorphic with the ring inclusion $K^p[[x_1^p, \ldots, x_n^p]] \subseteq K[[x_1, \ldots, x_n]]$. Letting $K^p = k$, we may factor this map as $k[[x_1^p, \ldots, x_n^p]] \subseteq k[[x_1, \ldots, x_n]] \subseteq K[[x_1, \ldots, x_n]]$. The first extension is free on the monomials $x_1^{h_1} \cdots x_n^{h_n}$ with $0 \le h_i < p$ for all i. The flatness of the second map (for any field inclusion $k \subseteq K$) may be seen as follows: since K is flat (in fact, free) over k, $K[x_1, \ldots, x_n]$ is flat over $k[x_1, \ldots, x_n]$. This is preserved when we localize at the maximal ideal generated by the x's in the larger ring and its contraction (also generated by the x's) to the smaller ring. Finally, it is further preserved when we complete both local rings. □

Recall that for an ideal I of R and element $u \in R$, $I : u = \{r \in R : ur \in I\}$. This may thought of as the annihilator in R of the image of u in R/I.

FACT 3.4. *If $f : R \to S$ is flat, $I \subseteq R$ and $u \in R$, then $IS :_S f(u) = (I :_R u)S$.*

To see why, note the exact sequence $(I : u)/I \to R/I \to R/I$ where the map is multiplication by u. Applying $S \otimes_R$ preserves exactness, from which the stated result follows.

COROLLARY 3.5. *If R is regular of positive characteristic, I is any ideal, and $u \in R$, then $I^{[q]} : u^q = (I : u)^{[q]}$ for all $q = p^e$.*

The point is that since $F : R \to R$ is flat, so is its e-th iterate F^e. If S denotes R viewed as an R-algebra via F^e then $IS = I^{[p^e]}$ when we "remember" that S is R. With this observation, Corollary 3.5 follows from Fact 3.4.

PROOF OF THEOREM 3.1. We can reduce to the case where R is a domain. If $c \neq 0$ and $cu^q \in I^{[q]}$ for all $q = p^e$, then $c \in \bigcap_q I^{[q]} : u^q = \bigcap_q (I : u)^{[q]} \subseteq \bigcap_q (I : u)^q$. Since the intersection is not 0, we must have that $I : u = R$, i.e., that $u \in R$.   □

We also mention here the very useful fact that tight closure captures contracted extensions from module-finite extensions.

THEOREM 3.6. *Let S be a domain module-finite over R and let I be an ideal of R. Then $IS \cap R \subseteq I^*$.*

PROOF. S can be embedded in a finitely generated free R-module. One of the projection maps back to R will be nonzero on the identity element of S. That is, there is an R-linear map $f : S \to R$ that sends $1 \in S$ to $c \in R - \{0\}$. If $u \in IS \cap R$, then $u^q \in I^{[q]}S$ for all q. Applying f to both sides yields that $cu^q \in I^q$.   □

Although we have not yet given the definitions the analogous fact holds for sub-modules of free modules, and can even be formulated for arbitrary submodules of arbitrary modules.

4. Direct Summands of Regular Rings are Cohen–Macaulay

Elements $x_1, \ldots, x_n$ in a ring R are called a *regular sequence* on an R-module M if $(x_1, \ldots, x_n)M \neq M$ and x_{i+1} is not a zerodivisor on $M/(x_1, \ldots, x_i)M$, $0 \leq i < n$. A sequence of indeterminates in a polynomial or formal power series ring R, with $M = R$ (or a nonzero free R-module) is an example. We shall make use of the following fact:

FACT 4.1. *Let A be a polynomial ring over a field K, say $A = K[x_1, \ldots, x_d]$ or let A be a regular local ring in which $x_1, \ldots, x_d$ is a minimal set of generators of the maximal ideal. Then a finitely generated nonzero A-module M (assumed graded in the first case) is A-free if and only if $x_1, \ldots, x_d$ is a regular sequence on M. Thus, a module-finite extension ring R (graded if A is a polynomial ring) of A is Cohen–Macaulay if and only if $x_1, \ldots, x_d$ is a regular sequence on R.*

The following two lemmas make the connection between tight closure and the Cohen–Macaulay property.

PROPOSITION 4.2. *Let S be a module-finite domain extension of the domain R (torsion-free is sufficient) and let $x_1, \ldots, x_d$ be a regular sequence in R. Suppose $0 \leq k < d$ and let $I = (x_1, \ldots, x_k)R$. Then $IS :_S x_{k+1} \subseteq (IS)^*$ in S.*

Thus, if every ideal of S is tightly closed, and $x_1, \ldots, x_d$ is a regular sequence in R, it is a regular sequence in S.

PROOF. Because S is a torsion-free R-module there is an an element c of $R-\{0\}$ that multiplies S into an R-free submodule $G \cong R^h$ of S. (This is really all we need about S.) Suppose that $ux_{k+1} \in IS$. Raise both sides to the $q = p^e$ power to get $u^q x_{k+1}^q \in I^{[q]}S$. Multiply by c to get $(cu^q)x_{k+1}^q \in I^{[q]}G$. Because the x_j form a regular sequence on G, so do their q-th powers, and we find that $cu^q \in I^{[q]}S = (IS)^{[q]}$. Since this holds for all $q = p^e$, we are done.   □

PROPOSITION 4.3. *Let R be a domain module-finite over a regular local ring A or $\mathbb{N}$-graded and module-finite over a polynomial ring A. Suppose that R is a direct summand of a regular ring S as an R-module. Then R is Cohen–Macaulay (i.e., A-free).*

PROOF. Let $x_1, \ldots, x_d$ be as in Fact 4.1. The result comes down to the assertion that $x_1, \ldots, x_d$ is a regular sequence on R. By Proposition 4.2, it suffices to show that every ideal of R is tightly closed. But if J is an ideal of R and $u \in R$ is in J^*, then it is clear that $u \in (JS)^* = JS$, since S is regular, and so $u \in JS \cap R = J$, because R is a direct summand of S.   □

Pushing this idea a bit further, one gets a full proof of Theorem 1.1. We need to extend the notion of tight closure to equal characteristic 0, however. This is tackled in Section 6.

5. The Ein–Lazarsfeld–Smith Comparison Theorem

We give here the characteristic-p proof of Theorem 1.3, and we shall even allow radical ideals, with h taken to be the largest height of any minimal prime. For a prime ideal P, $P^{(N)}$, the N-th symbolic power, is the contraction of $P^N R_P$ to R. When I is a radical ideal with minimal primes $P_1, \ldots, P_k$ and $W = R - \bigcup_j P_j$, we may define $P^{(N)}$ either as $\bigcap_j P_j^{(N)}$ or as the contraction of $I^N(W^{-1}R)$ to R.

Suppose that $I \neq (0)$ is radical ideal. If $u \in I^{(hn)}$, then for every $q = p^e$ we can write $q = an + r$ where $a \geq 0$ and $0 \leq r \leq n-1$ are integers. Then $u^a \in I^{(han)}$ and $I^{hn}u^a \subseteq I^{hr}u^a \subseteq I^{(han+hr)} = I^{(hq)}$. We now come to a key point: we can show that

$$I^{(hq)} \subseteq I^{[q]}. \tag{$*$}$$

To see this, note that because the Frobenius endomorphism is flat for regular rings, $I^{[q]}$ has no associated primes other than the minimal primes of I, and it suffices to check $(*)$ after localizing at each minimal prime P of I. But after localization, I has at most h generators, and so each monomial of degree hq in these generators is a multiple of the q-th power of at least one of the generators. This completes the proof of $(*)$. Taking n-th powers gives that $I^{hn^2}u^{an} \subseteq (I^{[q]})^n = (I^n)^{[q]}$, and since $q \geq an$, we have $I^{hn^2}u^q \subseteq (I^n)^{[q]}$ for fixed h and n and all q. Let d be any nonzero element of I^{hn^2}. The condition that $du^q \in (I^n)^{[q]}$

for all q says precisely that u is in the tight closure of I^n in R. But in a regular ring, every ideal is tightly closed, and so $u \in I^n$, as required.

6. Extending the Theory to Affine Algebras in Characteristic 0

In this section we discuss briefly how to extend the results of tight closure theory to finitely generated algebras over a field K of characteristic zero. There is a good theory with essentially the same properties as in positive characteristic. See [Hochster and Huneke 1999; Hochster 1996].

Suppose that we have a finitely generated K-algebra R. We may think of R as having the form $K[x_1, \ldots, x_n]/(f_1, \ldots, f_m)$ for finitely many polynomials f_j. An ideal $I \subseteq R$ can be given by specifying finitely many polynomials $g_j \in T = K[x_1, \ldots, x_n]$ that generate it, and an element u of R can be specified by giving a polynomial h that maps to. We can then choose a finitely generated $\mathbb{Z}$-subalgebra B of K that contains all of the coefficients of the f_j, the g_j and of h. We can form a ring $R_B = B[x_1, \ldots, x_n]/(f_1, \ldots, f_m)$ and we can consider the ideal I_B of the R_B generated by the images of the g_j in R_B. It turns out that after localizing B at one nonzero element we can make other pleasant assumptions: that $I_B \subseteq R_B \subseteq R$, that R_B and R_B/I_B are B-free (the lemma of generic freeness), and that tensoring with K over B converts $I_B \subseteq R_B$ to $I \subseteq R$. Moreover, h has an image in $R_B \subseteq R$ that we may identify with u.

We then define u to be in the tight closure of I in R provided that for all maximal Q in a dense open subset of the maximal spectrum of B, with $\kappa = B/Q$, the image of u in $R_\kappa = \kappa \otimes_B R_B$ is in the characteristic-p tight closure of $I_\kappa = IR_\kappa$ — this makes sense because B/Q will be a finite field.

This definition turns out to be independent of the choices of B R_B, I_B, etc.

Here is one very simple example. Let $R = K[x, y, z]/(x^3 + y^3 + z^3)$ where K is any field of characteristic 0, e.g., the complex numbers, let $I = (x, y)$ and u be the image of z^2. In this case we may take $B = \mathbb{Z}$, $R_{\mathbb{Z}} = \mathbb{Z}[x, y, z]/(x^3 + y^3 + z^3)$ and $I_{\mathbb{Z}} = (x, y)R_{\mathbb{Z}}$. Then z^2 is in the characteristic 0 tight closure of $(x, y)R$ because for every prime integer $p \neq 3$ (these correspond to the maximal ideals of $\mathbb{Z}$, the image of z^2 is in the characteristic-p tight closure of $(x, y)(\mathbb{Z}/p\mathbb{Z})[x, y, z]/(x^3 + y^3 + z^3)$). Take $c = x$, for example. One can check that $c(z^2)^q \in (x^q, y^q)(R_{\mathbb{Z}}/pR_{\mathbb{Z}})$ for all $q = p^e$. Write $2q = 3k + a$, $a \in \{1, 2\}$, and use that $xz^{2q} = \pm x(x^3 + y^3)^k z^a$. Each term in $x(x^3 + y^3)^k$ has the form $x^{3i+1}y^{3j}$ where $3i + 3j = 3k \geq 2q - 2$. Since $(3i + 1) + 3j \geq 2q - 1$, at least one of the exponents is $\geq q$.

7. Test Elements

In this section we again study the case of rings of characteristic $p > 0$. Let R be a Noetherian domain. We shall say that an element $c \in R - \{0\}$ is a *test element* if for every ideal I of R, $cI^* \subseteq I$. An equivalent condition is that for every ideal I and element u of R, $u \in I^*$ if and only if $cu^q \in I^{[q]}$ for every $q = p^e \geq 1$. The reason that this holds is the easily verified fact that if $u \in I^*$, then $u^q \subseteq (I^{[q]})^*$ for all q. Thus, an element that is known to be a test element can be used in all tight closure tests. A priori the element used in tight closure tests for whether $u \in I^*$ in the definition of tight closure can vary with both I and u. The test elements together with 0 form an ideal called the *test ideal*.

Test elements are known to exist for domains finitely generated over a field. Any element $d \neq 0$ such that R_d is regular turns out to have a power that is a test element. We won't prove this here.

We will explain, however, why the Jacobian ideal of a domain finitely generated over an algebraically closed field is contained in the test ideal, which is one of the ingredients of Theorem 1.4. The discussion of the results on test elements needed for Theorem 1.4 is continued in the next section.

Here is a useful result that leads to existence theorems for test elements.

THEOREM 7.1. *Let R be a Noetherian domain module-finite over a regular domain A of characteristic $p > 0$, and suppose that the extension of fraction fields is separable. Then*:

(a) *There are elements $d \in A - \{0\}$ such that $dR^{1/p} \subseteq R[A^{1/p}]$.*
(b) *For any d as in part (a), the element $c = d^2$ satisfies*

$$cR^{1/q} \subseteq R[A^{1/q}] \quad \text{for all } q. \tag{\dagger}$$

Let $R_q = R[A^{1/q}]$.
(c) *Any element $c \neq 0$ of R that satisfies condition (†) is a test element for R.*

Thus, R has test elements.

PROOF. If we localize at all nonzero elements of A we are in the case where A is a field and R is a separable field extension. This is well-known and is left as an exercise for the reader. It follows that $R^{1/p}/R[A^{1/p}]$, which we may think of as a finitely generated $A^{1/p}$-module, is a torsion module. But then it is killed by an element of $A^{1/p} - \{0\}$ and, hence, by an element of $A - \{0\}$.

For part (b) note we note that since $dR^{1/p} \subseteq R[A^{1/p}]$, we have $d^{1/q}R^{1/pq} \subseteq R^{1/q}[A^{1/pq}]$ for all $q = p^e$. Thus,

$$d^{1+1/p}R^{1/p^2} \subseteq d(d^{1/p}R^{1/p^2}) \subseteq d(R^{1/p}[A^{1/p^2}]) \subseteq R[A^{1/p}][A^{1/p^2}] = R[A^{1/p^2}].$$

Continuing in this way, one concludes easily by induction that

$$d^{1+1/p+\cdots+1/p^{e-1}}R^{1/p^e} \subseteq R[A^{1/p^e}].$$

Since $2 > 1 + 1/p + \cdots 1/p^{e-1}$ for all $p \geq 2$, we obtain the desired result.

Finally, suppose that c satisfies condition (†). It suffices to show that for all I and $u \in I^*$, that $cu \in I$. But if $u \in I^*$ we can choose $a \in A - \{0\}$ (all nonzero elements of R have nonzero multiples in A) such that $au^q \in I^{[q]}$ for all $q = p^e$.

Taking q-th roots gives $a^{1/q}u \in IR^{1/q}$ for all q. Multiplying by c gives that $a^{1/q}cu \in IR[A^{1/q}] = IR_q$ for all q, and so $a^{1/q} \in IR_q :_{R_q} cu$ for all q. It is not hard to show that $R \otimes_A A^{1/q} \cong R[A^{1/q}]$ here. (The obvious map is onto, and since R is torsion-free over A and $A^{1/q}$ is A-flat, $R \otimes_A A^{1/q}$ is torsion-free over, so that we can check injectivity after localizing at $A - \{0\}$, and we thus reduce to the case where A is a field and R is a finite separable extension field, where the result is the well-known linear disjointness of separable and purely inseparable field extensions.) The flatness of Frobenius for A means precisely that $A^{1/q}$ is flat over A, so that R_q is flat over R; this is simply a base change. Thus, $IR_q :_{R_q} cu = (I :_R cu)R_q \subseteq (I :_R cu)R^{1/q}$. Hence, for all $q = p^e$, $a^{1/q} \in JR^{1/q}$, where $J = I :_R cu$. This shows that $a \in J^{[q]}$ for all q. Since $a \neq 0$, we must have that J is the unit ideal, i.e., that $cu \in I$.

The same argument works essentially without change when I is a submodule of a free module instead of an ideal. $\qquad\square$

8. Test Elements Using the Lipman–Sathaye Theorem

This section describes material from [Hochster and Huneke 1999, Section 1.4].

For the moment, we do not make any assumption on the characteristic. Let $T \subseteq R$ be a module-finite extension, where T is a Noetherian domain, R is torsion-free as a T-module and the extension is generically smooth. Thus, if $\mathcal{K}$ is the fraction field of T and $\mathcal{L} = \mathcal{K} \otimes_T R$ is the total quotient ring of R then $\mathcal{K} \to \mathcal{L}$ is a finite product of separable field extensions of $\mathcal{K}$. The *Jacobian ideal* $\mathcal{J}(R/T)$ is defined as the 0-th Fitting ideal of the R-module of Kähler R-differentials $\Omega_{R/T}$, and may be calculated as follows: write $R \cong T[X_1, \ldots, X_n]/P$ and then $\mathcal{J}(R/T)$ is the ideal generated in R by the images of all the Jacobian determinants $\partial(g_1, \ldots, g_n)/\partial(X_1, \ldots, X_n)$ for n-tuples $g_1, \ldots, g_n$ of elements of P. Moreover, to generate $\mathcal{J}(R/T)$ it suffices to take all the n-tuples of g_i from a fixed set of generators of P.

Now suppose in addition that T is regular. Let R' be the integral closure of R in $\mathcal{L}$, which is well known to be module-finite over T (the usual way to argue is that any discriminant multiplies it into a finitely generated free T-module). Let $J = \mathcal{J}(R/T)$ and $J' = \mathcal{J}(R'/T)$. The result of Lipman and Sathaye [1981, Theorem 2, p. 200] may be stated as follows:

THEOREM 8.1 (LIPMAN–SATHAYE). *With notation as above (in particular, there is no assumption about the characteristic, and T is regular), suppose also that R is an integral domain. If $u \in \mathcal{L}$ is such that $uJ' \subseteq R'$ then $uJR' \subseteq R$. In particular, we may take $u = 1$, and so $JR' \subseteq R$.* $\qquad\square$

This property of "capturing the normalization" will enable us to produce test elements.

COROLLARY 8.2 (EXISTENCE OF TEST ELEMENTS VIA THE LIPMAN–SATHAYE THEOREM). *If R is a domain module-finite over a regular domain A of character-*

istic p such that the extension of fraction fields is separable, then every element c of $J = \mathscr{J}(R/A)$ is such that $cR^{1/q} \subseteq A^{1/q}[R]$ for all q, and, in particular, $cR^{\infty} \subseteq A^{\infty}[R]$. Thus, if $c \in J \cap (R - \{0\})$, it is a test element.

PROOF. Since $A^{1/q}[R] \cong A^{1/q} \otimes_A R$, the image of c is in $\mathscr{J}(A^{1/q}[R]/A^{1/q})$, and so the Lipman–Sathaye theorem implies that c multiplies the normalization S of $A^{1/q}[R]$ into $A^{1/q}[R]$. Thus, it suffices to see that $R^{1/q}$ is contained in S. Since it is clearly integral over $A^{1/q}[R]$ (it is obviously integral over R), we need only see that the elements of $R^{1/q}$ are in the total quotient ring of $A^{1/q}[R]$, and for this purpose we may localize at $A^{\circ} = A - \{0\}$. Thus, we may replace A by its fraction field and assume that A is a field, and then R is replaced by $(A^{\circ})^{-1}R$, which is a separable field extensions. Thus, we come down to the fact that if $A \subseteq R$ is a finite separable field extension, then the injection $A^{1/q} \otimes_A R \to R^{1/q}$ (the map is an injection because separable and purely inseparable field extensions are linearly disjoint) is an isomorphism, which is immediate by a degree argument.

$\square$

COROLLARY 8.3 (MORE TEST ELEMENTS VIA LIPMAN–SATHAYE). *Let K be a field of characteristic p and let R be a d-dimensional geometrically reduced (i.e., the ring stays reduced even when one tensors with an inseparable extension of K — this is automatic if K is perfect) domain over K that is finitely generated as a K-algebra. Let $R = K[x_1, \ldots, x_n]/(g_1, \ldots, g_r)$ be a presentation of R as a homomorphic image of a polynomial ring. Then the $(n - d) \times (n - d)$ minors of the Jacobian matrix $(\partial g_i / \partial x_j)$ are contained in the test ideal of R, and remain so after localization and completion. Thus, any element of the Jacobian ideal generated by all these minors that is in $R - \{0\}$ is a test element.*

PROOF. We pass to $K(t) \otimes_K R$, if necessary, where $K(t)$ is a simple transcendental extension of K, to guarantee that the field is infinite. Our hypothesis remains the same, the Jacobian matrix does not change, and, since $K(t) \otimes_K R$ is faithfully flat over R, it suffices to consider the latter ring. Thus, we may assume without loss of generality that K is infinite. The calculation of the Jacobian ideal is independent of the choice of indeterminates. We are therefore free to make a linear change of coordinates, which corresponds to choosing an element of $G = GL(n, K) \subseteq K^{n^2}$ to act on the one-forms of $K[x_1, \ldots, x_n]$. For a dense Zariski open set U of $G \subseteq K^{n^2}$, if we make a change of coordinates corresponding to an element $\gamma \in U \subseteq G$ then, for every choice of d of the (new) indeterminates, if A denotes the K-subalgebra of R that these d new indeterminates generate, the two conditions listed below will hold:

(1) R will be module-finite over A (and the d chosen indeterminates will then, perforce, be algebraically independent).

(2) R will be generically smooth over A.

We may consider these two statements separately, for if each holds for a dense Zariski open subset of G we may intersect the two subsets. The first statement follows from the standard "linear change of variable" proofs of the Noether normalization theorem for affine K-algebras (these may be used whenever the ring contains an infinite field). For the second, we want each d element subset, say, after renumbering, $x_1, \ldots, x_d$, of the variables to be a separating transcendence basis for the fraction field L of R over K. (The fact that R is geometrically reduced over K implies that L is separably generated over K.) By [Kunz 1986, Theorem 5.10(d)], for example, a necessary and sufficient condition for $x_1, \ldots, x_d$ to be a separating transcendence basis is that the differentials of these elements $dx_1, \ldots, dx_d$ in $\Omega_{L/K} \cong L^d$ be a basis for $\Omega_{L/K}$ as an L-vector space. Since the differentials of the original variables span $\Omega_{L/K}$ over L, it is clear that the set of elements of G for which all d element subsets of the new variables have differentials that span $\Omega_{L/K}$ contains a Zariski dense open set.

Now suppose that a suitable change of coordinates has been made, and, as above, let A be the ring generated over K by some set of d of the elements x_i. The $n-d$ size minors of $(\partial g_i/\partial x_j)$ involving the $n-d$ columns of $(\partial g_i/\partial x_j)$ corresponding to variables not chosen as generators of A precisely generate $\mathcal{J}(R/A)$. R is module-finite over A by the general position argument, and since it is equidimensional and reduced, it is likewise torsion-free over A, which is a regular domain. It is generically smooth likewise, because of the general position of the variables. The result is now immediate from Corollary 8.2: as we vary the set of d variables, every $n-d$ size minor occurs as a generator of some $\mathcal{J}(R/A)$. $\square$

9. Tight Closure for Submodules

We make some brief remarks on how to extend the theory of tight closure to submodules of arbitrary modules.

Let R be a Noetherian ring of positive prime characteristic p and let G be a free R-module with a specified free basis u_j, which we allow to be infinite. Then we may define an action of the Frobenius endomorphism F and its iterates on G very simply as follows: if $g = \sum_{i=1}^{t} r_i u_{j_i}$ (where the j_i are distinct) we let $F^e(g)$, which we also denote g^{p^e}, be $\sum_{i=1}^{t} r_i^{p^e} u_{j_i}$. Thus, we are simply letting F act (as it does on the ring) on all the coefficients that occur in the representation of an element of G in terms of the free basis. If $N \subseteq G$ is a submodule, we let $N^{[p^e]}$ denote the submodule of G spanned by all the elements g^{p^e} for $g \in N$. We then define an element $x \in G$ to be in N^* if there exists $c \in R^\circ$ such that $cx^{p^e} \in N^{[p^e]}$ for all $e \gg 0$.

More generally, if M is any R-module, N is a submodule, and we want to determine whether $x \in M$ is in the tight closure N^* of N in M, we can proceed by mapping a free module G onto M, taking an element $g \in G$ that maps to x, letting H be the inverse image of N in G, and letting x be in N_M^* precisely when $g \in H_G^*$, where we are using subscripts to indicate the ambient module. This definition turns out to be independent of the choice of free module G mapping onto M, and of the choice of free basis for G.

I believe that there are many important questions about the behavior of tight closure for modules that are not finitely generated over the ring, especially for Artinian modules over local rings. See question 3 in the next section.

However, for the rest of this section we restrict attention to the case of finitely generated modules. The theory of test elements for tight closure of ideals extends without change to the generality of modules.

In order to prove the result of Theorem 1.4 one may make use of a version of the phantom acyclicity theorem. We first recall the result of [Buchsbaum and Eisenbud 1973] concerning when a finite free complex over a Noetherian ring R is acyclic. Suppose that the complex is

$$0 \to R^{b_n} \to \cdots \to R^{b_0} \to 0$$

and that r_i is the (determinantal) rank of the matrix α_i giving the map from $R^{b_i} \to R^{b_i-1}$, $0 \le i \le n+1$, where b_{n+1} is defined to be 0. The result of [Buchsbaum and Eisenbud 1973] is that the complex is acyclic if and only if

(1) for $0 \le i \le n$, $b_i = r_{i+1} + r_i$, and
(2) for $1 \le i \le n$, the depth of the ideal J_i generated by the r_i size minors of α_i is at least i (this is automatic if the ideal generated by the minors is the unit ideal; by convention, the unit ideal has depth $+\infty$).

A complex $0 \to G_n \to \cdots \to G_0 \to 0$ is said to be *phantom acyclic* if for all $i \ge 1$, one has that the kernel Z_i of $G_i \to G_{i-1}$ is in the tight closure of the module of boundaries B_i (the image of G_{i+1} in G_i) in G_i. Note that this implies that Z_i/B_i is killed by the test ideal.

Consider the following weakening of condition (2) above:

(2°) for $1 \le i \le n$, the height of the ideal J_i generated by the r_i size minors of α_i is at least i (this is automatic if the ideal generated by the minors is the unit ideal; by convention, the unit ideal has height $+\infty$).

Then:

THEOREM 9.1 (PHANTOM ACYCLICITY CRITERION). *Let R be a reduced biequidimensional Noetherian ring of positive characteristic. A finite free complex as above is phantom acyclic provided that conditions (1) and (2°) hold.*

See [Hochster and Huneke 1990] and [Hochster and Huneke 1993] for detailed treatments where the result is established in much greater generality and a partial converse is proved, and to [Aberbach 1994] for the further development of the closely related notion of *finite phantom projective dimension*.

Note that in a domain, condition (2°) simply says that every J_i has height at least i: this replaces the subtle and difficult notion of "depth" by the much more tractable notion of "height" (or "codimension").

Theorem 1.4 is simply the result of applying the phantom acyclicity criterion to a Koszul complex. Conditions (1) and (2) are easy to verify. Therefore, the higher homology is killed by the test ideal, which contains the Jacobian ideal.

There is another point of view that is very helpful in understanding the phantom acyclicity theorem. It involves the main result of [Hochster and Huneke 1991a]. If R is a domain, let R^+ denote the integral closure of R in an algebraic closure of its fraction field, which is a maximal integral extension of R that is a domain. It is unique up to non-unique isomorphism. The theorem of [Hochster and Huneke 1991a] is that every system of parameters of R is a regular sequence in R^+: thus, R^+ is a big Cohen–Macaulay algebra for R (and for any module-finite extension domain of R, all of which are embeddable in R^+. Suppose that one has a complex that satisfies the hypothesis of the phantom acyclicity criterion. When one tensors with R^+ it actually becomes acyclic: heights become depths in R^+, and one may apply a generalization to the non-Noetherian case of the acyclicity criterion of [Buchsbaum and Eisenbud 1973] presented in great detail in [Northcott 1976]. One may use this to see that any cycle becomes a boundary after tensoring with a sufficiently large but module-finite extension of R. The fact that the cycles are in the tight closure of the boundaries is now analogous to the fact that when an ideal $I \subseteq R$ is expanded and then contracted from a module-finite extension S of R, we have $IS \cap R \subseteq I^*$: compare Theorem 3.6.

Finally, we mention the vanishing theorem for maps of Tor. Let $A \subseteq R \to S$ be maps of rings of characteristic p, where A is regular, R is module-finite and torsion-free over A, and S is any regular ring. The map $R \to S$ is arbitrary here: it need not be injective nor surjective. Let M be any R-module.

THEOREM 9.2 (VANISHING THEOREM FOR MAPS OF TOR). *With assumptions as just above, the maps* $\mathrm{Tor}_i^A(M, R) \to Tor_i^A(M, S)$ *are* 0 *for all* $i \geq 1$.

SKETCH OF PROOF. One may easily reduce to the case where S is complete local and then to the case where A is complete local. By a direct limit argument one may reduce to the case where M is finitely generated over A. Then M has a finite free resolution over A, which satisfies the hypothesis of the characterization of acyclic complexes given in [Buchsbaum and Eisenbud 1973]. When we tensor with R over A we get a free complex over R that satisfies the phantom acyclicity theorem: every cycle is in the tight closure of the boundaries. Taking its homology gives the $\mathrm{Tor}_i^A(M, R)$. Now when we tensor S, every module is tightly closed, so the cycles coming from the complex over R are now boundaries, which gives the desired result.   □

See [Hochster and Huneke 1990; 1993], the discussion in [Hochster and Huneke 1995], and [Ranganathan 2000]. This is an open question in mixed characteristic. This vanishing result is amazingly powerful. In the case where S is simply a field, it implies the direct summand conjecture, i.e., that regular rings are direct summands of their module-finite extensions. In the case where S is regular and R is a direct summand of S it implies that R is Cohen–Macaulay. Both questions are open in mixed characteristic. The details of these implications are given in [Hochster and Huneke 1995]. In [Ranganathan 2000], it is shown, somewhat surprisingly, that the vanishing theorem for maps of Tor is actually equivalent to the following question about splitting: let R be a regular local ring, let S be a module-finite extension, and suppose that P is a height one prime ideal of S that contracts to xR, where x is a regular parameter in R. Then xR is a direct summand of S as an R-module.

10. Further Thoughts and Questions

What we have said about tight closure so far is only the tip of an iceberg. Here are some major open questions.

1. Does tight closure commute with localization under mild assumptions on the ring? This is not known to be true even for finitely generated algebras over a field. Aspects of the problem are discussed in [Aberbach et al. 1993; Hochster and Huneke 2000; Vraciu 2000].

2. Under mild conditions, if a ring has the property that every ideal is tightly closed, does that continue to hold when one localizes? This is not known for finitely generated algebras over a field, nor for complete local rings. An affirmative answer to 1. would imply an affirmative answer to 2.

Rings such that every ideal is tightly closed are called *weakly F-regular*. The word "weakly" is omitted if this property also holds for all localizations of the ring. Weakly F-regular rings are Cohen–Macaulay and normal under very mild conditions — this holds even if one only assumes that ideals generated by parameters are tightly closed (this weaker property is called *F-rationality* and is closely related to the notion of rational singularities; see [Hara 1998; Smith 1997; Vélez 1995; Enescu 2000]). Both of the conditions of weak F-regularity and F-rationality tend to imply that the singularities of the ring are in some sense good. However, the theory is complicated [Hara and Watanabe 2002]. It is worth noting that weak F-regularity does not deform [Singh 1999], and that direct summands of F-rational rings are not necessarily F-rational [Watanabe 1997]. See also [Hara et al. 2002a; 2002b]. Weak F-regularity is established for some important classes of rings (those defined by the vanishing of the minors of fixed size of a matrix of indeterminates, and homogeneous coordinate rings of Grassmannians) in [Hochster and Huneke 1994b, Theorem 7.14].

3. Let M be an Artinian module over, say, a complete reduced local ring with a perfect residue field. Let N be a submodule of M, Is it true that $u \in N_M^*$ if and only if there exists Q with $N \subseteq Q \subseteq M$ with Q/N of finite length such that $u \in N_Q^*$? This is true in a graded version and for isolated singularities [Lyubeznik and Smith 1999; 2001]; other cases are established in [Elitzur 2003].

For any domain R, let R^+ denote the integral closure of R in an algebraic closure of its fraction field. This is unique up to non-unique isomorphism, and may be thought of as a "largest" domain extension of R that is integral over R.

4. For an excellent local domain R, is an element $r \in R$ in the tight closure of I if and only if it is in IS for some module-finite extension domain of R? It is equivalent to assert that for such a local domain R, $I^* = IR^+ \cap R$. This is known for ideals generated by part of a system of parameters [Smith 1994]. It is known that $IR^+ \cap R \subseteq I^*$. For some results on homogeneous coordinate rings of elliptic curves, see the remarks following the next question.

It is known in characteristic p that for a complete local domain R, and element $u \in R$ is in I^* if and only if it is in $IB \cap R$ for some big Cohen–Macaulay algebra extension ring B of R: see [Hochster 1994a, Section 11].

It is worth mentioning that there is an intimate connection between tight closure and the existence of big Cohen–Macaulay algebras B over local rings (R, m), i.e., algebras B such that $mB \neq B$ and every system of parameters for R is a regular sequence on B. Tight closure ideas led to the proof in [Hochster and Huneke 1992] that if R is an excellent local domain of characteristic p then R^+ is a big Cohen–Macaulay algebra. Moreover, for complete local rings R, it is known [Hochster 1994a, Section 11] that $u \in I^*$ if and only if R has a big Cohen–Macaulay algebra B such that $u \in IB$.

5. Is there an effective way to compute tight closures? The answer is not known even for ideals of cubical cones, i.e., of rings of the form $K[X, Y, Z]/(X^3 + Y^3 + Z^3)$ in positive characteristic different from 3. However, in cones over elliptic curves, tight closure agrees with plus closure (i.e., with $IR^+ \cap R$) for *homogeneous ideals I primary to the homogeneous maximal ideal*: see [Brenner 2003b; 2002]. For ideals that are not homogeneous, the question raised in 4. is open even for such rings. When the characteristic of K is congruent to 2 mod 3, it is even possible that tight closure agrees with Frobenius closure in these rings. See [McDermott 2000; Vraciu 2002].

6. How can one extend tight closure to mixed characteristic? By far the most intriguing result along these lines is due to Ray Heitmann [2002], who has proved that if (R, m) is a complete local domain of dimension 3 and mixed characteristic p, then every Koszul relation on parameters in R^+ is annihilated by multiplication by arbitrarily small positive rational powers of p (that is, by $p^{1/N}$ for arbitrarily large integers N). This implies that regular local rings of dimension 3 are direct summands of their module-finite extension rings. Heitmann's result can be used to prove the existence of big Cohen–Macaulay algebras in dimension 3: see [Hochster 2002]. Other possibilities are explored in [Hochster 2003] and [Hochster and Vélez 2004].

Appendix: Some Examples in Tight Closure

BY GRAHAM J. LEUSCHKE

Tight closure and related methods in the study of rings of prime character-istic have taken on central importance in commutative algebra, leading to both new results and improvements on old ones. Unfortunately, tight closure has a reputation for inaccessibility to novices, with what can seem a bewildering array of F- prefixes and other terminology. The very definition of tight closure is less than immediately illuminating:

DEFINITION A.1. Let R be a Noetherian ring of prime characteristic p. Let I be an ideal of R. An element $x \in R$ is said to be in the *tight closure* of I if there exists an element c, not in any minimal prime of R, so that for all large enough $q = p^e$, $cx^q \in I^{[q]}$, where $I^{[q]}$ is the ideal generated by the q-th powers of the elements of I. In this case we write $x \in I^*$.

This appendix is based on an hour-long help session about tight closure that I gave at MSRI following the series of lectures by Mel Hochster that constitute the bulk of this article. The help session itself was quite informal, driven mostly by questions from the audience, with the goal of presenting enough examples of computations to give a feeling for how the definition is used. The reader will quickly see that the methods are largely *ad hoc*; in fact, at this time there is no useful algorithm for determining that a given element is or is not in a certain tight closure.[2] Still, certain patterns will arise that indicate how problems of this sort are generally solved.

We first discuss the examples. Examples A.1 and A.2 are drawn from [Huneke 1998]. Example A.3 was shown me by Moira McDermott, whom I thank here for her help and insight into some of these computations.

After that, I present a few auxiliary results on tight closure, including the Strong Vanishing Theorem for hypersurfaces and some material on test elements. This section serves several purposes: In addition to putting the examples in context, the results address some of the audience questions raised during the help session and make this appendix relatively self-contained. I am grateful to Sean Sather-Wagstaff for his notes from the help session on this material.

Throughout, we work with Noetherian rings containing a field k of positive characteristic p, and write q for a varying power of p. Variables will be repre-sented by capital letters, which we routinely decapitalize to indicate their images in a quotient ring.

---

[2]See, however, [Sullivant 2002] for a procedure for calculating tight closures of monomial ideals in Fermat rings. Also, there is an algorithm due to Hochster for countable affine rings which involves enumerating all module-finite algebras over the ring. It is effective whenever tight closure is known to be the same as *plus closure*, but is impractical to implement.

The examples. We begin with the canonical first example of tight closure. It involves the "cubical cone" or "Fermat cubic" ring

$$R = k[X, Y, Z]/(X^3 + Y^3 + Z^3),$$

which is in some sense the first nontrivial ring from the point of view of tight closure.

EXAMPLE A.1. Let $R = k[X, Y, Z]/(X^3 + Y^3 + Z^3)$, where k is a field of characteristic $p \neq 3$, and let $I = (y, z)$. Then $I^* = (x^2, y, z)$.

We know from the example at the end of Section 6 (page 192) that $x^2 \in (y, z)^*$; we will reproduce the argument here, since it has a flavor to which we should become accustomed. We will take $c = z$ in the definition of tight closure, so we will show that $z(x^2)^q \in (y, z)^{[q]} = (y^q, z^q)$ for all $q = p^e$. For a general q, write $2q = 3u + i$, where i is 1 or 2. Expand $z(x^2)^q$:

$$z(x^2)^q = zx^{3u+i} = zx^i(x^3)^u = (-1)^u zx^i(y^3 + z^3)^u$$

$$= (-1)^u zx^i \sum_{j=0}^{u} \binom{u}{j} y^{3j} z^{3(u-j)}.$$

Consider a monomial $x^i y^{3j} z^{3(u-j)+1}$ in this sum. If we have both $3j \leq q-1$ and $3(u-j)+1 \leq q-1$, then $3u+1 \leq 2q-2$, so $2q \geq 3u+3$, a contradiction. Therefore each monomial in the expansion of $z(x^2)^q$ has degree at least q in either y or z, that is, each monomial is in (y^q, z^q), as desired.

Now we need only show that $x \notin I^*$. This argument is due to Mordechai Katzman, by way of [Huneke 1998]. We take for granted that z^N is a test element for some large N, that is, z^N can be used as c in any and all tight closure tests (see Definition A.2 and Theorem A.9). Then $x \in I^*$ if and only if $z^N x^q \in (y^q, z^q)$ for all q. Choose q to be larger than N and let $J = (X^3 + Y^3 + Z^3, Y^q, Z^q) \subseteq k[X, Y, Z]$.

Let $>$ be the reverse lexicographic term order on $k[X, Y, Z]$ with $X > Y > Z$. Then the initial ideal $\mathrm{in}_>(J)$ is (X^3, Y^q, Z^q). Write $q = 3u + i$, where i is either 1 or 2. Then $X^q = X^{3u+i} \equiv (-1)^u(Y^3 + Z^3)^u X^i$ modulo J. We also have

$$\mathrm{in}_>(Z^N X^q) = \mathrm{in}_>(Z^N(-1)^u(Y^3 + Z^3)^u X^i) = X^i Y^{3u} Z^N.$$

Since $N < q$, this last is not in $\mathrm{in}_>(J) = (X^3, Y^q, Z^q)$, and we see that $Z^N X^q$ is not in J. Thus $z^N x^q \notin (y^q, z^q)$ in $k[X, Y, Z]/(X^3 + Y^3 + Z^3)$. The same holds in R since $k[X, Y, Z]/J$ has finite length.

At this point in the help session, an audience member asked, "What difference do the numbers make?" That is, are the exponents $(3, 3, 3$ in the case of the Fermat cubic) vital to the outcome of the example? The next example, a side-by-side comparison, shows that they are indeed.

EXAMPLE A.2. Let $S = k[X, Y, Z]$, where the characteristic of k is greater than 7, and define two polynomials: $f_1 = X^2 + Y^3 + Z^5$, and $f_2 = X^2 + Y^3 + Z^7$. Let

$R_1 = S/(f_1)$ and $R_2 = S/(f_2)$, and put $I_i = (y, z)R_i$ for $i = 1, 2$. Then $I_1^* = I_1$, whereas $I_2^* = (x, y, z)R_2$.

We take for granted that for some large N, x^N is a test element for ideals of both rings, that is, we may take $c = x^N$ in the definition of tight closure (see Theorem A.9).

For the first assertion, it suffices to show that $x^N x^q \notin (y^q, z^q)$ for some $q = p^e$. Since $p > 5$, p is relatively prime to 30, and after possibly increasing N slightly, we can find a power q of p so that $q = 30u - N + 2$ for some u. Expand $x^N x^q$:

$$x^{N+q} = x^{30u+2} = (x^2)^{15u+1} = \pm(y^3 + z^5)^{15u+1} = \pm \sum_{j=0}^{15u+1} y^{3j} z^{5(15u-j+1)}.$$

To show that $x^{q+1} \notin (y^q, z^q)$, we just need to find j such that $3j < q$ and $5(15u - j + 1) < q$. Taking $j = 10u$ fills the bill. It remains only to show that the coefficient of $y^{3(10u)} z^{5(5u+1)}$ is nonzero modulo p, that is, that the binomial coefficient $\binom{15u+1}{10u}$ is not divisible by p. We must show that if a power of p divides the numerator of the fraction giving $\binom{15u+1}{10u}$, then it also divides the denominator. So suppose that p^a divides $15u + 1 - j$ for some $j \leq 5k$. Then $2p^a$ divides $30u + 2 - j$. Since $q = p^e = 30u + 1$, we see that $2p^a$ divides $p^e - (2j - 1)$. It follows that p^a divides $2j - 1$, which is a factor of the denominator, and we are done.

To see that $I_2^* = (x, y, z)R_2$, fix q and write $N + q = 2u$, again after increasing N if necessary. Then $x^N x^q = x^{2u} = (-1)^u (y^3 + z^7)^u$. Each monomial in the binomial expansion of the right-hand side is of the form $y^{3j} z^{7(u-j)}$. If both $3j < q$ and $7(u-j) < q$ for some j, then $21j + 21(u-j) < 7q + 3q = 10q$, forcing $21u < 10q$, or $21u < 20u - 10$, which is absurd. Thus, for each j, either $3j \geq q$ or $7(u-j) \geq q$, which implies $x^N x^q \in (y^q, z^q)$, so $x \in (y, z)^*$.

In fact, R_1 is *weakly F-regular,* which means that every ideal is tightly closed. On the other hand, we have shown above that R_2 is not weakly F-regular.

The next example is due to M. McDermott. In addition to showing that the Strong Vanishing Theorem for hypersurfaces (Theorem A.8) is sharp, it illustrates the occasionally mysterious nature of tight closure computations: Sometimes the numbers just work out, especially when p is small.

EXAMPLE A.3. Let $R = k[A, B, C, D, E]/(A^4 + B^4 + C^4 + D^4 + E^4)$, where k is a field of characteristic p, and let $I = (a^4, b^4, c^4, d^4)$. By the Strong Vanishing Theorem for hypersurfaces, for $p > 8$ we have $I^* = I + R_{\geq 16} = I$ is tightly closed. For smaller p, though, I need not be tightly closed. In particular, when $p = 7$ we have $a^3 b^3 c^3 d^3 e^3 \in I^*$.

To see this, let $w = a^3 b^3 c^3 d^3 e^3$. Then

$$w^7 = a^{21} b^{21} c^{21} d^{21} e^{21} = -a^{21} b^{21} c^{21} d^{21} e(a^4 + b^4 + c^4 + d^4)^5.$$

Every monomial of $(a^4 + b^4 + c^4 + d^4)^5$ has at least one variable to the eighth power, so $w^7 \in (a^{29}, b^{29}, c^{29}, d^{29}) \subseteq I^{[7]}$. So in fact $w^p \in I^{[p]}$ and w is in the

Frobenius closure of I. In particular, taking $c = 1$ shows that w is in the tight closure of I.

The next example is the most involved we will consider. It is due originally to Anurag Singh [1998], though the proof we will present is due to Holger Brenner, with improvements by Singh and Huneke. It returns to the Fermat cubic of Example A.1.

EXAMPLE A.4. Let $R = k[[X, Y, Z]]/(X^3 + Y^3 + Z^3)$, where k is a field of characteristic $p > 3$. Then $xyz \in (x^2, y^2, z^2)^*$.

We will need the following lemma, which is an easy consequence of colon-capturing [Huneke 1998, Theorem 2.3].

LEMMA A.5. *Let R be a complete equidimensional local ring with a test element c. Let $x_1, \ldots, x_n, y$ be part of a system of parameters, and set $I = (x_1, \ldots, x_n)$. Then for any ideal J and any element $h \in R$, $hy \in (I + yJ)^*$ if and only if $h \in (I + J)^*$.*

PROOF. Assume first that $hy \in (I + yJ)^*$. This happens if and only if for every q we have $c(yh)^q \in (I + yJ)^{[q]}$, which is equal to $I^{[q]} + y^q J^{[q]}$. So this happens if and only if there exists some $a_q \in J^{[q]}$ such that $y^q(ch^q - a_q) \in I^{[q]}$. Now, by colon-capturing [Huneke 1998, Theorem 2.3], $I^{[q]} : y^q \subseteq (I^{[q]})^*$, so $c(ch^q - a_q) \in I^{[q]}$. Unraveling this one more time gives $c^2 h^q \in I^{[q]} + J^{[q]}$, as desired. The converse follows by retracing these steps. $\square$

Returning to the example, we see by the Lemma that the claim is equivalent to showing that $xy^2 z \in (x^2, y^3, yz^2)^* = (x^2, z^3, yz^2)^*$. This in turn is equivalent to showing that $xy^2 \in (x^2, z^2, yz)^*$.

For $q = p^e$, write $q = 3u + i$, where i is 1 or 2. We will take $c = x^{3-i} y^{6-2i}$ in the definition of tight closure. First, expand:

$$y^{6-2i} y^{2q} = y^{6-2i} y^{6u+2i} = y^3 y^{3(2u+1)} = y^3 (-1)^{2u+1} (x^3 + z^3)^{2u+1}$$

$$= y^3 (-1)^{2u+1} \sum_{j=0}^{2u+1} \binom{2u+1}{j} x^{3(2u+1-j)} z^{3j}.$$

We separate this sum into one part with a factor of x^q and one with a factor of z^q, writing the preceding expression as $y^3 (-1)^{2u+1}$ times Q, where

$$Q = \left(\sum_{j=0}^{u} \binom{2u+1}{j} x^{3(2u+1-j)} z^{3j} + \sum_{j=u+1}^{2u+1} \binom{2u+1}{j} x^{3(2u+1-j)} z^{3j} \right)$$

$$= \left(x^{3u+3} \sum_{j=0}^{u} \binom{2u+1}{j} x^{3(u-j)} z^{3j} + z^{3u+3} \sum_{j=u+1}^{2u+1} \binom{2u+1}{j} x^{3(2u+1-j)} z^{3(j-u-1)} \right)$$

$$= \left(x^q x^{3-i} \sum_{j=0}^{u} \binom{2u+1}{j} x^{3(u-j)} z^{3j} + z^q z^{3-i} \sum_{j=0}^{u} \binom{2u+1}{j+u+1} x^{3(u-j)} z^{3j} \right).$$

Thus $y^{6-2i}y^{2q} = Fx^q + Gy^3z^q$, where

$$F = y^3(-1)^{2u+1}x^{3-i}\sum_{j=0}^{u}\binom{2u+1}{j}x^{3(u-j)}z^{3j},$$

$$G = (-1)^{2u+1}z^{3-i}\sum_{j=0}^{u}\binom{2u+1}{j+u+1}x^{3(u-j)}z^{3j}.$$

We are trying to show that the element

$$(x^{3-i}y^{6-2i})x^qy^{2q} = x^{3-i}Fx^{2q} + Gx^{3-i+q}y^3z^q$$

is in the ideal $(x^2, z^2, yz)^{[q]}$. The term involving Fx^{2q} is taken care of, so it suffices to show that $Gx^{3-i+q} \in (y,z)^{[q]}$. Write Gx^{3-i+q} solely in terms of y and z (recall that $q = 3u+i$):

$$Gx^{3-i+q} = x^{3+3u}(-1)^{2u+1}z^{3-i}\sum_{j=0}^{u}\binom{2u+1}{j+u+1}x^{3(u-j)}z^{3j}$$

$$= (-1)^{3u+2}(y^3+z^3)^{u+1}z^{3-i}\sum_{j=0}^{u}\binom{2u+1}{j+u+1}(-1)^{u-j}(y^3+z^3)^{u-j}z^{3j}.$$

Each monomial in this sum has degree $3(u+1)+(3-i)+3u = 6u+6-i \geq 6u+4 \geq 2q$. Since the sum involves only y and z, each monomial must have degree at least q in either y or z, as desired.

One might reasonably ask why we chose to show the equivalent statement that $xy^2 \in (x^2, z^2, yz)^*$, rather than the originally claimed inclusion. The glib answer is that it works. A more considered and satisfying reply might point to the fact that we reduced the problem in the end to showing that $Gx^{3-i+q} \in (y, z)^{[q]}$, which was quite easy, and in the original formulation there was simply *too much* symmetry to make a similar reduction.

Brenner has recently used powerful geometric methods ([Brenner 2004] and [Brenner 2003a]) to prove results like the following, which vastly generalizes the example above.

THEOREM A.6 [Brenner 2004, Corollary 9.3]. *Let k denote an algebraically closed field of characteristic 0 and let $F \in k[x, y, z]$ denote a homogeneous polynomial of degree δ such that $R = k[x, y, z]/(F)$ is a normal domain. Let $f_1, f_2, f_3 \in R$ denote R_+-primary homogeneous elements of degree d_1, d_2, d_3. Suppose that the sheaf of relations $\mathcal{R}$ is indecomposable on the curve $Y = \mathrm{Proj}\, R$. Then:*

(i) $R_m \subseteq (f_1, f_2, f_3)^*$ *for* $m \geq \frac{d_1+d_2+d_3}{2} + \frac{\delta-3}{2}$.
(ii) *For* $m < \frac{d_1+d_2+d_3}{2} - \frac{\delta+3}{2}$ *we have* $(f_1, f_2, f_3)^* \cap R_m = (f_1, f_2, f_3) \cap R_m$.

This follows from a more general fact involving semistability of vector bundles:

THEOREM A.7 [Brenner 2004, Theorem 8.1]. *Let k denote an algebraically closed field of characteristic 0 and let R be a normal two-dimensional standard-graded k-algebra. Set $t = \lceil \frac{d_1 + \ldots + d_n}{n-1} \rceil$. Suppose that the sheaf of relations $\mathscr{R}(m)$ for ideal generators $f_1, \ldots, f_n$ is semistable. Then*

$$(f_1, \ldots, f_n)^* = (f_1, \ldots, f_n) + R_{\geq t}.$$

Auxiliary results. We now mention a few results that were prepared for the help session, but were not presented. They are included here both as examples of "what might have been" and to make this appendix relatively self-contained (more so than the help session on which it is based). Complete proofs are given in [Huneke 1998].

The first result is the Strong Vanishing Theorem for hypersurfaces, which was mentioned in Example A.3.

THEOREM A.8. *Let $R = k[X_0, \ldots, X_d]/(f)$ be a quasi-homogeneous graded hypersurface over a field k of characteristic $p > 0$. Assume that R is an isolated singularity, and that the partials $\frac{\partial f}{\partial X_1}, \ldots, \frac{\partial f}{\partial X_d}$ form a system of parameters for R. If $p > (d-1)(\deg f) - \sum_{i=1}^{d} \deg X_i$, then for parameters $y_1, \ldots, y_d$ of degrees $a_1, \ldots, a_d$,*

$$(y_1, \ldots, y_d)^* = (y_1, \ldots, y_d) + R_{\geq a_1 + \ldots + a_d}.$$

This theorem is particularly well-suited for computations; see [Sullivant 2002].

The audience at MSRI was interested in the theory of *test elements*, specifically when they are known to exist.

DEFINITION A.2. An element c of R, not in any minimal prime, is called a *test element* for ideals of R if $xI^* \subseteq I$ for every ideal $I \subseteq R$. Equivalently, c can be used for all tight closure tests: $x \in R$ is in I^* if and only if $cx^q \in I^{[q]}$ for all $q = p^e$.

The most obvious immediate benefit of the existence of test elements is in showing that elements are *not* in tight closures. If c is known to be a test element, and it can be shown that $cx^q \notin I^{[q]}$ for any one q, then $x \notin I^*$. We saw this principle in action in Example A.2.

The theorem below is not the most general result on the existence of test elements, but suffices for many applications. We say that a ring R of characteristic p is *F-finite* provided the ring of p-th roots $R^{1/p}$ is a finitely generated R-module. A complete local ring $(R, \mathfrak{m}, k)$ such that $[k : k^p] < \infty$ is always F-finite, but there are many examples of rings, even fields, that are not.

THEOREM A.9 [Hochster and Huneke 1994a, Prop. 6.23]. *Let R be reduced and F-finite or reduced and essentially of finite type over an excellent local ring. Let c be an element of R not in any minimal prime. If the localization R_c is Gorenstein and weakly F-regular, then c has a power which is a test element. In particular, if R_c is regular, then c has a power which is a test element.*

The second case of Theorem A.9, in which R is assumed to be essentially of finite type over an excellent local ring, is deduced from the F-finite case by means of the "Γ-construction" [Hochster and Huneke 1994a], which shows that such a ring has a faithfully flat extension R^Γ which is F-finite, and such that R_c^Γ is still weakly F-regular and Gorenstein. Then some power of c is a test element in R^Γ, and that property descends automatically from faithfully flat extensions.

Other, similar, sources of abundant test elements are the theorem of Lipman–Sathaye and its consequences (see Sections 7 and 8 above).

References

[Aberbach 1994] I. M. Aberbach, "Finite phantom projective dimension", *Amer. J. Math.* **116**:2 (1994), 447–477.

[Aberbach et al. 1993] I. M. Aberbach, M. Hochster, and C. Huneke, "Localization of tight closure and modules of finite phantom projective dimension", *J. Reine Angew. Math.* **434** (1993), 67–114.

[Boutot 1987] J.-F. Boutot, "Singularités rationnelles et quotients par les groupes réductifs", *Invent. Math.* **88**:1 (1987), 65–68.

[Brenner 2002] H. Brenner, "Tight closure and plus closure for cones over elliptic curves", 2002. Available at math.AC/0209302.

[Brenner 2003a] H. Brenner, *The theory of tight closure from the viewpoint of vector bundles*, Habilitationsschrift, Universität Bochum, 2003. Available at www.arxiv.org/math.AG/0307161.

[Brenner 2003b] H. Brenner, "Tight closure and projective bundles", *J. Algebra* **265**:1 (2003), 45–78.

[Brenner 2004] H. Brenner, "Slopes of vector bundles on projective curves and applications to tight closure problems", *Trans. Amer. Math. Soc.* **356**:1 (2004), 371–392 (electronic).

[Bruns 1996] W. Bruns, "Tight closure", *Bull. Amer. Math. Soc. (N.S.)* **33**:4 (1996), 447–457.

[Buchsbaum and Eisenbud 1973] D. A. Buchsbaum and D. Eisenbud, "What makes a complex exact?", *J. Algebra* **25** (1973), 259–268.

[Ein et al. 2001] L. Ein, R. Lazarsfeld, and K. E. Smith, "Uniform bounds and symbolic powers on smooth varieties", *Invent. Math.* **144**:2 (2001), 241–252.

[Elitzur 2003] H. Elitzur, *Tight closure in Artinian modules*, Ph.D. thesis, University of Michigan, Ann Arbor, 2003.

[Enescu 2000] F. Enescu, "On the behavior of F-rational rings under flat base change", *J. Algebra* **233**:2 (2000), 543–566.

[Hara 1998] N. Hara, "A characterization of rational singularities in terms of injectivity of Frobenius maps", *Amer. J. Math.* **120**:5 (1998), 981–996.

[Hara 2001] N. Hara, "Geometric interpretation of tight closure and test ideals", *Trans. Am. Math. Soc.* **353**:5 (2001), 1885–1906.

[Hara and Watanabe 2002] N. Hara and K. Watanabe, "F-regular and F-pure rings vs. log terminal and log canonical singularities", *J. Algebraic Geom.* **11**:2 (2002), 363–392.

[Hara and Yoshida 2003] N. Hara and K.-I. Yoshida, "A generalization of tight closure and multiplier ideals", *Trans. Amer. Math. Soc.* **355**:8 (2003), 3143–3174.

[Hara et al. 2002a] N. Hara, K. Watanabe, and K.-i. Yoshida, "F-rationality of Rees algebras", *J. Algebra* **247**:1 (2002), 153–190.

[Hara et al. 2002b] N. Hara, K. Watanabe, and K.-i. Yoshida, "Rees algebras of F-regular type", *J. Algebra* **247**:1 (2002), 191–218.

[Heitmann 2002] R. C. Heitmann, "The direct summand conjecture in dimension three", *Ann. of Math.* (2) **156**:2 (2002), 695–712.

[Hochster 1994a] M. Hochster, "Solid closure", pp. 103–172 in *Commutative algebra: syzygies, multiplicities, and birational algebra* (South Hadley, MA, 1992), edited by W. J. Heinzer et al., Contemp. Math. **159**, Amer. Math. Soc., Providence, RI, 1994.

[Hochster 1994b] M. Hochster, "Tight closure in equal characteristic, big Cohen-Macaulay algebras, and solid closure", pp. 173–196 in *Commutative algebra: syzygies, multiplicities, and birational algebra* (South Hadley, MA, 1992), edited by W. J. Heinzer et al., Contemp. Math. **159**, Amer. Math. Soc., Providence, RI, 1994.

[Hochster 1996] M. Hochster, "The notion of tight closure in equal characteristic zero", pp. 94–106 (Appendix 1) in *Tight closure and its applications,* by Craig Huneke, CBMS Regional Conference Series in Mathematics **88**, American Math. Society, Providence, 1996.

[Hochster 2002] M. Hochster, "Big Cohen-Macaulay algebras in dimension three via Heitmann's theorem", *J. Algebra* **254**:2 (2002), 395–408.

[Hochster 2003] M. Hochster, "Parameter-like sequences and extensions of tight closure", pp. 267–287 in *Commutative ring theory and applications* (Fez, 2001), edited by M. Fontana et al., Lecture Notes in Pure and Appl. Math. **231**, Dekker, New York, 2003.

[Hochster and Huneke 1988] M. Hochster and C. Huneke, "Tightly closed ideals", *Bull. Amer. Math. Soc.* (*N.S.*) **18**:1 (1988), 45–48.

[Hochster and Huneke 1989a] M. Hochster and C. Huneke, "Tight closure", pp. 305–324 in *Commutative algebra* (Berkeley, 1987), edited by M. Hochster et al., Math. Sci. Res. Inst. Publ. **15**, Springer, New York, 1989.

[Hochster and Huneke 1989b] M. Hochster and C. Huneke, *Tight closure and strong F-regularity* (Orsay, 1987), Mém. Soc. Math. France (N.S.) **38**, 1989.

[Hochster and Huneke 1990] M. Hochster and C. Huneke, "Tight closure, invariant theory, and the Briançon–Skoda theorem", *J. Amer. Math. Soc.* **3**:1 (1990), 31–116.

[Hochster and Huneke 1991a] M. Hochster and C. Huneke, "Absolute integral closures are big Cohen-Macaulay algebras in characteristic p", *Bull. Amer. Math. Soc.* (*N.S.*) **24**:1 (1991), 137–143.

[Hochster and Huneke 1991b] M. Hochster and C. Huneke, "Tight closure and elements of small order in integral extensions", *J. Pure Appl. Algebra* **71**:2-3 (1991), 233–247.

[Hochster and Huneke 1992] M. Hochster and C. Huneke, "Infinite integral extensions and big Cohen-Macaulay algebras", *Ann. of Math.* (2) **135**:1 (1992), 53–89.

[Hochster and Huneke 1993] M. Hochster and C. Huneke, *Phantom homology,* Mem. Amer. Math. Soc. **490**, 1993.

[Hochster and Huneke 1994a] M. Hochster and C. Huneke, "F-regularity, test elements, and smooth base change", *Trans. Amer. Math. Soc.* **346**:1 (1994), 1–62.

[Hochster and Huneke 1994b] M. Hochster and C. Huneke, "Tight closure of parameter ideals and splitting in module-finite extensions", *J. Algebraic Geom.* **3**:4 (1994), 599–670.

[Hochster and Huneke 1995] M. Hochster and C. Huneke, "Applications of the existence of big Cohen-Macaulay algebras", *Adv. Math.* **113**:1 (1995), 45–117.

[Hochster and Huneke 1999] M. Hochster and C. Huneke, "Tight closure in equal characteristic zero", preprint, 1999. Available at http://www.math.lsa.umich.edu/~hochster/msr.html.

[Hochster and Huneke 2000] M. Hochster and C. Huneke, "Localization and test exponents for tight closure", *Michigan Math. J.* **48** (2000), 305–329.

[Hochster and Huneke 2002] M. Hochster and C. Huneke, "Comparison of symbolic and ordinary powers of ideals", *Invent. Math.* **147**:2 (2002), 349–369.

[Hochster and Roberts 1974] M. Hochster and J. L. Roberts, "Rings of invariants of reductive groups acting on regular rings are Cohen-Macaulay", *Advances in Math.* **13** (1974), 115–175.

[Hochster and Vélez 2004] M. Hochster and J. D. Vélez, "Diamond closure", pp. 511–523 in *Algebra, arithmetic and geometry with applications* (West Lafayette, IN, 2000), edited by C. Christensen et al., Springer, Berlin, 2004.

[Huneke 1996] C. Huneke, *Tight closure and its applications*, CBMS Regional Conference Series in Mathematics **88**, American Math. Society, Providence, 1996.

[Huneke 1998] C. Huneke, "Tight closure, parameter ideals, and geometry", pp. 187–239 in *Six lectures on commutative algebra* (Bellaterra, 1996), edited by J. Elias et al., Progr. Math. **166**, Birkhäuser, Basel, 1998.

[Huneke and Smith 1997] C. Huneke and K. E. Smith, "Tight closure and the Kodaira vanishing theorem", *J. Reine Angew. Math.* **484** (1997), 127–152.

[Kunz 1986] E. Kunz, *Kähler differentials*, Advanced Lectures in Mathematics, Vieweg, Braunschweig, 1986.

[Lipman and Sathaye 1981] J. Lipman and A. Sathaye, "Jacobian ideals and a theorem of Briançon-Skoda", *Michigan Math. J.* **28**:2 (1981), 199–222.

[Lipman and Teissier 1981] J. Lipman and B. Teissier, "Pseudorational local rings and a theorem of Briançon–Skoda about integral closures of ideals", *Michigan Math. J.* **28**:1 (1981), 97–116.

[Lyubeznik and Smith 1999] G. Lyubeznik and K. E. Smith, "Strong and weak F-regularity are equivalent for graded rings", *Amer. J. Math.* **121**:6 (1999), 1279–1290.

[Lyubeznik and Smith 2001] G. Lyubeznik and K. E. Smith, "On the commutation of the test ideal with localization and completion", *Trans. Amer. Math. Soc.* **353**:8 (2001), 3149–3180 (electronic).

[McDermott 2000] M. A. McDermott, "Tight closure, plus closure and Frobenius closure in cubical cones", *Trans. Amer. Math. Soc.* **352**:1 (2000), 95–114.

[Northcott 1976] D. G. Northcott, *Finite free resolutions*, Cambridge Tracts in Mathematics **71**, Cambridge University Press, Cambridge, 1976.

[Ranganathan 2000] N. Ranganathan, *Splitting in module-finite extensions and the vanishing conjecture for maps of Tor*, Ph.D. thesis, University of Michigan, Ann Arbor, 2000.

[Singh 1998] A. K. Singh, "A computation of tight closure in diagonal hypersurfaces", *J. Algebra* **203**:2 (1998), 579–589.

[Singh 1999] A. K. Singh, "*F*-regularity does not deform", *Amer. J. Math.* **121**:4 (1999), 919–929.

[Skoda 1972] H. Skoda, "Application des techniques L^2 à la théorie des idéaux d'une algèbre de fonctions holomorphes avec poids", *Ann. Sci. École Norm. Sup.* (4) **5** (1972), 545–579.

[Skoda and Briançon 1974] H. Skoda and J. Briançon, "Sur la clôture intégrale d'un idéal de germes de fonctions holomorphes en un point de $\mathbf{C}^n$", *C. R. Acad. Sci. Paris Sér. A* **278** (1974), 949–951.

[Smith 1994] K. E. Smith, "Tight closure of parameter ideals", *Invent. Math.* **115**:1 (1994), 41–60.

[Smith 1997] K. E. Smith, "*F*-rational rings have rational singularities", *Amer. J. Math.* **119**:1 (1997), 159–180.

[Smith 2000] K. E. Smith, "The multiplier ideal is a universal test ideal", *Commun. Algebra* **28**:12 (2000), 5915–5929.

[Sullivant 2002] S. Sullivant, "Tight closure of monomial ideals in Fermat rings", preprint, 2002. Available at http://math.berkeley.edu/~seths/papers.html.

[Vélez 1995] J. D. Vélez, "Openness of the F-rational locus and smooth base change", *J. Algebra* **172**:2 (1995), 425–453.

[Vraciu 2000] A. Vraciu, "Local cohomology of Frobenius images over graded affine algebras", *J. Algebra* **228**:1 (2000), 347–356.

[Vraciu 2002] A. Vraciu, "∗-independence and special tight closure", *J. Algebra* **249**:2 (2002), 544–565.

[Watanabe 1997] K. Watanabe, "*F*-rationality of certain Rees algebras and counterexamples to "Boutot's theorem" for *F*-rational rings", *J. Pure Appl. Algebra* **122**:3 (1997), 323–328.

MELVIN HOCHSTER
MATHEMATICS DEPARTMENT
UNIVERSITY OF MICHIGAN
EAST HALL
525 EAST UNIVERSITY
ANN ARBOR, MI 48109-1109
UNITED STATES
    hochster@umich.edu

GRAHAM J. LEUSCHKE
DEPARTMENT OF MATHEMATICS
UNIVERSITY OF KANSAS
LAWRENCE, KS 66045
UNITED STATES
    gjleusch@syr.edu, http://www.leuschke.org/

Trends in Commutative Algebra
MSRI Publications
Volume **51**, 2004

Monomial Ideals, Binomial Ideals, Polynomial Ideals

BERNARD TEISSIER

ABSTRACT. These lectures provide a glimpse of the applications of toric geometry to singularity theory. They illustrate some ideas and results of commutative algebra by showing the form which they take for very simple ideals of polynomial rings: monomial or binomial ideals, which can be understood combinatorially. Some combinatorial facts are the expression for monomial or binomial ideals of general results of commutative algebra or algebraic geometry such as resolution of singularities or the Briançon–Skoda theorem. In the opposite direction, there are methods that allow one to prove results about fairly general ideals by continuously specializing them to monomial or binomial ideals.

CONTENTS

1. Introduction

Let k be a field. We denote by $k[u_1, \ldots, u_d]$ the polynomial ring in d variables, and by $k[[u_1, \ldots, u_d]]$ the power series ring.

If $d = 1$, given two monomials u^m, u^n, one divides the other, so that if $m > n$, say, a binomial $u^m - \lambda u^n = u^n(u^{m-n} - \lambda)$ with $\lambda \in k^*$ is, viewed now in $k[[u]]$, a monomial times a unit. For the same reason any series $\sum_i f_i u^i \in k[[u]]$ is the product of a monomial u^n, $n \geq 0$, by a unit of $k[[u]]$. Staying in $k[u]$, we can also view our binomial as the product of a monomial and a cyclic polynomial $u^{m-n} - \lambda$.

For $d = 2$, working in $k[[u_1, u_2]]$, we meet a serious difficulty: a series in two variables does not necessarily have a dominant term (a term that divides all others). The simplest example is the binomial $u_1^a - cu_2^b$ with $c \in k^*$. As we shall see, if we allow enough transformations, this is essentially the only example in dimension 2. So the behavior of a series $f(u_1, u_2)$ near the origin does not reduce to that of the product of a monomial $u_1^a u_2^b$ by a unit.

In general, for $d > 1$ and given $f(u_1, \ldots, u_d) \in k[[u_1, \ldots, u_d]]$, say $f = \sum_m f_m u^m$, where $m \in \mathbf{Z}_{\geq 0}^d$ and $u^m = u_1^{m_1} \ldots u_d^{m_d}$, we can try to measure how far f is from a monomial times a unit by considering the ideal of $k[[u_1, \ldots, u_d]]$ or $k[u_1, \ldots, u_d]$ generated by the monomials $\{u^m : f_m \neq 0\}$ that actually appear in f. Since both rings are noetherian, this ideal is finitely generated in both cases, and we are faced with the following problem:

PROBLEM. *Given an ideal generated by finitely many monomials (a monomial ideal) in* $k[[u_1, \ldots, u_d]]$ *or* $k[u_1, \ldots, u_d]$, *study how far it is from being principal.*

We shall also meet a property of finitely generated ideals that is stronger than principality, namely that given any pair of generators, one divides the other. This implies principality (exercise), but is stronger in general: take an ideal in a principal ideal domain such as $\mathbf{Z}$, or a nonmonomial ideal in $k[u]$. I shall call this property *strong principality*. Integral domains in which every finitely generated ideal is strongly principal are known as *valuation rings*. Most are not noetherian.

Here we reach a bifurcation point in methodology:

– One approach is to generalize the notion of divisibility by studying all linear relations, with coefficients in the ambient ring, between our monomials. This leads to the construction of syzygies for the generators of our monomial ideal $\mathscr{M}$, or free resolutions for the quotient of the ambient ring by $\mathscr{M}$. There are many beautiful results in this direction; see [Eisenbud and Sidman 2004] in this volume and [Sturmfels 1996]. One is also led to try and compare monomials using *monomial orders* to produce Gröbner bases, since as soon as the ideal is not principal, deciding whether a given element belongs to it becomes arduous in general.

– Another approach is to try and force the ideal $\mathscr{M}$ to become principal after a change of variables. This is the subject of the next section.

2. Strong Principalization of Monomial Ideals by Toric Maps

In order to make a monomial ideal principal by changes of variables, the first thing to try is changes of variables that transform monomials into monomials, that is, which are themselves described by monomial functions:

$$u_1 = y_1{}^{a_1^1} \cdots \cdot y_d{}^{a_1^d},$$

$$u_2 = y_1{}^{a_2^1} \cdots \cdot y_d{}^{a_2^d},$$

$$\cdots\cdots\cdots\cdots\cdots\cdots$$

$$u_d = y_1{}^{a_d^1} \cdots \cdot y_d{}^{a_d^d},$$

where we can consider the exponents of y_i appearing in the expressions of $u_1, \ldots, u_d$ as the coordinates of a vector a^i with integral coordinates. These expressions decribe a monomial, or *toric*, map of d-dimensional affine spaces

$$\pi(a^1, \ldots, a^d) : \mathbf{A}^d(k) \to \mathbf{A}^d(k)$$

in the coordinates (y_i) for the first affine space and (u_i) for the second.

If we compute the effect of the change of variables on a monomial u^m, we see that

$$u^m \mapsto y_1^{\langle a^1, m\rangle} \cdots y_d^{\langle a^d, m\rangle}.$$

EXERCISE. Show that the degree of the fraction field extension $k(u_1, \ldots, u_d) \to k(y_1, \ldots, y_d)$ determined by $\pi(a^1, \ldots, a^d)$ is the absolute value of the determinant of the vectors $(a^1, \ldots, a^d)$. In particular, it is equal to one — that is, our map $\pi(a^1, \ldots, a^d)$ is birational — if and only if the determinant of the vectors $(a^1, \ldots, a^d)$ is ± 1, that is, $(a^1, \ldots, a^d)$ is a basis of the integral lattice $\mathbf{Z}^d$.

In view of the form of the transformation on monomials by our change of variables, it makes sense to introduce a copy of $\mathbf{Z}^d$ where the exponents of our monomials dwell, and which we will denote by M, and a copy of $\mathbf{Z}^d$ in which our vectors a^j dwell, which we will call the *weight space* and denote by W. The lattices M and W are dual and we consider W as the integral lattice of the vector space $\check{\mathbf{R}}^d$ dual to the vector space $\mathbf{R}^d$ in which our monomial exponents live. In this manner, we think of $m \mapsto \langle a^i, m\rangle$ as the linear form on M corresponding to $a^i \in W$.

Given two monomials u^m and u^n, the necessary and sufficient condition for the transform of u^n to divide the transform of u^m in $k[y_1, \ldots, y_d]$ is that $\langle a^i, m\rangle \geq \langle a^i, n\rangle$ for all i with $1 \leq i \leq d$. If we read this as $\langle a^i, m-n\rangle \geq 0$ for all i, $1 \leq i \leq d$, and seek a symmetric formulation, we are led to introduce the rational hyperplane H_{m-n} in $\check{\mathbf{R}}^d$ dual to the vector $m - n \in M$, and obtain the following elementary but fundamental fact, where the transform of a monomial is just its composition with the map $\pi(a^1, \ldots, a^d)$ in the coordinates $(y_1, \ldots, y_d)$:

LEMMA 2.1. *A necessary and sufficient condition for the transform of one of the monomials u^m, u^n by the map $\pi(a^1, \ldots, a^d)$ to divide the transform of the other* *in $k[y_1, \ldots, y_d]$ is that all the vectors a^j lie on the same side of the hyperplane H_{m-n} in $\check{\mathbf{R}}_{\geq 0}^d$.*

The condition is nonvacuous if and only if one of the monomials u^m, u^n does not already divide the other in $k[u_1, \ldots, u_d]$, because to say that such divisibility does not occur is to say that the equation of the hyperplane H_{m-n} does not have all its coefficients of the same sign, and therefore separates into two regions the first quadrant $\check{\mathbf{R}}_{\geq 0}^d$ where our vectors a^j live.

To force one monomial to divide the other in the affine space $\mathbf{A}^d(k)$ with coordinates (y_i) is nice, but not terribly useful, since it provides information on the original monomials only in the image of the map $\pi(a^1, \ldots, a^d)$ in the affine space $\mathbf{A}^d(k)$ with coordinates (u_i), which is a constructible subset different from $\mathbf{A}^d(k)$. It is much more useful to find a proper and birational (hence surjective) map $\pi : Z \to \mathbf{A}^d(k)$ of algebraic varieties over k such that the compositions with π of our monomials generate a sheaf of ideals in Z which is *locally* principal; if you prefer, Z should be covered by affine charts U such that if our monomial ideal $\mathscr{M}$ is generated by $u^{m^1}, \ldots, u^{m^q}$, the ideal $(u^{m^1} \circ \pi, \ldots, u^{m^q} \circ \pi)|_U$ is principal or strongly principal.

Toric geometry provides a way to do this. To set the stage, we need a few definitions (see [Ewald 1996]):

A *cone* σ in $\mathbf{R}^d$ (or $\check{\mathbf{R}}^d$) is a set closed under multiplication by nonnegative numbers. A cone is *strictly convex* if it contains no positive-dimensional vector subspace. Cones contained in the first quadrant are strictly convex. The *convex dual* of σ is the set

$$\check{\sigma} = \{m \in \mathbf{R}^d : \langle m, a \rangle \geq 0 \text{ for all } a \in \sigma\}.$$

This is also a cone. A cone is strictly convex if and only if its convex dual has nonempty interior.

A *rational* convex cone is one bounded by finitely many hyperplanes whose equations have rational (or equivalently, integral) coefficients. An equivalent definition is that a rational convex cone is the cone positively generated by finitely many vectors with integral coordinates.

A *rational fan* with support $\check{\mathbf{R}}_{\geq 0}^d$ is a finite collection Σ of rational strictly convex cones $(\sigma_\alpha)_{\alpha \in A}$ with the following properties:

(1) The union of all the $(\sigma_\alpha)_{\alpha \in A}$ is the closed first quadrant $\check{\mathbf{R}}_{\geq 0}^d$ of $\check{\mathbf{R}}^d$.
(2) Each face of a $\sigma_\alpha \in \Sigma$ is in Σ; in particular $\{0\} \in \Sigma$.
(3) Each intersection $\sigma_\alpha \cap \sigma_\beta$ is a face of σ_α and of σ_β.

In general, the *support* of a fan Σ is defined as $\bigcup_{\alpha \in A} \sigma_\alpha$.

A fan is *regular* if each of its k-dimensional cones is generated by k integral vectors (a simplicial cone) that form part of a basis of the integral lattice. If $k = d$ this means that their determinant is ± 1.

If we go back to our monomial map, assuming that the determinant of the vectors $(a^1, \ldots, a^d)$ is ± 1, we can express the y_j as monomials in the u_i; the matrix of exponents will then be the inverse of the matrix $(a^1, \ldots, a^d)$, and will have some negative entries. Monomials with possibly negative exponents will be called *Laurent monomials* here.

If $\sigma = \langle a^1, \ldots, a^d \rangle$, the cone positively generated by the vectors $a^1 \ldots, a^d$, then the monomials in $y_1, \ldots, y_d$, viewed as Laurent monomials in $u_1, \ldots u_d$ via the expression of the y_j as Laurent monomials in the u_i, correspond to the integral points of the convex dual cone of σ, that is, those points $m \in \mathbf{Z}^d$ such that $\langle a^i, m \rangle \geq 0$ for all $1 \leq i \leq d$. So we can identify the polynomial algebra $k[y_1, \ldots, y_d]$ with the algebra $k[\check{\sigma} \cap M]$ of the semigroup $\check{\sigma} \cap M$ with coefficients in k. Since σ is contained in the first quadrant of $\check{\mathbf{R}}^d$, its convex dual $\check{\sigma}$ contains the first quadrant of $\mathbf{R}^d$, so we have a graded inclusion of algebras

$$k[\mathbf{R}^d_{\geq 0} \cap M] = k[u_1, \ldots, u_d] \subset k[\check{\sigma} \cap M] = k[y_1, \ldots, y_d],$$

the inclusion being described by sending each variable u_i to a monomial in $y_1, \ldots, y_d$ as we did in the beginning.

This slightly more abstract formulation has the following use: Given a fan in $\check{\mathbf{R}}^d$, to each of its cones σ we can associate the algebra $k[\check{\sigma} \cap M]$, even if the strictly convex cone σ is not generated by d vectors with determinant ± 1.

By a lemma of Gordan [Kempf et al. 1973], the algebra $k[\check{\sigma} \cap M]$ is finitely generated, so it corresponds to an affine algebraic variety $X_\sigma = \mathrm{Spec}\ k[\check{\sigma} \cap M]$. This variety is a d-dimensional affine space if and only if the cone $\check{\sigma}$ (or σ) is d-dimensional and generated by vectors that form a basis of the integral lattice of $\check{\mathbf{R}}^d$. It is, however, always normal and has rational singularities only [Kempf et al. 1973]; moreover it is rational, which means that the field of fractions of $k[\check{\sigma} \cap M]$ is $k(u_1, \ldots, u_d)$.

If two cones σ_α and σ_β have a common face $\tau_{\alpha\beta}$, the affine varieties X_{σ_α} and X_{σ_β} can be glued up along the open set corresponding to the shared $X_{\tau_{\alpha\beta}}$. By this process, the fan Σ gives rise to an algebraic variety $Z(\Sigma)$ proper over $\mathbf{A}^d(k)$:

$$\pi(\Sigma) : Z(\Sigma) \to \mathbf{A}^d(k).$$

The variety $Z(\Sigma)$ is covered by affine charts corresponding to the d-dimensional cones σ of Σ, and in each of these charts the map $\pi(\Sigma)$ corresponds to the inclusion of algebras $k[u_1, \ldots, u_d] \subset k[\check{\sigma} \cap M]$. If σ is generated by d vectors forming a basis of the integral lattice (determinant ± 1), the latter algebra is a polynomial ring $k[y_1, \ldots, y_d]$ and the inclusion is given by the monomial expression we started from.

DEFINITION. A convex polyhedral cone σ is *compatible* with a convex polyhedral cone σ' if $\sigma \cap \sigma'$ is a face of each. A fan is compatible with a polyhedral cone if each of its cones is.

Remember that $\{0\}$ is a face of every strictly convex cone.

LEMMA 2.2. *Given two monomials u^m, u^n, if we can find a fan Σ compatible with the hyperplane H_{m-n} in the weight space, then in each chart of $Z(\Sigma)$ the transform of one of our monomials will divide the other.*

PROOF. This follows from Lemma 2.1.   □

EXAMPLE. In dimension $d = 2$, let's try to make one of the two monomials (u_1, u_2) divide the other after a monomial transformation. The hyperplane in the weight space is $w_1 = w_2$; its intersection with the first quadrant defines a fan whose cones are σ_1 generated by $a^1 = (0,1)$, $a^2 = (1,1)$ and σ_2 generated by $b^1 = (1,1)$, $b^2 = (0,1)$, together with and their faces. The semigroup of integral points of $\check{\sigma}_1 \cap M$ is generated by $(1,0)$, $(-1,1)$, which correspond respectively to the monomials $y_1 = u_1$, $y_2 = u_1^{-1}u_2$. The semigroup of integral points of $\check{\sigma}_2 \cap M$ is generated by $(0,1)$, $(1,-1)$, which correspond to $y_2' = u_2$, $y_1' = u_1 u_2^{-1}$. There is a natural isomorphism of the open sets where $u_1 \neq 0$ and $u_2 \neq 0$, and gluing gives the two-dimensional subvariety of $\mathbf{A}^2(k) \times \mathbf{P}^1(k)$ defined by $t_2 u_1 - t_1 u_2 = 0$, where $(t_1 : t_2)$ are the homogeneous coordinates on $\mathbf{P}^1(k)$, with its natural projection to $\mathbf{A}^2(k)$: it is the blowing-up of the origin in $\mathbf{A}^2(k)$.

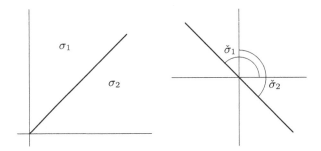

Now if we have a finite number of distinct monomials $\neq 1$, say $u^{m^1}, \ldots, u^{m^q}$, and if we can find a fan Σ with support $\check{\mathbf{R}}^d_{\geq 0}$ and compatible with *all* the hyperplanes $H_{m^s - m^t}$, $1 \leq s, t \leq q$, $s \neq t$, this will give us an algebraic (toric) variety $Z(\Sigma)$, possibly singular and endowed with a *proper surjective map* $\pi(\Sigma) : Z(\Sigma) \to \mathbf{A}^d(k)$ such that the pullback by $\pi(\Sigma)$ of the ideal $\mathscr{M}$ generated by our monomials is strongly principal in each chart. Properness and surjectivity are ensured (see [Kempf et al. 1973]) by the fact that the support of Σ is $\mathbf{R}^d_{\geq 0}$.

Our collection of hyperplanes $H_{m^s - m^t}$, $1 \leq s, t \leq q$, $s \neq t$ through the origin in fact defines a fan $\Sigma_0(F)$ that depends only upon the finite set $F = \{m^1, \ldots, m^q\}$ of elements of $\mathbf{Z}^d$: take as cones the closures of the connected components of the complement in $\check{\mathbf{R}}^d_{\geq 0}$ of the union of all the hyperplanes. They are strictly convex rational cones because they lie in the first quadrant and are bounded by hyperplanes whose equations have integral coefficients. Add all the faces of these cones, and we have a fan, of course not regular in general. To say that a monomial ideal generated by monomials in the generators of the algebra $k[\check{\sigma} \cap M]$ is locally strongly principal is not nearly as useful when these generators do not form a system of coordinates as when they do. However, note that we first make our ideal $\mathscr{M}$ locally strongly principal via the map $\pi(\Sigma_0) : Z(\Sigma_0) \to \mathbf{A}^d(k)$, and then resolve the singularities of $Z(\Sigma_0)$ using a toric map.

The second step corresponds to a refinement of Σ_0 into a regular fan Σ, where refinement means that each cone of the second fan in contained in a cone of the original.

This is always possible in view of a result of Kempf, Knudsen, Mumford and St. Donat:

THEOREM 2.3 [Kempf et al. 1973]. *A rational fan can always be refined into a regular fan.*

From this follows:

THEOREM 2.4. *Let k be a field. Given a monomial ideal $\mathscr{M}$ in $k[u_1, \ldots, u_d]$, there exists a fan Σ_0 with support $\check{\mathbf{R}}^d_{\geq 0}$ such that, given any regular refinement Σ of Σ_0, the associated proper birational toric map of nonsingular toric varieties*

$$\pi(\Sigma) : Z(\Sigma) \to \mathbf{A}^d(k)$$

has the property that the transform of $\mathscr{M}$ is strongly principal in each chart.

REMARK. By construction, for each chart $Z(\sigma)$ of $Z(\Sigma)$ there is an element of $\mathscr{M}$ whose transform generates the ideal $\mathscr{M}\mathcal{O}_{Z(\sigma)}$. This element cannot be the same for all charts unless $\mathscr{M}$ is already principal.

To see this, assume that there is a monomial u^n whose transform generates $\mathscr{M}\mathcal{O}_{Z(\Sigma)}$ in every chart. This means that every simplicial cone σ of our fan with support $\check{\mathbf{R}}^d_{\geq 0}$ is on the positive side of all the hyperplanes H_{m-n} for all other monomials u^m generating $\mathscr{M}$. But this is possible only if none of these hyperplanes meets the positive quadrant outside $\{0\}$, which means that u^n divides all the other u^m.

REMARK (STRONG PRINCIPALIZATION AND BLOWING-UP). Given a finitely generated ideal I in a commutative integral domain R, there is a proper birational map $\pi : B(I) \to \operatorname{Spec} R$, unique up to unique isomorphism, with the property that the ideal sheaf $I\mathcal{O}_{B(I)}$ generated by the compositions with π of the elements of I is locally principal and generated by a nonzero divisor (that is, it's an *invertible* ideal), and that any map $W \to \operatorname{Spec} R$ with the same property factors uniquely through π. The map π is called the *blowing-up* of I in R, or in $\operatorname{Spec} R$.

The blowing-up is independent of the choice of generators of I. Since a product of ideals is invertible if and only if each ideal is, for $I = (f_1, \ldots, f_s)R$ the blowing-up in $\operatorname{Spec} R$ of the ideal $J = \prod_{i<j}(f_i, f_j)R$ will make I strongly principal.

If I is a monomial ideal in $k[u_1, \ldots, u_d]$, according to [Kempf et al. 1973], the blowing-up of I *followed by normalization* is the equivariant map associated to the fan dual to the Newton polyhedron of I. (The Newton polyhedron is defined in the Appendix.) The reader is encouraged to check that the fan just mentioned admits the fan Σ_0 introduced above as a refinement, illustrating the general fact that a strong principalization map factors through the blowing-up.

Strong principalization is stressed in these lectures because it is directly linked with the resolution of singularities of binomial ideals explained in Section 6.

EXERCISE. Check that one can in all statements and proofs in this section replace the positive quadrant of $\check{\mathbf{R}}^d$ by any strictly convex rational cone $\sigma_0 \subset \check{\mathbf{R}}^d$. The affine space $\mathbf{A}^d(k)$ is then replaced by the affine toric variety X_{σ_0}.

3. The Integral Closure of Ideals

Given a finite set $F = \{m^1, \ldots, m^q\}$ of elements of $\mathbf{Z}^d$, define its *support function* as the function $\mathrm{h}_F : \check{\mathbf{R}}^d \to \mathbf{R}$ defined by

$$\mathrm{h}_F(\ell) = \min_{1 \leq s \leq q} \ell(m^s).$$

For reasons that will become apparent, I denote the convex hull of F by $\bar{F}$. It is a classical result that

$$\bar{F} = \{n \in \mathbf{R}^d : \ell(n) \geq \mathrm{h}_F(\ell) \text{ for all } \ell \in \check{\mathbf{R}}^d\};$$

in words, the convex hull of a set is the intersection of the half-spaces containing that set (or, as often stated in books on convexity, a convex set is the intersection of the half-spaces determined by its support hyperplanes). The proof of this statement also shows that the "positive convex hull" is defined by the same inequalities, restricted to the linear forms lying in the positive quadrant of $\check{\mathbf{R}}^d$:

$$\overline{\bigcup_{1 \leq s \leq q} (m^s + \mathbf{R}_{\geq 0}^d)} = \{n \in \mathbf{R}^d : \ell(n) \geq \mathrm{h}_F(\ell) \text{ for all } \ell \in \check{\mathbf{R}}_{\geq 0}^d\}.$$

LEMMA 3.1. *The support function h_F is linear in each cone of the fan $\Sigma_0(F)$ introduced in Section 2.*

PROOF. This follows directly from the definitions.   □

Choose a strongly principalizing map $\pi(\Sigma) : Z(\Sigma) \to \mathbf{A}^d(k)$ with Σ a refinement of $\Sigma_0(F)$, as in Theorem 2.4. Then $Z(\Sigma)$ is normal by [Kempf et al. 1973] (it is regular if Σ is regular), and $\pi(\Sigma)$ is proper and birational. Let u^n be a monomial in $k[u_1, \ldots, u_d]$. Given a chart X_σ of $Z(\Sigma)$, corresponding to $\sigma \in \Sigma$, a necessary and sufficient condition for $u^n \circ \pi(\Sigma)$ to belong in $k[\check{\sigma} \cap M]$ to the ideal generated by the transforms of the generators of $\mathscr{M}$ is that we have $\ell(n) \geq \mathrm{h}_F(\ell)$ for all $\ell \in \sigma$: by Lemma 3.1, we have for some $t \in \{1, \ldots, q\}$ that $\mathrm{h}_F(\ell) = \ell(m^t)$ for all $\ell \in \sigma$, and then by the definition of $\check{\sigma}$ our inequality means that the quotient of the transform of u^n by the transform of u^{m^t} is in $k[\check{\sigma} \cap M]$, which means that $u^n k[\check{\sigma} \cap M] \subset \mathscr{M} k[\check{\sigma} \cap M]$. For this to be true in all charts it is necessary and sufficient, as we saw, that n should be in the convex hull of $\bigcup_{1 \leq s \leq q} (m^s + \mathbf{R}_{\geq 0}^d)$. So we have finally, using a little sheaf-theoretic language (in particular, $u^n \mathcal{O}_{Z(\Sigma)} = u^n \circ \pi(\Sigma)$ viewed as a global section of the sheaf $\mathcal{O}_{Z(\Sigma)}$):

LEMMA 3.2. $u^n \mathcal{O}_{Z(\Sigma)} \in \mathcal{M} \mathcal{O}_{Z(\Sigma)}$ if and only if n is in the convex hull of $\bigcup_{1 \le s \le q}(m^s + \mathbf{R}^d_{\ge 0})$.

Now one defines integral dependance over an ideal (a concept which goes back to Prüfer or even Dedekind) as follows:

DEFINITION. An element h of a commutative ring R is integral over an ideal I of R if it satisfies an algebraic relation

$$h^r + a_1 h^{r-1} + \cdots + a_r = 0, \quad \text{with } a_i \in I^i \text{ for } 1 \le i \le r.$$

It is not difficult to see that the set of elements integral over I is an ideal $\bar{I}$ containing I and contained in $\sqrt{I}$; it is the *integral closure* of I. We have the following characterization in algebraic geometry, which follows from the Riemann extension theorem:

PROPOSITION 3.3 [Lipman and Teissier 1981]. *Let k be a field and R a localization of a finitely generated reduced k-algebra. Let I be an ideal of R and $h \in R$. The element h is integral over I if and only if there exists a proper and birational morphism $t : Z \to \operatorname{Spec} R$ such that $h \circ t \in I \mathcal{O}_Z$ (i.e., $h \mathcal{O}_Z \in I \mathcal{O}_Z$), and then this inclusion holds for any such morphism such that Z is normal and $I \mathcal{O}_Z$ is invertible.*

From this follows the interpretation of Lemma 3.2:

PROPOSITION 3.4. *The integral closure in $k[u_1, \ldots, u_d]$ of a monomial ideal generated by the monomials $u^{m^1}, \ldots, u^{m^q}$ is the monomial ideal generated by the monomials with exponents in the convex hull $\bar{E}$ of $E = \bigcup_{1 \le s \le q}(m^s + \mathbf{R}^d_{\ge 0})$.*

EXAMPLE. In the ring $k[u_1, \ldots, u_d]$, for each integer $n \ge 1$ the integral closure of the ideal generated by $u_1^n, \ldots, u_d^n$ is $(u_1, \ldots, u_d)^n$.

EXERCISE. Check that in the preceding subsection, one can in all statements and proofs replace the positive quadrant of $\mathbf{R}^d$ by any strictly convex rational cone $\sigma_0 \subset \mathbf{R}^d$ and let $\mathcal{M}$ denote the ideal generated by monomials $u^{m^1}, \ldots, u^{m^q}$ of the normal toric algebra $k[\check{\sigma}_0 \cap M]$; its integral closure $\bar{\mathcal{M}}$ in that algebra is generated by the monomials with exponents in the convex hull in $\check{\sigma}_0$ of $\bigcup_{1 \le s \le q}(m^s + \check{\sigma}_0)$.

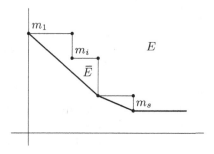

4. The Monomial Briançon–Skoda Theorem

THEOREM 4.1 (CARATHÉODORY). *Let E be a subset of $\mathbf{R}^d$; every point of the convex hull of E is in the convex hull of $d + 1$ points of E.*

PROOF. For the reader's convenience, here is a sketch of the proof, according to [Grünbaum 1967]. First one checks that the convex hull of E, defined as the intersection of all convex subsets of $\mathbf{R}^d$ containing E, coincides with the set of points of $\mathbf{R}^d$ which are in the convex hull of a finite number of points of E:

Given a finite set F of points of E, its convex hull $\bar{F}$ is contained in the convex hull $\bar{E}$ of E. Now for two finite sets F and F' we have $\bar{F} \cup \bar{F}' \subseteq \overline{F \cup F'}$, so that $\bigcup_F \bar{F}$ is convex. It contains E and so has to be equal to $\bar{E}$, which proves the assertion.

Given a point x of the convex hull of E, let p be the smallest integer such that x is in the convex hull of $p + 1$ points of E, i.e., that $x = \sum_{i=0}^{p} \alpha_i x_i$, with $\alpha_i \geq 0$, $\sum_{i=0}^{p} \alpha_i = 1$ and $x_i \in E$; we must prove that $p \leq d$. Assume that $p > d$. Then the points x_i must be affinely dependent: there is a relation $\sum_{i=0}^{p} \beta_i x_i = 0$ with $\beta_i \in \mathbf{R}$, where not all the β_i are zero and $\sum_{i=0}^{p} \beta_i = 0$. We may choose the β_i so that at least one is > 0 and renumber the points x_i so that $\beta_p > 0$ and for each index i such that $\beta_i > 0$ we have $\alpha_i/\beta_i \geq \alpha_p/\beta_p$. For $0 \leq i \leq p - 1$ set $\gamma_i = \alpha_i - \alpha_p/\beta_p \beta_i$, and $\gamma_p = 0$. Now we have

$$\sum_{i=0}^{p-1} \gamma_i x_i = \sum_{i=0}^{p} \gamma_i x_i = \sum_{i=0}^{p} \alpha_i x_i - \frac{\alpha_p}{\beta_p} \sum_{i=0}^{p} \beta_i x_i = x,$$

and moreover

$$\sum_{i=0}^{p-1} \gamma_i = \sum_{i=0}^{p} \gamma_i = \sum_{i=0}^{p} \alpha_i - \frac{\alpha_p}{\beta_p} \sum_{i=0}^{p} \beta_i = 1.$$

Finally, each γ_i is indeed ≥ 0 since if $\beta_i \leq 0$ we have $\gamma_i \geq \alpha_i \geq 0$ and if $\beta_i > 0$ then by our choice of numbering we have $\gamma_i = \beta_i(\alpha_i/\beta_i - \alpha_p/\beta_p) \geq 0$. Assuming that $p > d$ we have expressed x as the barycenter of the p points $x_0, \ldots, x_{p-1}$ of E with coefficients γ_i, which contradicts the definition of p and thus proves the theorem. $\square$

Taking for E the set consisting of $d + 1$ affinely independent points of $\mathbf{R}^d$ shows that the bound of the theorem is optimal. However, the following result means that this is essentially the only case where $d + 1$ points are necessary:

PROPOSITION 4.2 [Fenchel 1929; Hanner and Rådström 1951]. *Let $E \subset \mathbf{R}^d$ be a subset having at most d connected components. Every point of the convex hull of E is in the convex hull of d points of E.*

PROOF. We follow [Hanner and Rådström 1951]. Assume that a point m of the convex hull is *not* in the convex hull of any subset of d points of E. By Caratheodory's theorem, m is in the convex hull $\tau \subset \mathbf{R}^d$ of $d + 1$ points of E; if these $d + 1$ points were not linearly independent, the point m would be in the convex hull of the intersection of E with a hyperplane and we could apply Caratheodory's theorem in a space of dimension $d - 1$ and contradict our assumption, so the convex hull τ of the $d + 1$ points is a d-simplex. Choose m as origin, and denote by $(q_0, \ldots, q_d)$ the vertices of τ. We have therefore $0 = \sum_0^d r_i q_i$ with $r_i \geq 0$ and $\sum_0^d r_i = 1$. Our assumption that 0 is not the barycenter of d points implies that 0 is in the interior of τ, that is, $r_i > 0$ for $0 \leq i \leq d$. Consider the simplex $-\tau$ and the cones with vertex 0 drawn on the faces of $-\tau$. Since the r_i are > 0, we can reinterpret the expression of 0 as a barycenter of the q_i to mean that each q_i is in the cone with vertex 0 generated by the vectors $-q_j$ for $j \neq i$; thus each of these cones drawn from 0 on the faces of $-\tau$ contains a point of E, namely one of the q_i. The union of their closures is $\mathbf{R}^d$ because $-\tau$ is a d-simplex, and no point of E can be on the boundary of one of these cones; if such was the case, this point, together with $d - 1$ of the vertices of τ, would define a $(d - 1)$-simplex with vertices in E and containing 0, a possibility which we have excluded. Therefore these $d + 1$ cones divide E into $d + 1$ disjoint nonempty parts, and E must have at least $d + 1$ connected components. $\qquad\square$

We remark that, given finitely many points $m^1, \ldots, m^q$ in the positive quadrant $\mathbf{R}^d_{\geq 0}$, the set $E = \bigcup_{s=1}^q (m^s + \mathbf{R}^d_{\geq 0})$ is connected. Indeed, by definition, each point of this set is connected by a line to at least one of the points m^s, and any point of $\mathbf{R}^d_{\geq 0}$ having each of its coordinates larger than the maximum over $s \in \{1, \ldots, q\}$ of the corresponding coordinate of the m^s is in E and connected by lines to all the points m^s, so that any two of the points m^s are connected in E.

Now let σ be a strictly convex rational cone in $\check{\mathbf{R}}^d$ and $\check{\sigma} \subset \mathbf{R}^d$ its dual. We need not assume that σ is regular, or even simplicial. Let $m^1, \ldots, m^q$ be integral points in $\check{\sigma}$, corresponding to monomials $u^{m^1}, \ldots, u^{m^q}$ in the algebra $k[\check{\sigma} \cap M]$. The integral closure $\overline{\mathcal{M}}$ in $k[\check{\sigma} \cap M]$ of the ideal $\mathcal{M}$ generated by the monomials u^{m^s} is the ideal generated by the monomials u^n such that n is in the convex hull of the set $E = \bigcup_{s=1}^q (m^s + \check{\sigma})$. What we have just said about the connectedness of E extends immediately.

THEOREM 4.3 (MONOMIAL BRIANÇON–SKODA THEOREM). *Let k be a field and let σ be a strictly convex rational cone in $\mathbf{R}^d$. Given a monomial ideal $\mathcal{M}$ in $k[\check{\sigma} \cap M]$, we have the inclusion of ideals*

$$\overline{\mathcal{M}^d} \subset \mathcal{M}.$$

PROOF. (Compare with [Teissier 1988].) Let $y_1, \ldots, y_N$ be a system of homogeneous generators of the graded k-algebra $k[\check{\sigma} \cap M]$ and let $y^{m^1}, \ldots, y^{m^q}$ be generators of $\mathcal{M}$ in $k[\check{\sigma} \cap M]$. Set $E = \bigcup_{1 \leq s \leq q} (m^s + \check{\sigma}) \subset \check{\sigma}$. Thanks to Proposition 4.2 and the fact that E is connected, it suffices to show that any point $n \in \check{\sigma} \cap M$ which is the barycenter of d points $x_1, \ldots, x_d$, each of which is the sum of d points of E, is in E. But then n/d is also, as a barycenter of barycenters of points of E, in the convex hull of E, and therefore, by Proposition 4.2, the barycenter of d points of E. Write $n/d = \sum_{i=1}^{d} r_i e_i$ with $e_i \in E$, $r_i \geq 0$ and $\sum_{i=1}^{d} r_i = 1$. At least one of the r_i, say r_1, must be at least $1/d$, so that $n \in e_1 + \check{\sigma} \subset E$, which proves the result. $\square$

EXERCISE. Prove by the same method that for each integer $\lambda \geq 1$ we have

$$\overline{\mathcal{M}^{d+\lambda-1}} \subset \mathcal{M}^{\lambda}.$$

REMARK. It is not difficult to check that

$$\overline{E} = \lim_{n \to \infty} \frac{nE}{n} = \bigcup_{n \in \mathbf{N}} \frac{nE}{n},$$

where nE is the Minkowski multiple (the set of all sums of n elements of E) and division by n means a homothety of ratio $1/n$. In fact, the inclusion $\bigcup_{n \in \mathbf{N}}(nE/n) \subset \overline{E}$ is clear, and the first set is also clearly convex, so they are equal. The combinatorial avatar of the weak form of the Briançon–Skoda theorem, which states that $x \in \overline{\mathcal{M}}$ implies $x^d \in \mathcal{M}$, is the existence of a uniform bound for the n such that $x \in \overline{E}$ implies $nx \in E$, namely $n = d$.

The Briançon–Skoda theorem is the statement $\overline{\mathcal{M}^d} \subset \mathcal{M}$ for an ideal in a d-dimensional regular local ring. The rings $k[\check{\sigma} \cap M]$ are not regular in general, nor are they local, but the monomial Briançon–Skoda theorem for ideals in their localizations $k[\check{\sigma} \cap M]_m$ follows from the results of [Lipman and Teissier 1981] in the case where $\mathcal{M}$ contains an ideal generated by a regular sequence and with the same integral closure, since $k[\check{\sigma} \cap M]$ has only rational singularities (see [Kempf et al. 1973]) and hence $k[\check{\sigma} \cap M]_m$ is a pseudorational local ring.

The Briançon–Skoda theorem was originally proved [Skoda and Briançon 1974] by analytic methods for ideals of $\mathbf{C}\{z_1, \ldots, z_d\}$, and has been the subject of many algebraic proofs and generalizations. The first algebraic proof was given in [Lipman and Teissier 1981], albeit for a restricted class of ideals in an extended class of rings (pseudorational ones). See [Hochster 2004] and [Blickle and Lazarsfeld 2004] in this volume for references and recent developments.

5. Polynomial Ideals and Nondegeneracy

The hypothesis of nondegeneracy of a polynomial with respect to its Newton polyhedron has a fairly ancient history in the sense that it was made more or less implicitly by authors trying to compute various invariants of a projective hypersurface from its Newton polyhedron. In the nineteenth century one may mention Minding and Elliott, and in the twentieth Baker (1905) and Hodge (1930). The last three were interested in computing the geometric genus of a projective curve or surface with isolated singularities from its Newton polygon or polyhedron. This is a special case of computation of a multiplier ideal. See

[Merle and Teissier 1980], and compare its Theorem of Hodge 2.3.1 with the recent work of J. Howald expounded in [Blickle and Lazarsfeld 2004]; see also [Howald 2001].

The modern approach to nondegeneracy was initiated essentially by Kushnirenko [1976] and Khovanskii, who made the nondegeneracy condition explicit and computed from the Newton polyhedron invariants of a similar nature. In particular Khovanskii gave the general form of Hodge's result. The essential facts behind the classical computations turned out to be that nondegenerate singularities have embedded toric (pseudo-)resolutions which depend only on their Newton polyhedron and from which one can read combinatorially various interesting invariants, and that in the spaces of coefficients of all those functions or systems of functions having given polyhedra, those which are nondegenerate are Zariski-dense.

Let $f = \sum_p f_p u^p$ be an arbitrary polynomial or power series in d variables with coefficients in the field k. Let $\operatorname{Supp} f = \{p \in \mathbf{R}^d : f_p \neq 0\}$ be its *support*. The affine Newton polyhedron of f in the coordinates $(u_1, \ldots, u_d)$ is the boundary $\mathcal{N}(f)$ of the convex hull in $\mathbf{R}^d_{\geq 0}$ of the support of f. The local Newton polyhedron is the boundary $\mathcal{N}_+(f)$ of

$$\mathcal{P}_+(f) = \text{convex hull of } (\operatorname{Supp} f + \mathbf{R}^d_{\geq 0}).$$

It has finitely many compact faces and its noncompact faces of dimension at most $d - 1$ are parallel to coordinate hyperplanes. Both polyhedra depend not only on f but also on the choice of coordinates. Remark also that the local Newton polyhedron is of no interest if f has a nonzero constant term.

We can define the affine and the local support functions associated with the function f as follows (in the affine case, this is the same definition as before, applied to the set of monomials appearing in f):

For the affine Newton polyhedron it is the function defined on $\check{\mathbf{R}}^d$ by

$$\mathrm{h}_{\mathcal{N}(f)}(\ell) = \min_{p \in \mathcal{N}(f)} \ell(p),$$

and for the local Newton polyhedron it is defined on the first quadrant $\check{\mathbf{R}}^d_{\geq 0}$ by

$$\mathrm{h}_{\mathcal{N}_+(f)}(\ell) = \min_{p \in \mathcal{N}_+(f)} \ell(p).$$

Both functions are piecewise linear in their domain of definition, meaning that there is a decomposition of the domain of definition into convex cones such that the function is linear in each cone. These collections of cones are actually fans, in $\check{\mathbf{R}}^d$ and $\check{\mathbf{R}}^d_{\geq 0}$ respectively. These fans are "dual" to the Newton polyhedra in the following sense:

Consider, say for the local polyhedron, the following equivalence relation between linear functions:

$$\ell \equiv \ell' \iff \{p \in \mathcal{N}_+(f) : \ell(p) = \mathrm{h}_{\mathcal{N}_+(f)}(\ell)\} = \{p \in \mathcal{N}_+(f) : \ell'(p) = \mathrm{h}_{\mathcal{N}_+(f)}(\ell')\}.$$

Its equivalence classes form a decomposition of the first quadrant into strictly convex rational cones, and by definition the support function is linear in each of them, given by $\ell \mapsto \ell(p)$ for any p in the set $\{p \in \mathcal{N}_+(f) : \ell(p) = \mathrm{h}_{\mathcal{N}_+(f)}(\ell)\}$. These sets are *faces* of the Newton polyhedron, and the collection of the cones constitutes a fan $\Sigma_{\mathcal{N}}$ in $\mathbf{R}_{\geq 0}^d$, called the *dual fan* of the Newton polyhedron. This establishes a *one-to-one decreasing correspondence between the cones of the dual fan of a Newton polyhedron and the faces of all dimensions of that Newton polyhedron*. Corresponding to noncompact faces of the Newton polyhedron meet coordinate hyperplanes outside the origin.

We have now associated to each polynomial $f = \sum_p f_p u^p$ a dual fan in $\check{\mathbf{R}}^d$ corresponding to the global Newton polyhedron, and another in $\check{\mathbf{R}}_{\geq 0}^d$ corresponding to the local Newton polyhedron. The local polyhedron is also defined for a series $f = \sum_p f_p u^p$, and the combinatorial constructions are the same. For the moment, let's restrict our attention to the local polyhedron, assuming that $f_0 = 0$, and let's choose a regular refinement Σ of the fan associated to it.

By the definition just given, this means that for each cone $\sigma = \langle a^1, \ldots, a^k \rangle$ of the fan Σ, the primitive vectors a^i form part of a basis of the integral lattice, and all the linear forms $p \mapsto \langle a^i, p \rangle$, when restricted to the set $\{p : f_p \neq 0\}$, take their minimum value on the same subset, which is a face, of the (local) Newton polyhedron of $f = \sum_p f_p u^p$. This face may or may not be compact.

We examine the behavior of f under the map $\pi(\sigma) : Z(\sigma) \to \mathbf{A}^d(k)$ corresponding to a cone $\sigma = \langle a^1, \ldots, a^d \rangle \subset \check{\mathbf{R}}_{\geq 0}^d$ of a regular fan which is a subdivision of the fan associated to the local polyhedron of f. If we write h for $\mathrm{h}_{\mathcal{N}_+(f)}$ we get

$$f \circ \pi(\sigma) = \sum_p f_p y_1^{\langle a^1, p \rangle} \cdots y_d^{\langle a^d, p \rangle}$$

$$= y_1^{h(a^1)} \cdots y_d^{h(a^d)} \sum_p f_p y_1^{\langle a^1, p \rangle - h(a^1)} \cdots y_d^{\langle a^d, p \rangle - h(a^d)}.$$

The last sum is by definition the *strict transform* of f by $\pi(\sigma)$.

EXERCISES. Check that:

(a) In each chart $Z(\sigma)$ the exceptional divisor consists (set-theoretically) of the union of those hyperplanes $y_j = 0$ such that a^j is not a basis vector of $\check{\mathbf{Z}}^d$.

(b) Provided that no monomial in the u_i divides f, the hypersurface

$$\sum_p f_p y_1^{\langle a^1, p \rangle - h(a^1)} \cdots y_d^{\langle a^d, p \rangle - h(a^d)} = 0$$

is indeed the strict transform by the map $\pi(\sigma) : Z(\sigma) \to \mathbf{A}^d(k)$ of the hypersurface $X \subset \mathbf{A}^d(k)$ defined by $f(u_1, \ldots, u_d) = 0$, in the sense that it is the closure in $Z(\sigma)$ of the image of $X \cap (k^*)^d$ by the isomorphism induced by $\pi(\sigma)$ on the tori of the two toric varieties $Z(\sigma)$ and $\mathbf{A}^d(k)$ as well as in the sense that it is obtained from $f \circ \pi(\sigma)$ by factoring out as many times as possible the defining functions of the components of the exceptional divisor.

Denote by $\tilde{f}$ the strict transform of f and note that by construction it has a nonzero constant term: the cone σ is of maximal dimension, which means that there is a unique exponent p such that $\langle a, p \rangle = h(a)$ for $a \in \sigma$.

The map $\pi(\tau)$ associated to a face τ of σ coincides with the restriction of $\pi(\sigma)$ to an open set $Z(\tau) \subset Z(\sigma)$ which is of the form $y_j \neq 0$ for $j \in J$, where J depends on $\tau \subset \sigma$.

Now we can, for each cone σ of our regular fan, stratify the space $Z(\sigma)$ in such a way that $\pi(\sigma)^{-1}(0)$ is a union of strata. Let I be a subset of $\{1, 2, \ldots, d\}$ and define S_I to be the constructible subset of $Z(\sigma)$ defined by $y_i = 0$ for $i \in I$, $y_i \neq 0$ for $i \notin I$. The S_I for $I \subset \{1, 2, \ldots, d\}$ constitute a partition of $Z(\sigma)$ into nonsingular varieties, constructible in $Z(\sigma)$, which we call the *natural stratification* of $Z(\sigma)$. If we glue up two charts $Z(\sigma)$ and $Z(\sigma')$ along $Z(\sigma \cap \sigma')$, the natural stratifications glue up as well.

If we restrict the strict transform

$$\tilde{f}(y_1, \ldots, y_d) = \sum_p f_p y_1^{\langle a^1, p \rangle - h(a^1)} \cdots y_d^{\langle a^d, p \rangle - h(a^d)}$$

to a stratum S_I, we see that in the sum representing $\tilde{f}(y_1, \ldots, y_d)$ only the terms $f_p y_1^{\langle a^1, p \rangle - h(a^1)} \cdots y_d^{\langle a^d, p \rangle - h(a^d)}$ such that $\langle a^i, p \rangle - h(a^i) = 0$ for $i \in I$ survive. These equalities define a unique face γ_I of the Newton polyhedron of f, since our fan is a subdivision of its dual fan. Given a series $f = \sum_p f_p u^p$ and a weight vector $\check{a} \in \mathbf{R}_{\geq 0}^d$, the set

$$\{p \in \mathbf{Z}_{\geq 0}^d : f_p \neq 0 \text{ and } \langle \check{a}, p \rangle = h(\check{a})\}$$

is a face of the local Newton polyhedron of f, corresponding to the cone of the dual fan which contains $\check{a}$ in its relative interior. If all the coordinates of the vector $\check{a}$ are positive, this face is compact.

Moreover, if we define

$$f_{\gamma_I} = \sum_{p \in \gamma_I} f_p u^p$$

to be the partial polynomial associated to the face γ_I, which is nothing but the sum of the terms of f whose exponent is in the face γ_I, we see that we have the fundamental equality

$$\tilde{f}|_{S_I} = \tilde{f}_{\gamma_I}|_{S_I}$$

and we remark moreover that $\tilde{f}_{\gamma_I}$ is a function on $Z(\sigma)$ which is independent of the coordinates y_i for $i \in I$, so that it is determined by its restriction to S_I.

Now, to say that the strict transform $\tilde{f} = 0$ in $Z(\sigma)$ of the hypersurface $f = 0$ is transversal to the stratum S_I and is nonsingular in a neighborhood of its intersection with it is equivalent to saying that the restriction $\tilde{f}|_{S_I}$ of the function $\tilde{f}$ defines, by the equation $\tilde{f}|_{S_I} = 0$, a nonsingular hypersurface of S_I.

By the definition of S_I and what we have just seen, this in turn is equivalent to saying that the equation $\widetilde{f_{\gamma_I}} = 0$ defines a nonsingular hypersurface in the torus $(k^*)^d = \{u : \prod_1^d u_j \neq 0\}$ of $Z(\sigma)$, and this finally is equivalent to saying that $f_{\gamma_I} = 0$ defines a nonsingular hypersurface in the torus $(k^*)^d$ of the affine space $\mathbf{A}^d(k)$ since $\pi(\sigma)$ induces an isomorphism of the two tori.

This motivates the definition:

DEFINITION. The series $f = \sum_p f_p u^p$ is *nondegenerate* with respect to its Newton polyhedron in the coordinates $(u_1, \ldots, u_d)$ if for every compact face γ of $\mathscr{N}_+(f)$ the polynomial f_γ defines a nonsingular hypersurface of the torus $(k^*)^d$.

REMARK. By definition of the faces of the Newton polyhedron and of the dual fan, in each chart $Z(\sigma)$ of a regular fan refining the dual fan of $\mathscr{N}_+(f)$, the compact faces γ_I correspond to strata S_I of the canonical stratification which are contained in $\pi(\sigma)^{-1}(0)$. Each stratum S_I which is not contained in $\pi(\sigma)^{-1}(0)$ contains in its closure strata which are.

PROPOSITION 5.1. *If the germ of hypersurface X is defined by the vanishing of a series f which is nondegenerate, there is a neighborhood U of 0 in $\mathbf{A}^d(k)$ (a formal neighborhood if the series f does not converge) such that the strict transform of $X \cap U$ by the toric map*

$$\pi(\Sigma) : Z(\Sigma) \to \mathbf{A}^d(k)$$

associated to a regular fan refining the dual fan of its Newton polyhedron is nonsingular and transversal in each chart to the strata of the canonical stratification.

PROOF. By the fundamental equality seen above, the restriction of the strict transform to one of the strata contained in $\pi(\sigma)^{-1}(0)$, say S_I, has the same behavior as the restriction of f_{γ_I}, where γ_I is a compact face of the Newton polyhedron of f, to the torus $(k^*)^d$. As we saw, this implies that the strict transform of $X \cap U$ is nonsingular and transversal to S_I. By openness of transversality the same transversality holds, whithin a neighborhood of each point of $\pi(\Sigma)^{-1}(0)$, for all strata.

Since the map $\pi(\Sigma) : Z(\Sigma) \to \mathbf{A}^d(k)$ is proper, there is a neighborhood U of 0 in $\mathbf{A}^d(k)$ such that the strict transform by $\pi(\Sigma)$ of the hypersurface $X \subset \mathbf{A}^d(k)$ is nonsingular in $\pi(\Sigma)^{-1}(U)$ and transversal in each chart $Z(\sigma)$ to all the strata of the canonical stratification. □

The definition and properties of nondegeneracy extend to systems of functions as follows. Let $f_1, \ldots, f_k$ be series in the variables $u_1, \ldots, u_d$ defining a subspace $X \subset \mathbf{A}^d(k)$ in a neighborhood of 0. For each $j = 1, \ldots, k$ we have a local Newton polyhedron $\mathscr{N}_+(f_j)$. Choose a regular fan Σ of $\mathbf{R}^d_{\geq 0}$ compatible with all the fans dual to the polyhedra $\mathscr{N}_+(f_j)$ for $j = 1, \ldots, k$. We have for each j the same correspondence as above between the strata S_I of each chart $Z(\sigma)$ for $\sigma \in \Sigma$ and the faces of $\mathscr{N}_+(f_j)$, the strata contained in $\pi(\sigma)^{-1}(0)$ corresponding to compact faces.

For each vector $\breve{a} \in \check{\mathbf{R}}^d_{\geq 0}$ we get as above a system of equations $f_{1,\breve{a}}, \ldots, f_{k,\breve{a}}$, where

$$f_{j,\breve{a}} = \sum_{\{p: \langle \breve{a}, p \rangle = h(\breve{a})\}} f_{jp} u^p.$$

DEFINITION. The system of equations $f_1, \ldots, f_k$ is said to be nondegenerate of rank c with respect to the Newton polyhedra of the f_j in the coordinates $(u_1, \ldots, u_d)$ if for each vector $\breve{a} \in \mathbf{R}^d_{>0}$ the ideal of $k[u_1^{\pm 1}, \ldots, u_d^{\pm 1}]$ generated by the polynomials $f_{1,\breve{a}}, \ldots, f_{k,\breve{a}}$ defines a nonsingular subvariety of dimension $d - c$ of the torus $(k^*)^d$.

EXERCISE. Check that, since we took $\breve{a} \in \mathbf{R}^d_{>0}$ in the definition, it is equivalent to say that for each choice of a compact face γ_j in each $\mathscr{N}_+(f_j)$, the ideal generated by the polynomials $f_{1\gamma_1}, \ldots, f_{k\gamma_k}$ defines a nonsingular subvariety of dimension $d - c$ of the torus $(k^*)^d$.

Exactly as in the case of hypersurfaces, one then has:

PROPOSITION 5.2. *If the system of equations $f_1, \ldots, f_k$ is nondegenerate of rank c, for any regular fan Σ of $\mathbf{R}^d_{\geq 0}$ compatible with the dual fans of the polyhedra $\mathscr{N}_+(f_j)$, there is a neighborhood U of 0 in $\mathbf{A}^d(k)$ (a formal neighborhood if all the series f_j do not converge) such that the strict transform $X' \subset Z(\Sigma)$ by the toric map $\pi(\Sigma) : Z(\Sigma) \to \mathbf{A}^d(k)$ of the subvariety $X \cap U$ defined in U by the ideal generated by $f_1, \ldots, f_k$ is nonsingular and of dimension $d - c$ and transversal to the strata of the natural stratification in $\pi(\Sigma)^{-1}(U)$.*

PROOF. The same as that of Proposition 5.1. □

There is a difference, however, between the birational map $X' \to X \cap U$ induced by $\pi(\Sigma)$ and a resolution of singularities; this map is not necessarily an isomorphism outside of the singular locus; it is therefore only a *pseudoresolution* in the sense of [Goldin and Teissier 2000]. In fact, even in the nondegenerate case, and even for a hypersurface, the Newton polyhedron contains in general far too little information about the singular locus of X near 0. Kushnirenko introduced, for isolated hypersurface singularities, the notion of being *convenient* with respect to a coordinate system. It means that the Newton polyhedron meets all the coordinate axis of $\mathbf{R}^d_{\geq 0}$. For a hypersurface with isolated singularity, it implies that a toric pseudoresolution associated to the Newton polyhedron is a resolution. This was extended and generalized by M. Oka for complete intersections. The reader is referred to [Oka 1997, Ch. III] (especially Theorem 3.4) and we will only quote here the following fact, which is also a consequence of the existence of a toric pseudoresolution:

THEOREM [Oka 1997, Ch. III, Lemma 2.2]. *If k is a field and $(X, 0) \subset \mathbf{A}^d(k)$ is a germ of a complete intersection with equations $f_1 = \cdots = f_c = 0$, which is nondegenerate with respect to the Newton polyhedra of its equations in the coordinates $u_1, \ldots, u_d$, then there is a (possibly formal) neighborhood U of $0 \in$*

$\mathbf{A}^d(k)$ such that the intersection of X and the torus $(k^*)^d$ has no singularities in U.

In the formal case this should be understood as saying that formal germ at 0 of the singular locus of X is contained in the union of the coordinate hyperplanes.

Finally, it seems that the following coordinate-free definition of nondegeneracy is appropriate:

DEFINITION. An algebraic or formal subscheme X of an affine space $\mathbf{A}^d(k)$ is nondegenerate at a point $x \in X$ if there exist local coordinates $u_1, \ldots, u_d$ centered at x and an open (étale or formal) neighborhood U of x in $\mathbf{A}^d(k)$ such that there is a proper birational toric map $\pi : Z \to U$ in the coordinates $u_1, \ldots, u_d$ with Z nonsingular and such that the strict transform X' of $X \cap U$ by π is nonsingular and transversal to the exceptional divisor at every point of $\pi^{-1}(x) \cap X'$.

If X admits a system of equations which in some coordinates is nondegenerate with respect to its Newton polyhedra, it is also nondegenerate in this sense as we saw. The converse will not be discussed here.

QUESTION [Teissier 2003]. Given a reduced and equidimensional algebraic or formal space X over an algebraically closed field k, is it true that for every point $x \in X$ there is a local formal embedding of X into an affine space $\mathbf{A}^N(k)$ such that X is nondegenerate in $\mathbf{A}^N(k)$ at the point x?

A subsequent problem is to give a geometric interpretation of the systems of coordinates in which an embedded toric resolution for X exists.

For branches (analytically irreducible curve singularities), the question is answered positively, and the problem settled in Section 7 below. Recent work of P. González Pérez [2003] also settles question and problem for irreducible quasi-ordinary hypersurface singularities.

In [Teissier 2003] one finds an approach to the simpler problem where the nondegeneracy is requested only with respect to a valuation of the local ring of X at x.

In a given coordinate system, and for given Newton polyhedra, "almost all" systems of polynomials having these given Newton polyhedra are nondegenerate with respect to them. In this sense there are many nondegenerate singularities. However, nondegenerate singularities are very special from the viewpoint of the classification of singularities. A plane complex branch is nondegenerate in some coordinate system if and only if it has only one characteristic pair, which means that its equation can be written in some coordinate system as

$$u_1^p - u_0^q + \sum_{i/q+j/p>1} a_{ij} u_0^i u_1^j = 0,$$

where $a_{ij} \in k$ and the integers p, q are coprime. The curve

$$(u_1^2 - u_0^3)^2 - u_0^5 u_1 = 0$$

is degenerate in any coordinate system since it has two characteristic pairs [Smith 1873; Brieskorn and Knörrer 1986].

6. Resolution of Binomial Varieties

This section presents what is in a way the simplest class of nondegenerate singularities, according to the results in [González Pérez and Teissier 2002]:

Let k be a field. Binomial varieties over k are irreducible varieties of the affine space $\mathbf{A}^d(k)$ which can, in a suitable coordinate system, be defined by the vanishing of binomials in these coordinates, which is to say expressions of the form $u^m - \lambda_{mn} u^n$ with $\lambda_{mn} \in k^*$. An ideal generated by such binomial expressions is called a *binomial ideal*. These affine varieties defined by prime binomial ideals are also the irreducible affine varieties on which a torus of the same dimension acts algebraically with a dense orbit (see [Sturmfels 1996]); they are the (not necessarily normal) *affine toric varieties*.

Binomial ideals were studied in [Eisenbud and Sturmfels 1996]; these authors showed in particular that if k is algebraically closed their geometry is determined by the lattice generated by the differences $m - n$ of the exponents of the generating binomials. If the field k is not algebraically closed, the study becomes more complicated. Here I will assume throughout that k is algebraically closed. It is natural to study the behavior of binomial ideals under toric maps.

Let $\sigma = \langle a^1, \ldots, a^d \rangle$ be a regular cone in $\check{\mathbf{R}}_{\geq 0}^d$. The image of a binomial

$$u^m - \lambda_{mn} u^n$$

under the map $k[u_1, \ldots, u_d] \to k[y_1, \ldots, y_d]$ determined by $u_i \mapsto y_1^{a_i^1} \ldots y_d^{a_i^d}$ is given by

$$u^m - \lambda_{mn} u^n \mapsto y_1^{\langle a^1, m \rangle} \ldots y_d^{\langle a^d, m \rangle} - \lambda_{mn} y_1^{\langle a^1, n \rangle} \ldots y_d^{\langle a^d, n \rangle}.$$

In general this only tells us that the transform of a binomial is a binomial, which is no news since by definition of a toric map the transform of a monomial is a monomial.

However, something interesting happens if we assume that the cone σ is *compatible* with the hyperplane H_{m-n} which is the dual in the space of weights of the vector $m - n$ of the space of exponents, in the sense of definition on page 215, where we remember that the origin $\{0\}$ is a face of any polyhedral cone. Note that the Newton polyhedron of a binomial has only one compact face, which is a segment, so that for a cone in $\check{\mathbf{R}}_{\geq 0}^d$, being compatible with the hyperplane H_{m-n} is the same as being compatible with the dual fan of the Newton polyhedron of our binomial.

Let us assume that the binomial hypersurface $u^m - \lambda_{mn}u^n = 0$ is irreducible; this means that no variable u_j appears in both monomials, and the vector $m - n$ is primitive. In the sequel, I will tacitly assume this and also that our binomial is really singular, that is, not of the form $u_1 - \lambda u^m$.

If the convex cone σ of dimension d is compatible with the hyperplane H_{m-n}, it is contained in one of the closed half-spaces determined by H_{m-n}. This means that all the nonzero $\langle a^i, m-n \rangle$ have the same sign, say positive. It also means that, if we renumber the vectors a^i in such a way that $\langle a^i, m-n \rangle = 0$ for $1 \leq i \leq t$ and $\langle a^i, m-n \rangle > 0$ for $t+1 \leq i \leq d$, we can write the transform of our binomial as

$$u^m - \lambda_{mn}u^n \mapsto y_1^{\langle a^1, n \rangle} \cdots y_d^{\langle a^d, n \rangle} \left(y_{t+1}^{\langle a^{t+1}, m-n \rangle} \cdots y_d^{\langle a^d, m-n \rangle} - \lambda_{mn} \right).$$

And we can see that the strict transform $y_{t+1}^{\langle a^{t+1}, m-n \rangle} \cdots y_d^{\langle a^d, m-n \rangle} - \lambda_{mn} = 0$ of our hypersurface in the chart $Z(\sigma)$ is nonsingular!

It is also irreducible in view of the results of [Eisenbud and Sturmfels 1996] because we assumed that the vector $m - n$ is primitive and the matrix (a_j^i) is unimodular. This implies that the vector $(0, \ldots, 0, \langle a^{t+1}, m-n \rangle, \ldots, \langle a^d, m-n \rangle)$ is also primitive, and the strict transform irreducible. Moreover, in the chart $Z(\sigma)$ with $\sigma = \langle a^1, \ldots, a^d \rangle$, the strict transform meets the hyperplane $y_j = 0$ if and only if $\langle a^j, m-n \rangle = 0$. Unless our binomial is nonsingular, a case we excluded, this implies that a^j is not a vector of the canonical basis of W, so that $y_j = 0$ is a component of the exceptional divisor. So we see that the strict transform meets the exceptional divisor only in those charts such that $\sigma \cap H_{m-n} \neq \{0\}$, and then meets it transversally.

So we have in this very special case achieved that the total transform of our irreducible binomial hypersurface defines in each chart a *divisor with normal crossings* that is, a divisor locally at every point defined in suitable local coordinates by the vanishing of a monomial and whose irreducible components are nonsingular.

Now we consider a prime binomial ideal of $k[u_1, \ldots, u_d]$ generated by $(u^{m^\ell} - \lambda_\ell u^{n^\ell})_{\ell \in \{1, \ldots, L\}}$, $\lambda_\ell \in k^*$. Let us denote by $\mathcal{L}$ the sublattice of $\mathbf{Z}^d$ generated by the differences $m^\ell - n^\ell$. According to [Eisenbud and Sturmfels 1996], the dimension of the subvariety $X \subset \mathbf{A}^d(k)$ defined by the ideal is $d - r$ where r is the rank of the $\mathbf{Q}$-vector space $\mathcal{L} \otimes_{\mathbf{Z}} \mathbf{Q}$. To each binomial is associated a hyperplane $H_\ell \subset \check{\mathbf{R}}^d$, the dual of the vector $m^\ell - n^\ell \in \mathbf{R}^d$. The intersection $\mathcal{W}$ of the hyperplanes H_ℓ is the dual of the vector subspace $\mathcal{L} \otimes_{\mathbf{Z}} \mathbf{R}$ of $\mathbf{R}^d$ generated by the vectors $m^\ell - n^\ell$; its dimension is $d - r$.

Let Σ be a fan with support $\mathbf{R}_{\geq 0}^d$ which is compatible with each of the hyperplanes H_ℓ. Let us compute the transforms of the generators $u^{m^\ell} - \lambda_\ell u^{n^\ell}$ in a chart $Z(\sigma)$ associated to the cone $\sigma = \langle a^1, \ldots, a^d \rangle$: after renumbering the vectors a^j and possibly exchanging some m^ℓ, n^ℓ and replacing λ_ℓ by its inverse, we may assume that $a^1, \ldots, a^t$ are in $\mathcal{W}$, that the $\langle a^j, m^\ell - n^\ell \rangle$ are ≥ 0 for

$j = t + 1, \ldots, d$, and that moreover for each such index j there is an ℓ such $\langle a^j, m^\ell - n^\ell \rangle > 0$. The transforms of the binomials can be written

$$y_1^{\langle a^1, n^\ell \rangle} \cdots y_d^{\langle a^d, n^\ell \rangle} \left(y_{t+1}^{\langle a^{t+1}, m^\ell - n^\ell \rangle} \cdots y_d^{\langle a^d, m^\ell - n^\ell \rangle} - \lambda_\ell \right)$$

with that additional condition. If $\sigma \cap \mathscr{W} = \{0\}$, we have $t = 0$ and the subvariety defined by the equations just written (the strict transform of X in $Z(\sigma)$) does not meet any coordinate hyperplane; in particular it does not meet the exceptional divisor. In general, still assuming that none of the binomials is already in the form $u_j - \lambda u^m$, one sees that the additional condition implies that, just like in the case of hypersurfaces, the strict transform meets the hyperplane $y_j = 0$ if and only if a^j is in $\mathscr{W}$.

Now the claim is that in each chart $Z(\sigma)$ the strict transform is either empty or nonsingular and transversal to the exceptional divisor.

The $\mathbf{Q}$-vector space generated by the $m^\ell - n^\ell$ is of dimension r. Let us assume that $m^1 - n^1, \ldots, m^r - n^r$ generate it and let us denote by $\mathscr{L}_1$ the lattice which they generate in $\mathbf{Z}^d$. By construction, the quotient $\mathscr{L} / \mathscr{L}_1$ is a torsion $\mathbf{Z}$-module. Let us first show that the strict transform of the subspace $X_1 \subset X$ defined by the first r binomial equations is nonsingular and transversal to the exceptional divisor.

We consider then, for each cone $\sigma = \langle a^1, \ldots, a^d \rangle$, the equations

$$y_1^{\langle a^1, m^1 - n^1 \rangle} \cdots y_d^{\langle a^d, m^1 - n^1 \rangle} - \lambda_1 = 0$$
$$y_1^{\langle a^1, m^2 - n^2 \rangle} \cdots y_d^{\langle a^d, m^2 - n^2 \rangle} - \lambda_2 = 0$$
$$\cdots\cdots\cdots\cdots\cdots\cdots\cdots\cdots\cdots\cdots$$
$$y_1^{\langle a^1, m^r - n^r \rangle} \cdots y_d^{\langle a^d, m^r - n^r \rangle} - \lambda_r = 0$$

of the strict transform of X_1 in $Z(\sigma)$.

We can compute by logarithmic differentiation their jacobian matrix J, and find with the same definition of t as above an equality of $d \times r$ matrices:

$$y_{t+1} \cdots y_d J = y_{t+1}^{\sum_s \langle a^{t+1}, m^s - n^s \rangle} \cdots y_d^{\sum_s \langle a^d, m^s - n^s \rangle} \left(\langle a^j, m^s - n^s \rangle \right)_{1 \le j \le d, 1 \le s \le r}.$$

LEMMA 6.1. *Given an irreducible binomial variety $X \subset \mathbf{A}^d(k)$, with the notations just introduced, for any regular cone $\sigma = \langle a^1, \ldots, a^d \rangle$ compatible with the hyperplanes H_ℓ, the image in $\mathrm{Mat}_{d \times L}(k)$ of the matrix*

$$\left(\langle a^j, m^s - n^s \rangle \right)_{1 \le j \le d, 1 \le s \le L} \in \mathrm{Mat}_{d \times L}(\mathbf{Z})$$

has rank r.

PROOF. Since the vectors a^j form a basis of $\mathbf{Q}^d$, and the space $\mathscr{W} = \mathscr{L} \otimes_{\mathbf{Z}} \mathbf{R}$ generated by the $m^s - n^s$ is of dimension r, the rank of the matrix $\left(\langle a^j, m^s - n^s \rangle \right)$ is r, which proves the lemma if k is of characteristic zero. If the field k is of positive characteristic the proof is a little less direct; see [Teissier 2003, Ch. 6].

In particular, the rank of the image in $\mathrm{Mat}_{d\times r}(k)$ of the matrix $(\langle a^j, m^s - n^s\rangle)_{1\leq j\leq d, 1\leq s\leq r} \in \mathrm{Mat}_{d\times r}(\mathbf{Z})$ is r. $\hfill\square$

LEMMA 6.2. *The strict transform X_1' by $\pi(\Sigma)$ of the subspace $X \subset \mathbf{A}^d(k)$ defined by the ideal of $k[u_1, \ldots, u_d]$ generated by the binomials*

$$u^{m^1} - \lambda_1 u^{n^1}, \ldots, u^{m^r} - \lambda_r u^{n^r}$$

is regular and transversal to the exceptional divisor.

PROOF. Let σ be a cone of of maximal dimension in the fan Σ. In the chart $Z(\sigma)$, none of the coordinates $y_{t+1}, \ldots, y_d$ vanishes on the strict transform X_1' of X_1 and the equations of X_1' in $Z(\sigma)$ are independent of $y_1, \ldots, y_t$. Therefore to prove that the jacobian J of the equations has rank r at each point of this strict transform it suffices to show that the rank of the image in $\mathrm{Mat}_{d\times L}(k)$ of the matrix $(\langle a^j, m^s - n^s\rangle)_{1\leq j\leq d, s\in L} \in \mathrm{Mat}_{d\times L}(\mathbf{Z})$ is r, which follows from the lemma. $\hfill\square$

PROPOSITION 6.3. *If the regular fan Σ with support $\check{\mathbf{R}}_{\geq 0}^d$ is compatible with all the hyperplanes $H_{m^\ell - n^\ell}$, the strict transform X' under the map $\pi(\Sigma): Z(\Sigma) \to \mathbf{A}^d(k)$ of the subspace $X \subset \mathbf{A}^d(k)$ defined by the ideal of $k[u_1, \ldots, u_d]$ generated by the $(u^{m^\ell} - \lambda_\ell u^{n^\ell})_{\ell \in \{1, \ldots, L\}}$ is regular and transversal to the exceptional divisor; it is also irreducible in each chart.*

PROOF. The preceding discussion shows that the rank of J is r everywhere on the strict transform of X, and by Zariski's jacobian criterion this strict transform is smooth and transversal to the exceptional divisor. But it is not necessarily irreducible; we show that the strict transform of our binomial variety is one of its irreducible components. Since the differences of the exponents in the total transform and the strict transform of a binomial are the same, the lattice of exponents generated by the exponents of all the strict transforms of the binomials $(u^{m^\ell} - \lambda_{mn} u^{n^\ell})_{\ell \in \{1, \ldots, L\}}$ is the image $M(\sigma)\mathscr{L}$ of the lattice $\mathscr{L}$ by the linear map $\mathbf{Z}^d \to \mathbf{Z}^d$ corresponding to the matrix $M(\sigma)$ with rows $(a^1, \ldots, a^d)$. Similarly the exponents of the strict transforms of $u^{m^1} - \lambda_{m^1 n^1} u^{n^1}, \ldots, u^{m^r} - \lambda_{m^r n^r} u^{n^r}$ generate the lattice $M(\sigma)\mathscr{L}_1$. The lattice $M(\sigma)\mathscr{L}$ is the saturation of $M(\sigma)\mathscr{L}_1$, and so according to [Eisenbud and Sturmfels 1996], since we assume that k is algebraically closed, the strict transform of our binomial variety is one of the irreducible components of the binomial variety defined by the r equations displayed above.

The charts corresponding to regular cones $\sigma \in \Sigma$ of dimension $< d$ are open subsets of those which we have just studied, so they contribute nothing new. $\hfill\square$

In the case of binomial varieties one can show that the regular refinement Σ of the fan Σ_0 determined by the hyperplanes $H_{m^s - n^s}$ can be chosen in such a way that the restriction $X' \to X$ of the map $\pi(\Sigma)$ to the strict transform X' of X induces an isomorphism outside of the singular locus of X; it is therefore an embedded resolution of $X \subset \mathbf{A}^d(k)$ and not only a pseudoresolution; see [González Pérez and Teissier 2002] and [Teissier 2003, § 6.2].

REMARK. Since [Hironaka 1964], one usually seeks to achieve resolution of singularities by successions of blowing-ups with nonsingular centers, which moreover are "permissible". According to [De Concini and Procesi 1983; 1985], toric maps are dominated by finite successions of blowing-ups with nonsingular centers.

Now in view of the results of Section 5, we expect that if we deform a binomial variety by adding to each of its equations terms which do not affect the Newton polyhedron, the same toric map will resolve the deformed variety as well. However, it may be only a pseudoresolution, since the effect of the deformation on the singular locus is difficult to control. The next section shows that in a special case one can, conversely, present a singularity as a deformation of a toric variety, and thus obtain an embedded toric resolution.

7. Resolution of Singularities of Branches

This section is essentially an exposition of material in [Goldin and Teissier 2000] and [Teissier 2003]. The idea is to show that any analytically irreducible germ of curve is in a canonical way a deformation of a *monomial* curve, which is defined by *binomial* equations. In this terminological mishap, the *monomial* refers to the parametric presentation of the curve; the parametric presentation is more classical, but the binomial character of the equations is more suitable for resolution of singularities.

The deformation from the monomial curve to the curve is "equisingular", so that the toric map which resolves the singularties of the monomial curve according to Section 6 also resolves the singularities of our original curve once it is suitably embedded in the affine space where the monomial curve embeds. One interpretation of this is that *after a suitable reembedding, any analytically irreducible curve becomes nondegenerate.*

For example, in order to resolve the singularities at the origin of the plane curve C with equation

$$(u_1^2 - u_0^3)^2 - u_0^5 u_1 = 0,$$

a good method is to view it as the fiber for $v = 1$ of the family of curves C_v in $\mathbf{A}^3(k)$ defined by the equations

$$u_1^2 - u_0^3 - v u_2 = 0,$$
$$u_2^2 - u_0^5 u_1 = 0,$$

as one can see by eliminating u_2 between the two equations. The advantage is that the fiber for $v = 0$ is a binomial variety, which we know how to resolve, and its resolution also resolves all the fibers C_v. For $v \neq 0$, the fiber C_v is isomorphic to our original plane curve C, re-embedded in $\mathbf{A}^3(k)$ by the functions $u_0, u_1, u_1^2 - u_0^3$.

In more algebraic terms, it gives this:

Let R be a one dimensional excellent equicharacteristic local ring whose completion is an integral domain and whose residue field is algebraically closed. A basic example is $R = k[[x, y]]/(f)$ where k is algebraically closed and $f(x, y)$ is irreducible in $k[[x, y]]$. Then the normalization $\bar{R}$ of R is a (discrete) valuation ring because it is a one dimensional normal local ring. The maximal ideal of $\bar{R}$ is generated by a single element, say t, and each nonzero element of $\bar{R}$ can be written uniquely as ut^n, where u is invertible in $\bar{R}$ and $n \in \mathbf{N} \cup \{0\}$. The valuation $\nu(ut^n)$ of that element is n.

In our basic example, the inclusion $R \subset \bar{R}$ is $k[[x, y]]/(f) \subset k[[t]]$ given by $x \mapsto x(t)$, $y \mapsto y(t)$, where $x(t), y(t)$ is a parametrization of the plane curve with equation $f(x, y) = 0$.

Since R is a subalgebra of $\bar{R}$, the values taken by the valuation on the elements of R (except 0) form a semigroup Γ contained in $\mathbf{N}$. This semigroup has a finite complement in $\mathbf{N}$ and is finitely generated. Let us write it

$$\Gamma = \langle \gamma_0, \gamma_1, \ldots, \gamma_g \rangle.$$

The powers of the maximal ideal of $\bar{R}$ form a filtration

$$\bar{R} \supset t\bar{R} \supset t^2 \bar{R} \supset \cdots \supset t^n \bar{R} \supset \cdots$$

whose associated graded ring

$$\mathrm{gr}_\nu \bar{R} = \bigoplus_{n \in \mathbf{N} \cup \{0\}} t^n \bar{R}/t^{n+1} \bar{R}$$

is a k-algebra isomorphic to the polynomial ring $k[t]$ by the map $t \pmod{t^2 \bar{R}} \mapsto t$.

This filtration induces a filtration on the ring R itself, by the ideals $\mathscr{P}_n = R \cap t^n \bar{R}$, and one defines the corresponding associated graded ring

$$\mathrm{gr}_\nu R = \bigoplus_{n \in \mathbf{N} \cup \{0\}} \mathscr{P}_n/\mathscr{P}_{n+1} \subseteq \mathrm{gr}_\nu \bar{R} = k[t].$$

PROPOSITION 7.1 [Goldin and Teissier 2000]. *The subalgebra* $\mathrm{gr}_\nu R$ *of* $k[t]$ *is equal to the subalgebra generated by* $t^{\gamma_0}, t^{\gamma_1}, \ldots, t^{\gamma_g}$. *It is the semigroup algebra over* k *of the semigroup* Γ *of the valuation* ν *on* R; *it is also the affine algebra of the monomial curve in the affine space* $\mathbf{A}^{g+1}(k)$ *described parametrically by* $u_i = t^{\gamma_i}$ *for* $0 \leq i \leq g$.

There is a precise geometrical relationship between the original curve C with algebra R and the monomial curve C^Γ with algebra $\mathrm{gr}_\nu R$: according to a general principle of algebra, the ring R is a deformation of its associated graded ring. More precisely, assume that R contain a field of representatives of its residue field k, i.e., that we have a subfield $k \subset R$ such that the composed map $k \subset R \to R/m = k$ is the identity. This will be the case in particular, according to Cohen's theorem, if the local ring R is complete (and equicharacteristic of course).

Start from the filtration by the ideals $\mathscr{P}_n$ introduced above, set $\mathscr{P}_n = R$ for $n \leq 0$ and consider the algebra

$$\mathscr{A}_\nu(R) = \bigoplus_{n \in \mathbf{Z}} \mathscr{P}_n v^{-n} \subset R[v, v^{-1}].$$

It can be shown (see [Teissier 2003]) that it is generated as a $R[v]$-algebra by the $\xi_i v^{-\gamma_i}$, $0 \leq i \leq g$, where $\xi_i \in R$ is of t-adic order γ_i. Since $\mathscr{P}_n = R$ for $n \leq 0$ it contains as a graded subalgebra the polynomial algebra $R[v]$, and therefore also $k[v]$.

PROPOSITION 7.2 [Teissier 1975; Bourbaki 1983, Ch. VIII § 6, exerc. 2]; see also [Gerstenhaber 1964; 1966].

(a) *The composed map* $k[v] \to \mathscr{A}_\nu(R)$ *is faithfully flat.*
(b) *The map*

$$\sum x_n v^{-n} \mapsto \sum \overline{x}_n,$$

where $\overline{x}_n$ *is the image of* x_n *in the quotient* $\mathscr{P}_n / \mathscr{P}_{n+1}$, *induces an isomorphism*

$$\mathscr{A}_\nu(R)/v\mathscr{A}_\nu(R) \to \mathrm{gr}_\nu R.$$

(c) *For any* $v_0 \in k^*$ *the map*

$$\sum x_n v^{-n} \mapsto \sum x_n v_0^{-n}$$

induces an isomorphism of k-algebras

$$\mathscr{A}_\nu(R)/(v - v_0)\mathscr{A}_\nu(R) \to R.$$

PROOF. Since $k[v]$ is a principal ideal domain, to prove (a) it suffices by [Bourbaki 1968, Ch. I § 3.1] to prove that $\mathscr{A}_\nu(R)$ has no torsion as a $k[v]$-module and that for any $v_0 \in k$ we have $(v - v_0)\mathscr{A}_\nu(R) \neq \mathscr{A}_\nu(R)$. The second statement follows from (b) and (c), which are easy to verify, and the first follows from the fact that $\mathscr{A}_\nu(R)$ is a subalgebra of $R[v, v^{-1}]$. □

This proposition means that there is a one parameter flat family of algebras whose special fiber is the graded algebra and all other fibers are isomorphic to R. Geometrically, this gives us a flat family of curves whose special fiber is the monomial curve and all other fibers are isomorphic to our given curve. This deformation can be realized in the following way. I assume for simplicity that R is complete. Then by the definition of the semigroup Γ there are elements $\xi_0(t), \ldots, \xi_g(t)$ in $k[\![t]\!]$ that belong to R and are such that their t-adic valuations are the generators γ_i of the semigroup Γ. We may write $\xi_i(t) =$

$t^{\gamma_i} + \sum_{j>\gamma_i} b_{ij} t^j$ with $b_{ij} \in k$. Now introduce a parameter v and consider the family of parametrized curves in $\mathbf{A}^{g+1}(k)$ described as follows:

$$u_0 = \xi_0(vt)v^{-\gamma_0} = t^{\gamma_0} + \sum_{j>\gamma_0} b_{0j} v^{j-\gamma_0} t^j,$$

$$u_1 = \xi_1(vt)v^{-\gamma_1} = t^{\gamma_1} + \sum_{j>\gamma_1} b_{1j} v^{j-\gamma_1} t^j,$$

$$\cdots\cdots\cdots\cdots\cdots\cdots\cdots\cdots\cdots\cdots$$

$$u_g = \xi_g(vt)v^{-\gamma_g} = t^{\gamma_g} + \sum_{j>\gamma_g} b_{gj} v^{j-\gamma_g} t^j.$$

The parametrization shows that for $v = 0$ we obtain the monomial curve, and for any $v \neq 0$ a curve isomorphic to our given curve, as embedded in $\mathbf{A}^{g+1}(k)$ by the functions $\xi_0, \ldots, \xi_g$. This is a realisation of the family of Proposition 7.2. In order to get an equational representation of that family, we must begin by finding the equations of the monomial curve, which we will then proceed to deform.

The equations of the monomial curve C^Γ correspond to the relations between the generators γ_i of Γ. They are fairly simple in the case where Γ is the semigroup of a plane branch, and in that case C^Γ is a complete intersection. The general setup is as follows:

Consider the $\mathbf{Z}$-linear map $w : \mathbf{Z}^{g+1} \to \mathbf{Z}$ determined by sending the i-th base vector e_i to γ_i; the image of $\mathbf{Z}^{g+1}_{\geq 0}$ is Γ. It is not difficult to see that the kernel of w is generated by differences $m - m'$, where $m, m' \in \mathbf{Z}^{g+1}_{\geq 0}$ and $w(m) = w(m')$. The kernel of w is a lattice (free sub $\mathbf{Z}$-module) $\mathscr{L}$ in $\mathbf{Z}^{g+1}$, which must be finitely generated because $\mathbf{Z}^{g+1}$ is a noetherian $\mathbf{Z}$-module and $\mathbf{Z}$ is a principal ideal domain.

If we choose a basis $m^1 - n^1, \ldots, m^q - n^q$ for $\mathscr{L}$, such that all the m^j, n^j are in $\mathbf{Z}^{g+1}_{\geq 0}$, it follows from the very construction of semigroup algebras that C^Γ is defined in the space $\mathbf{A}^{g+1}(k)$ with coordinates $u_0, \ldots, u_g$ by the vanishing of the binomials $u^{m^1} - u^{n^1}, \ldots, u^{m^q} - u^{n^q}$.

Now the faithful flatness of the family of Proposition 7.2 implies that it can be defined in $\mathbf{A}^1(k) \times \mathbf{A}^{g+1}(k)$ by equations which are deformations of the equations of the monomial curve [Teissier 2003, §5, proof of 5.49]. Here I cheat a little by leaving out the fact that one in fact defines a formal space. Anyway, our family of (formal) curves is also defined by equations of the form

$$u^{m^1} - u^{n^1} + \sum_{w(r)>w(m^1)} c_r^{(1)}(v) u^r = 0,$$

$$u^{m^2} - u^{n^2} + \sum_{w(r)>w(m^2)} c_r^{(2)}(v) u^r = 0,$$

$$\cdots\cdots\cdots\cdots\cdots\cdots\cdots\cdots\cdots\cdots$$

$$u^{m^q} - u^{n^q} + \sum_{w(r)>w(m^q)} c_r^{(1)}(v) u^r = 0,$$

where the $c_r^{(j)}(v)$ are in $(v)k[\![v]\!]$, $w(r) = \sum_0^g \gamma_j r_j$ is the weight of the monomial u^r with respect to the weight vector $w = (\gamma_0, \ldots, \gamma_g)$, that is, $w(r) = \langle w, r \rangle$. Remember that by construction $w(m^i) = w(n^i)$ for $1 \le i \le q$. This means that we deform each binomial equations by adding terms of weight greater than that of the binomial. It is shown in [Teissier 2003] that the parametric representation and the equation representation both describe the deformation of Proposition 7.2. Up to completion with respect to the $(u_0, \ldots, u_g)$-adic topology, the algebra $\mathscr{A}_\nu(R)$ is the quotient of $k[v][\![u_0, \ldots, u_g]\!]$ by the ideal generated by the equations written above. It is also equal to the subalgebra $k[\![\xi_0(vt)v^{-\gamma_0}, \ldots, \xi_g(vt)v^{-\gamma_g}]\!]$ of $k[v][\![t]\!]$.

One may remark that, in the case where the $\xi_j(t)$ are polynomials, there is a close analogy with the SAGBI algebras bases for the subalgebra $k[\xi_0(t), \xi_1(t)] \subset k[t]$ (see [Sturmfels 1996]). This is developed in [Bravo 2004].

This equation description is the generalization of the example shown at the beginning of this section.

Now it should be more or less a computational exercise to check that a toric map $Z(\Sigma) \to \mathbf{A}^{g+1}$ which resolves the binomial variety C^Γ also resolves the "nearby fibers", which are all isomorphic to C re-embedded in $\mathbf{A}^{g+1}$. There is however a difficulty [Goldin and Teissier 2000] which requires the use of Zariski's main theorem.

The results of this section have been extended in [González Pérez 2003] to the much wider class of irreducible *quasi-ordinary* germs of hypersurface singularities, whose singularities are not isolated in general.

This shows that a toric resolution of binomial varieties can be used, by considering suitable deformations, to resolve singularities which are at first sight far from binomial.

Appendix: Multiplicities, Volumes and Nondegeneracy

Multiplicities and volumes. One of the interesting features of the Briançon–Skoda theorem is that it provides a way to pass from the integral closure of an ideal to the ideal itself, while it is much easier to check that a given element is in the integral closure of an ideal than to check that it is in the ideal. For this reason, the theorem has important applications in problems of effective commutative algebra motivated by transcendental number theory. In the same vein, this section deals, in the monomial case, with the problem of determining from numerical invariants whether two ideals have the same integral closure, which is much easier than to determine whether they are equal. The basic fact coming to light is that multiplicities in commutative algebra are like volumes in the theory of convex bodies, and indeed, for monomial ideals, they *are* volumes, up to a factor of $d!$ (compare with [Teissier 1988]). The same is true for degrees of invertible sheaves on algebraic varieties. Exactly as monomial ideals, and for the same reason, the degrees of equivariant invertible sheaves generated by their global sections on toric varieties *are* volumes of compact convex bodies multiplied by $d!$ [Teissier 1979].

In this appendix proofs are essentially replaced by references; for the next four paragraphs, see [Bourbaki 1983, Ch. VIII, § 4].

Let R be a noetherian ring and $\mathfrak{q}$ an ideal of R such that the R-module $R/\mathfrak{q}$ has finite length $\ell_R(R/\mathfrak{q}) = \ell_{R/\mathfrak{q}}(R/\mathfrak{q})$. Then the quotients $\mathfrak{q}^n/\mathfrak{q}^{n+1}$ have finite length as $R/\mathfrak{q}$-modules and one can define the Hilbert–Samuel series

$$H_{R,\mathfrak{q}} = \sum_{n=0}^{\infty} \ell_{R/\mathfrak{q}}(\mathfrak{q}^n/\mathfrak{q}^{n+1})T^n \in \mathbf{Z}[\![T]\!].$$

There exist an integer $d \geq 0$ and an element $P \in \mathbf{Z}[T, T^{-1}]$ such that $P(1) > 0$ and

$$H_{R,\mathfrak{q}} = (1 - T)^{-d}P.$$

From this follows:

PROPOSITION A.1 (SAMUEL). *Given R and $\mathfrak{q}$ as above, there exist an integer N_0 and a polynomial $Q(U)$ with rational coefficients such that for $n \geq N_0$ we have*

$$\ell_{R/\mathfrak{q}}(R/\mathfrak{q}^n) = Q(n).$$

If we assume that $\mathfrak{q}$ is primary for some maximal ideal m of R, i.e., $\mathfrak{q} \supset m^k$ for large enough k, the degree of the polynomial Q is the dimension d of the local ring R_m, and the highest degree term of $Q(U)$ can be written $e(\mathfrak{q}, R)U^d/d!$. In fact, $e(\mathfrak{q}, R) = P(1) \in \mathbf{N}$.

By definition, the integer $e(\mathfrak{q}, R)$ is the *multiplicity* of the ideal $\mathfrak{q}$ in R.

If R contains a field k such that $k = R/m$, we can replace $\ell_{R/\mathfrak{q}}(R/\mathfrak{q}^n)$ by its dimension $\dim_k(R/\mathfrak{q}^n)$ as a k-vector space.

Take $R = k[u_1, \ldots, u_d]$ and $\mathfrak{q} = (u^{m^1}, \ldots, u^{m^q})R$; the ideal $\mathfrak{q}$ is primary for the maximal ideal $m = (u_1, \ldots, u_d)R$ if and only if $\dim_k R/\mathfrak{q} < \infty$. Now one sees that the images of the monomials u^m such that m is not contained in $E = \bigcup_{i=1}^{q}(m^i + \mathbf{R}_{\geq 0}^d)$ constitute a basis of the k-vector space $R/\mathfrak{q}$:

$$\dim_k R/\mathfrak{q} = \#\mathbf{Z}^d \cap (\mathbf{R}_{\geq 0}^d \setminus E).$$

For the same reason we have for all $n \geq 1$, since $\mathfrak{q}^n$ is also monomial,

$$\dim_k R/\mathfrak{q}^n = \#\mathbf{Z}^d \cap (\mathbf{R}_{\geq 0}^d \setminus nE),$$

where nE is the set of sums of n elements of E.

From this follows, in view of the polynomial character of the first term of the equality:

COROLLARY A.2. *Given a subset $E = \bigcup_{s=1}^{q}(m^s + \mathbf{R}_{\geq 0}^d)$ whose complement in $\mathbf{R}_{\geq 0}^d$ has finite volume, there exists an integer N_0 and a polynomial $Q(n)$ of degree d with rational coefficients such that for $n \geq N_0$ we have*

$$\#\mathbf{Z}^d \cap (\mathbf{R}_{\geq 0}^d \setminus nE) = Q(n).$$

Therefore,

$$\lim_{n\to\infty} \frac{Q(n)}{n^d} = \lim_{n\to\infty} \frac{\#\mathbf{Z}^d \cap (\mathbf{R}^d_{\geq 0} \setminus nE)}{n^d} = \lim_{n\to\infty} \text{Covol} \frac{nE}{n} = \text{Covol}\,\bar{E},$$

where Covol A, the *covolume* of A, is the volume of the complement of A in $\mathbf{R}^d_{\geq 0}$. The last equality follows from the remark made in Section 4, and the previous one from the classical fact of calculus that as $n \to \infty$,

$$\text{Covol} \frac{nE}{n} = \frac{\#\mathbf{Z}^d \cap (\mathbf{R}^d_{\geq 0} \setminus nE)}{n^d} + o(1).$$

Since the limit as $n \to \infty$ of $Q(n)/n^d$ is $e(\mathfrak{q}, R)/d!$, we have immediately:

COROLLARY A.3. *For a monomial ideal* $\mathfrak{q} = (u^{m^1}, \ldots, u^{m^s})$ *in* $R = k[u_1, \ldots, u_d]$ *which is primary for* $m = (u_1, \ldots, u_d)$, *with the notations above, we have*

$$\dim_k(R/\mathfrak{q}) = \#\mathbf{Z}^d \cap (\mathbf{R}^d_{\geq 0} \setminus E),$$

$$e(\mathfrak{q}, R) = d!\,\text{Covol}\,\bar{E}.$$

COROLLARY A.4 (MONOMIAL REES THEOREM, an avatar of [Rees 1961]).

(a) *For a monomial primary ideal* $\mathfrak{q}$ *as above, me have*

$$e(\bar{\mathfrak{q}}, R) = e(\mathfrak{q}, R).$$

(b) *Given two such ideals* $\mathfrak{q}_1, \mathfrak{q}_2$ *such that* $\mathfrak{q}_1 \subseteq \mathfrak{q}_2$, *we have* $\bar{\mathfrak{q}}_1 = \bar{\mathfrak{q}}_2$ *if and only if* $e(\mathfrak{q}_1, R) = e(\mathfrak{q}_2, R)$.

These results hold for ideals containing a power of the maximal ideal in a noetherian local ring R whose completion is equidimensional [Rees 1961].

Now there is a well-known theorem in the theory of convex bodies, concerning the volume of the Minkowski sum of compact convex sets. Recall that for K_1, K_2 in $\mathbf{R}^d$, the *Minkowski sum* $K_1 + K_2$ is the set of sums $\{x_1 + x_2 : x_1 \in K_1, x_2 \in K_2\}$; also we set $\lambda K = \{\lambda x : x \in K\}$ for $\lambda \in \mathbf{R}$. Then:

THEOREM A.5 (MINKOWSKI). *Given* s *compact convex subsets* $K_1, \ldots, K_s$ *of* $\mathbf{R}^d$, *there is a homogeneous expression for the d-dimensional volume of the positive Minkowski linear combination of the* K_i, *with* $(\lambda_i)_{1 \leq i \leq s} \in \mathbf{R}^s_{\geq 0}$:

$$\text{Vol}_d(\lambda_1 K_1 + \cdots + \lambda_s K_s) = \sum_{\sum_1^s \alpha_i = d} \frac{d!}{\alpha_1! \ldots \alpha_s!} \text{Vol}(K_1^{[\alpha_1]}, \ldots, K_s^{[\alpha_s]}) \lambda_1^{\alpha_1} \ldots \lambda_s^{\alpha_s},$$

where the coefficients $\text{Vol}(K_1^{[\alpha_1]}, \ldots, K_s^{[\alpha_s]})$ *are nonnegative and are called the mixed volumes of the convex sets* K_i.

In particular, with $s = 2$,

$$\text{Vol}_d(\lambda_1 K_1 + \lambda_2 K_2) = \sum_{i=0}^d \binom{d}{i} \text{Vol}(K_1^{[i]}, K_2^{[d-i]}) \lambda_1^i \lambda_2^{d-i}.$$

The proof is obtained by approximating the convex bodies by polytopes, and using the Cauchy formula for the volume of polytopes. Exactly the same proof applies to the covolumes of *convex* subsets of $\mathbf{R}_{\geq 0}^d$ to give the corresponding theorem:

$$\operatorname{Covol}_d(\lambda_1 \bar{E}_1 + \cdots + \lambda_s \bar{E}_s) = \sum_{\sum_1^s \alpha_i = d} \frac{d!}{\alpha_1! \ldots \alpha_s!} \operatorname{Covol}(\bar{E}_1^{[\alpha_1]}, \ldots, \bar{E}_s^{[\alpha_s]}) \lambda_1^{\alpha_1} \ldots \lambda_s^{\alpha_s},$$

defining the *mixed covolumes* of such convex subsets.

There is an analogous formula in commutative algebra:

THEOREM A.6 [Teissier 1973]. *Given ideals* $\mathfrak{q}_1, \ldots, \mathfrak{q}_s$ *which are primary for a maximal ideal* m *in a noetherian ring* R *such that the localization* R_m *is a d-dimensional local ring and the residue field* R_m/mR_m *is infinite, there is for* $\lambda_1, \ldots, \lambda_s \in \mathbf{Z}_{\geq 0}^s$ *an expression*

$$e(\mathfrak{q}_1^{\lambda_1} \ldots \mathfrak{q}_s^{\lambda_s}, R) = \sum_{\sum_1^s \alpha_i = d} \frac{d!}{\alpha_1! \ldots \alpha_s!} e(\mathfrak{q}_1^{[\alpha_1]}, \ldots, \mathfrak{q}_s^{[\alpha_s]}; R) \lambda_1^{\alpha_1} \ldots \lambda_s^{\alpha_s},$$

where the coefficients $e(\mathfrak{q}_1^{[\alpha_1]}, \ldots, \mathfrak{q}_s^{[\alpha_s]}; R)$ *are nonnegative integers and are called the mixed multiplicities of the primary ideals* $\mathfrak{q}_i$. (*This name is justified by the fact that* $e(\mathfrak{q}_1^{[\alpha_1]}, \ldots, \mathfrak{q}_s^{[\alpha_s]}; R)$ *is the multiplicity of an ideal generated by* α_1 *elements of* $\mathfrak{q}_1$, ..., α_s *elements of* $\mathfrak{q}_s$, *chosen in a sufficiently general way.*) *Taking* $s = 2$ *gives*

$$e(\mathfrak{q}_1^{\lambda_1} \mathfrak{q}_2^{\lambda_2}, R) = \sum_{i=0}^{d} \binom{d}{i} e(\mathfrak{q}_1^{[i]}, \mathfrak{q}_2^{[d-i]}; R) \lambda_1^i \lambda_2^{d-i}.$$

From this and Corollary A.3 there follows immediately:

COROLLARY A.7. *Let* k *be an infinite field. Given monomial ideals* $\mathfrak{q}_1, \ldots, \mathfrak{q}_s$ *which are primary for the maximal ideal* $(u_1, \ldots, u_d)$ *in* $R = k[u_1, \ldots, u_d]$, *and denoting by* E_i *the corresponding subsets generated by their exponents, we have for all* $\alpha \in \mathbf{Z}_{\geq 0}^s$ *such* $\sum_1^s \alpha_i = d$ *the equality*

$$e(\mathfrak{q}_1^{[\alpha_1]}, \ldots, \mathfrak{q}_s^{[\alpha_s]}; R) = d! \operatorname{Covol}(\bar{E}_1^{[\alpha_1]}, \ldots, \bar{E}_s^{[\alpha_s]}).$$

In particular, the mixed multiplicities depend only on the integral closures of the ideals $\mathfrak{q}_i$. Now we have the well-known Alexandrov–Fenchel inequalities for the mixed volumes of two compact convex bodies:

THEOREM A.8 (ALEXANDROV AND FENCHEL; see [Gromov 1990]).

(a) *Let* K_1, K_2 *be compact convex bodies in* $\mathbf{R}^d$; *set* $v_i = \operatorname{Vol}(K_1^{[i]}, K_2^{[d-i]})$. *For all* $2 \leq i \leq d$,

$$v_{i-1}^2 \geq v_i v_{i-2}.$$

(b) *Equality holds in all these inequalities if and only if for some $\rho \in \mathbf{R}_+$ we have $K_1 = \rho K_2$ up to translation. If all the v_i are equal, then $K_1 = K_2$ up to translation, and conversely.*

Let $\mathbf{B}^d$ denote the d-dimensional unit ball, and A any subset of $\mathbf{R}^d$ which is tame enough for the volumes to exist.

The problem that inspired this theorem is to prove that in the isoperimetric inequality $\mathrm{Vol}_{d-1}(\partial A)^d \geq d^d \mathrm{Vol}_d(\mathbf{B}^d) \mathrm{Vol}_d(A)^{d-1}$, equality should hold only if A is a multiple of the unit ball, to which some "hairs" of a smaller dimension than ∂A have been added. In the case where A is convex, taking K_1 to be the unit ball and $K_2 = A$, one notices that $v_0 = \mathrm{Vol}_d(A)$ and $v_1 = d^{-1} \mathrm{Vol}_{d-1}(\partial A)$; the isoperimetric inequality then follows very quickly by an appropriate telescoping of the Alexandrov–Fenchel inequalities. From this telescoping follows the fact that if we have equality in the isoperimetric inequality for a convex subset A of $\mathbf{R}^d$, then we have equality in *all* the Alexandrov–Fenchel inequalities for A and the unit ball, so that A must be a ball. By the same type of telescoping, one proves the inequalities $v_i^d \geq v_0^{d-i} v_d^i$, which yields:

THEOREM A.9 (BRÜNN AND MINKOWSKI; see [Gromov 1990]). *For convex compact subsets K_1, K_2 of $\mathbf{R}^d$,*

$$\mathrm{Vol}_d(K_1 + K_2)^{1/d} \geq \mathrm{Vol}_d(K_1)^{1/d} + \mathrm{Vol}_d(K_2)^{1/d}.$$

Equality holds if and only if the two sets are homothetic up to translation, or one of them is a point, or $\mathrm{Vol}_d(K_1 + K_2)^{1/d} = 0$.

The same constructions and proof apply to covolumes, where the inequalities are reversed; they correspond to inequalities for the mixed multiplicities of monomial ideals, which are in fact true for primary ideals in formally equidimensional noetherian local rings:

THEOREM A.10 [Teissier 1977; 1978; Rees and Sharp 1978; Katz 1988]. *Let $\mathfrak{q}_1, \mathfrak{q}_2$ be primary ideals in the d-dimensional noetherian local ring R. Set*

$$w_i = e(\mathfrak{q}_1^{[i]}, \mathfrak{q}_2^{[d-i]}; R).$$

(a) *We have $w_{i-1}^2 \leq w_i w_{i-2}$ for $2 \leq i \leq d$.*

(b) *The inequalities $e(\mathfrak{q}_1 \mathfrak{q}_2, R)^{1/d} \leq e(\mathfrak{q}_1, R)^{1/d} + e(\mathfrak{q}_2, R)^{1/d}$ hold, with equality if and only if the inequalities of (a) are equalities.*

(c) *Assuming in addition that R is formally equidimensional (quasi-unmixed), equality holds in all these inequalities if and only if $\overline{\mathfrak{q}_1^a} = \overline{\mathfrak{q}_2^b}$ for some $a, b \in \mathbf{N}$. If all the w_i are equal, then $\overline{\mathfrak{q}}_1 = \overline{\mathfrak{q}}_2$, and conversely.*

So in this case again, the combinatorial inequalities appear as the avatar for monomial ideals of general inequalities of commutative algebra. One can see that if $\mathfrak{q}_1 \subseteq \mathfrak{q}_2$, we have $e(\mathfrak{q}_1, R) = w_d \geq w_i \geq w_0 = e(\mathfrak{q}_2, R)$, for $1 \leq i \leq d-1$. So this result implies Rees' Theorem, which is stated after Corollary A.4.

In fact the same happens for the Alexandrov–Fenchel inequalities, which are the avatars for toric varieties associated to polytopes of general inequalities of Kähler geometry known as the Hodge Index Theorem. This is because the mixed volumes of rational convex polytopes are equal, up to a $d!$ factor, to the mixed degrees of invertible sheaves (or of divisors) on certain toric varieties associated to the collection of polytopes, exactly as in Corollary A.7. This approach to Alexandrov–Fenchel inequalities was introduced by Khovanskii and the author; see [Gromov 1990] for an excellent exposition of this topic, and [Khovanskii 1979; Teissier 1979].

In all these cases, it is remarkable that, thanks to the positivity and convexity properties of volumes and of multiplicities, a *finite* number of equations on a pair (A_1, A_2) of objects in an *infinite dimensional* space (convex bodies modulo translation or integrally closed primary ideals) suffices to ensure that $A_1 = A_2$.

Newton nondegenerate ideals in $k[\![u_1, \ldots, u_d]\!]$ and multiplicities. Define the support $S(I)$ of an ideal I of $k[\![u_1, \ldots, u_d]\!]$ to be the set of the exponents m appearing as one of the exponents in at least one series belonging to the ideal I. Define the Newton polyhedron $\mathcal{N}_+(I)$ of I as the boundary of the convex hull $\mathcal{P}_+(I)$ of $\bigcup_{m \in S(I)}(m + \mathbf{R}_{\geq 0}^d)$.

According to [Bivià-Ausina et al. 2002], a primary ideal $\mathfrak{q}$ is said to be *nondegenerate* if it admits a system of generators $q_1, \ldots, q_t$ such that their restrictions to each compact face of $\mathcal{N}_+(I)$ have no common zero in the torus $(k^*)^d$. The following is part of what is proved in [Bivià-Ausina et al. 2002, § 3]:

THEOREM A.11. *For a primary ideal $\mathfrak{q}$ of $R = k[\![u_1, \ldots, u_d]\!]$, the following conditions are equivalent*:

(a) *The ideal $\mathfrak{q}$ is nondegenerate in the coordinates $u_1, \ldots, u_d$.*

(b) $e(\mathfrak{q}, R) = d! \operatorname{Covol} \mathcal{P}_+(I)$.

(c) *The integral closure $\bar{\mathfrak{q}}$ of $\mathfrak{q}$ is generated by monomials in $u_1, \ldots, u_d$.*

It follows from this that monomial ideals are nondegenerate, and that products of nondegenerate primary ideals are nondegenerate [Bivià-Ausina et al. 2002, Corollary 3.14]. Moreover, all the numerical facts mentioned above for monomial ideals with respect to their Newton polyhedron are valid for nondegenerate ideals (*loc. cit.*). Nondegenerate ideals behave as *reductions* of monomial ideals, which in fact they are. Here we can think of a reduction (in the sense of [Northcott and Rees 1954]; see also see [Rees 1984]) of an ideal $\mathcal{M} \subset k[\![u_1, \ldots, u_d]\!]$ as an ideal generated by d sufficiently general combinations of generators of $\mathcal{M}$. More precisely, it is an ideal $\mathcal{M}'$ contained in $\mathcal{M}$ and having the same integral closure.

There is a close connection between this nondegeneracy for ideals and the results of section 5; if the ideal $\mathfrak{q} = (q_1, \ldots, q_s)k[\![u_1, \ldots, u_d]\!]$ is nondegenerate, then a general linear combination $q = \sum_{i=1}^s \lambda_i q_i$ is nondegenerate with respect to its Newton polyhedron.

There are many other interesting consequences of the relationship between monomial ideals and combinatorics; I refer the reader to [Sturmfels 1996].

All the results of this appendix remain valid if $k[u_1, \ldots, u_d]$ and its completion $k[\![u_1, \ldots, u_d]\!]$ are replaced respectively by $k[\check{\sigma} \cap \mathbf{Z}^d]$ and its completion, for a strictly convex cone $\sigma \subset \mathbf{R}^d_{\geq 0}$.

There are also generalizations of mixed multiplicities to collections of not necessarily primary ideals [Rees 1986] and to the case where one of the ideals is replaced by a submodule of finite colength of a free R-module of finite type [Kleiman and Thorup 1996].

It would be interesting to determine how the results of this appendix extend to monomial submodules of a free $k[u_1, \ldots, u_d]$-module.

References

[Bivià-Ausina et al. 2002] C. Bivià-Ausina, T. Fukui, and M. J. Saia, "Newton filtrations, graded algebras and codimension of non-degenerate ideals", *Math. Proc. Cambridge Philos. Soc.* **133**:1 (2002), 55–75.

[Blickle and Lazarsfeld 2004] M. Blickle and R. Lazarsfeld, "An informal introduction to multiplier ideals", pp. 87–114 in *Trends in algebraic geometry*, edited by L. Avramov et al., Math. Sci. Res. Inst. Publ. **51**, Cambridge University Press, New York, 2004.

[Bourbaki 1968] N. Bourbaki, *Algèbre commutative, Ch. I à IV*, Masson, Paris, 1968.

[Bourbaki 1983] N. Bourbaki, *Algèbre commutative, Ch. VIII et IX*, Masson, Paris, 1983.

[Bravo 2004] A. Bravo, "Some facts about canonical subalgebra bases", pp. 247–254 in *Trends in algebraic geometry*, edited by L. Avramov et al., Math. Sci. Res. Inst. Publ. **51**, Cambridge University Press, New York, 2004.

[Brieskorn and Knörrer 1986] E. Brieskorn and H. Knörrer, *Plane algebraic curves*, Birkhäuser, Basel, 1986. German original, Birkhäuser, 1981.

[De Concini and Procesi 1983] C. De Concini and C. Procesi, "Complete symmetric varieties", pp. 1–44 in *Invariant theory* (Montecatini, 1982), edited by F. Gherardelli, Lecture Notes in Math. **996**, Springer, Berlin, 1983.

[De Concini and Procesi 1985] C. De Concini and C. Procesi, "Complete symmetric varieties, II: Intersection theory", pp. 481–513 in *Algebraic groups and related topics* (Kyoto/Nagoya, 1983), edited by R. Hotta, Adv. Stud. Pure Math. **6**, Kinokuniya, Tokyo, 1985.

[Eisenbud and Sidman 2004] D. Eisenbud and J. Sidman, "The geometry of syzygies", pp. 115–152 in *Trends in algebraic geometry*, edited by L. Avramov et al., Math. Sci. Res. Inst. Publ. **51**, Cambridge University Press, New York, 2004.

[Eisenbud and Sturmfels 1996] D. Eisenbud and B. Sturmfels, "Binomial ideals", *Duke Math. J.* **84**:1 (1996), 1–45.

[Ewald 1996] G. Ewald, *Combinatorial convexity and algebraic geometry*, Graduate Texts in Math. **168**, Springer, Paris, 1996.

[Fenchel 1929] W. Fenchel, "Über Krümmung und Windung geschlossener Raumkurven", *Math. Annalen* **101** (1929), 238–252.

[Gerstenhaber 1964] M. Gerstenhaber, "On the deformation of rings and algebras", *Ann. of Math.* (2) **79** (1964), 59–103.

[Gerstenhaber 1966] M. Gerstenhaber, "On the deformation of rings and algebras, II", *Ann. of Math.* (2) **84** (1966), 1–19.

[Goldin and Teissier 2000] R. Goldin and B. Teissier, "Resolving singularities of plane analytic branches with one toric morphism", pp. 315–340 in *Resolution of singularities* (Obergurgl, 1997), edited by H. Hauser et al., Progr. Math. **181**, Birkhäuser, Basel, 2000.

[González Pérez 2003] P. D. González Pérez, "Toric embedded resolutions of quasi-ordinary hypersurface singularities", *Ann. Inst. Fourier (Grenoble)* **53**:6 (2003), 1819–1881.

[González Pérez and Teissier 2002] P. D. González Pérez and B. Teissier, "Embedded resolutions of non necessarily normal affine toric varieties", *C. R. Math. Acad. Sci. Paris* **334**:5 (2002), 379–382.

[Gromov 1990] M. Gromov, "Convex sets and Kähler manifolds", pp. 1–38 in *Advances in differential geometry and topology*, edited by F. Tricerri, World Sci. Publishing, Teaneck, NJ, 1990.

[Grünbaum 1967] B. Grünbaum, *Convex polytopes*, Pure and Applied Mathematics **16**, Wiley/Interscience, New York, 1967. Second edition, Springer, 2003 (GTM **221**).

[Hanner and Rådström 1951] O. Hanner and H. Rådström, "A generalization of a theorem of Fenchel", *Proc. Amer. Math. Soc.* **2** (1951), 589–593.

[Hironaka 1964] H. Hironaka, "Resolution of singularities of an algebraic variety over a field of characteristic zero. I, II", *Ann. of Math.* (2) **79**:1–2 (1964), 109–203, 205–326.

[Hochster 2004] M. Hochster, "Tight closure theory and characteristic p methods", pp. 181–210 in *Trends in algebraic geometry*, edited by L. Avramov et al., Math. Sci. Res. Inst. Publ. **51**, Cambridge University Press, New York, 2004.

[Howald 2001] J. A. Howald, "Multiplier ideals of monomial ideals", *Trans. Amer. Math. Soc.* **353**:7 (2001), 2665–2671.

[Katz 1988] D. Katz, "Note on multiplicity", *Proc. Amer. Math. Soc.* **104**:4 (1988), 1021–1026.

[Kempf et al. 1973] G. Kempf, F. F. Knudsen, D. Mumford, and B. Saint-Donat, *Toroidal embeddings, I*, Lecture Notes in Math. **339**, Springer, Berlin, 1973.

[Khovanskii 1979] A. G. Khovanskii, "Geometry of convex bodies and algebraic geometry", *Uspekhi Mat. Nauk* **34**:4 (1979), 160–161.

[Kleiman and Thorup 1996] S. Kleiman and A. Thorup, "Mixed Buchsbaum–Rim multiplicities", *Amer. J. Math.* **118**:3 (1996), 529–569.

[Kouchnirenko 1976] A. G. Kouchnirenko, "Polyèdres de Newton et nombres de Milnor", *Invent. Math.* **32**:1 (1976), 1–31.

[Lipman and Teissier 1981] J. Lipman and B. Teissier, "Pseudorational local rings and a theorem of Briançon–Skoda about integral closures of ideals", *Michigan Math. J.* **28**:1 (1981), 97–116.

[Merle and Teissier 1980] M. Merle and B. Teissier, "Conditions d'adjonction, d'après Du Val", pp. 229–245 in *Séminaire sur les singularités des surfaces* (Palaiseau 1976/77), edited by M. Demazure et al., Lecture Notes in Math. **777**, Springer, 1980.

[Northcott and Rees 1954] D. G. Northcott and D. Rees, "Reductions of ideals in local rings", *Proc. Cambridge Philos. Soc.* **50** (1954), 145–158.

[Oka 1997] M. Oka, *Non-degenerate complete intersection singularity*, Actualités Mathématiques, Hermann, Paris, 1997.

[Rees 1961] D. Rees, "α-transforms of local rings and a theorem on multiplicities of ideals", *Proc. Cambridge Philos. Soc.* **57**:1 (1961), 8–17.

[Rees 1984] D. Rees, "Generalizations of reductions and mixed multiplicities", *J. London Math. Soc.* (2) **29**:3 (1984), 397–414.

[Rees 1986] D. Rees, "The general extension of a local ring and mixed multiplicities", pp. 339–360 in *Algebra, algebraic topology and their interactions* (Stockholm, 1983), edited by J.-E. Roos, Lecture Notes in Math. **1183**, Springer, Berlin, 1986.

[Rees and Sharp 1978] D. Rees and R. Y. Sharp, "On a theorem of B. Teissier on multiplicities of ideals in local rings", *J. London Math. Soc.* (2) **18**:3 (1978), 449–463.

[Skoda and Briançon 1974] H. Skoda and J. Briançon, "Sur la clôture intégrale d'un idéal de germes de fonctions holomorphes en un point de $\mathbf{C}^n$", *C. R. Acad. Sci. Paris Sér. A* **278** (1974), 949–951.

[Smith 1873] H. J. S. Smith, "On the higher singularities of plane curves", *Proc. London Math. Soc.* **6** (1873), 153–182.

[Sturmfels 1996] B. Sturmfels, *Gröbner bases and convex polytopes*, University Lecture Series **8**, Amer. Math. Soc., Providence, 1996.

[Teissier 1973] B. Teissier, "Cycles évanescents, sections planes et conditions de Whitney", pp. 285–362 in *Singularités à Cargèse* (Cargèse, 1972), Astérisque **7–8**, Soc. Math. France, Paris, 1973.

[Teissier 1975] B. Teissier, "Appendice", pp. 145–199 in *Le problème des modules pour les branches planes*, by O. Zariski, Centre de Mathématiques de l'École Polytechnique, Paris, 1975. Second edition, Hermann, 1986.

[Teissier 1977] B. Teissier, "Sur une inégalité à la Minkowski pour les multiplicités", appendix to D. Eisenbud and H. I. Levine, "An algebraic formula for the degree of a C^∞ map germ", *Ann. Math.* (2) **106**:1 (1977), 19–44.

[Teissier 1978] B. Teissier, "On a Minkowski-type inequality for multiplicities. II", pp. 347–361 in *C. P. Ramanujam, a tribute*, edited by K. G. Ramanathan, Tata Studies in Math. **8**, Springer, Berlin, 1978.

[Teissier 1979] B. Teissier, "Du théorème de l'index de Hodge aux inégalités isopérimétriques", *C. R. Acad. Sci. Paris Sér. A-B* **288**:4 (1979), A287–A289.

[Teissier 1988] B. Teissier, "Monômes, volumes et multiplicités", pp. 127–141 in *Introduction à la théorie des singularités, II*, edited by L. D. Tráng, Travaux en Cours **37**, Hermann, Paris, 1988.

[Teissier 2003] B. Teissier, "Valuations, deformations, and toric geometry", pp. 361–459 in *Valuation theory and its applications* (Saskatoon, SK, 1999), vol. 2, edited by F.-V. Kuhlmann et al., Fields Inst. Commun. **33**, Amer. Math. Soc., Providence, RI, 2003.

BERNARD TEISSIER
INSTITUT MATHÉMATIQUE DE JUSSIEU
UMR 7586 DU C.N.R.S.
ÉQUIPE "GEOMÉTRIE ET DYNAMIQUE"
BUREAU 8E18
175 RUE DU CHEVALERET
F 75013 PARIS
FRANCE
   teissier@math.jussieu.fr

.

Trends in Commutative Algebra
MSRI Publications
Volume **51**, 2004

Some Facts About Canonical Subalgebra Bases

ANA BRAVO

ABSTRACT. This is a brief exposition of canonical subalgebra bases, their uses and their computation.

CONTENTS

1. Introduction

Let k be a field, let $R \subset k[x_1, \ldots, x_n]$ be a finitely generated subalgebra, and let $>$ be a term ordering in $k[x_1, \ldots, x_n]$. A subset B of R is said to be a *canonical subalgebra basis*, or *SAGBI basis*, of R, if

$$\mathrm{in}_> B := \{\mathrm{in}_> f : f \in B\}$$

generates the subalgebra $\mathrm{in}_> R := \{\mathrm{in}_> g : g \in R\}$ of $k[x_1, \ldots, x_n]$. If B is a SAGBI basis for R, it generates R as a k-algebra.

The abbreviation SAGBI stands for "subalgebra analog to Gröbner basis for ideals"; as we will see in Section 2 there are several similarities between canonical subalgebra bases and Gröbner bases.

The simplest example occurs when $R \subset k[x_1, \ldots, x_n]$ is generated by a single element, in which case this same generator is also a canonical subalgebra basis.

The notion was introduced by Kapur and Madlener [1989] and independently by Robbiano and Sweedler [1990]. SAGBI bases are used to test subalgebra

membership. Algorithms for computing canonical subalgebra bases are presented
in both [Kapur and Madlener 1989] and [Robbiano and Sweedler 1990].

From the algebraic point of view, canonical bases are very interesting. For
instance, if $\text{in}_> R$ is finitely generated, the study of $\text{in}_> R$ is simpler than that of
R, and in many cases both algebras share the same properties. As an example,
in [Conca et al. 1996] it is shown that if $\text{in}_> R$ is normal, Cohen–Macaulay, and
has rational singularities, R has the same properties.

From the geometric perspective, SAGBI bases also offer interesting possibilities. When $\text{in}_> R$ is finitely generated, it can be regarded as the associated graded
ring of a suitable degree filtration of R. As a consequence $\text{in}_> R$ can be interpreted as the special fiber of a one-parameter family with R as a general fiber.
In this case the general fiber and the special fiber of the family share geometric
properties. See [Conca et al. 1996; Sturmfels 1996] and also Section 6 below for
discussion.

This philosophy appears, in the analytic case, in [Teissier 1975] and [Goldin
and Teissier 2000], as an approach to resolution of singularities of plane curves:
Given a suitable parametrization of a plane curve, construct a flat family of
curves in such a way that the general fiber is isomorphic to the original curve,
and the special fiber is a monomial curve. Then a toric resolution of singularities
of the special fiber induces a resolution of the generic fiber [Goldin and Teissier
2000, § 6].

Canonical subalgebra bases have also been studied for algebras over arbitrary
rings in [Miller 1996] and [Stillman and Tsai 1999]. For other applications and
examples, see [Göbel 2002; 2001; 2000; 1999c; 1999b; 1999a; 1998; Göbel and
Maier 2000; Miller 1998; Nordbeck 2002].

2. SABGI Bases Versus Gröbner Bases

As pointed out in the introduction, canonical subalgebra bases and Gröbner
bases play similar roles in two different contexts: The first are used to test
subalgebra membership while the second do the same work for ideals.

This similarity can be carried out one step further in two different directions:
The computational point of view and the geometric interpretation.

The Subduction Algorithm described in [Sturmfels 1996, Chapter 11] corresponds to the subalgebra analog of the Division Algorithm for ideals (which
produces, for any element f in an ideal I, an expression of f as a linear combination of a Gröbner basis of I):

ALGORITHM 2.1 (SUBDUCTION ALGORITHM FOR A CANONICAL BASIS $\mathscr{C}$).
Given a canonical basis $\mathscr{C}$ for a subalgebra $R \subset k[x_1, \ldots, x_n]$ and given $f \in$
$k[x_1, \ldots, x_n]$, the algorithm computes an expression for f as a polynomial in the
elements of $\mathscr{C}$, provided that $f \in R$.

Step 1. Find $f_1, \ldots, f_r \in \mathscr{C}$ and exponents $i_1, \ldots, i_r \in \mathbb{N}$ and $c \in k \setminus \{0\}$ such that

$$\mathrm{in}_> f = c \cdot \mathrm{in}_> f_1^{i_1} \cdots \mathrm{in}_> f_r^{i_r}. \tag{2-1}$$

Step 2. If it is not possible to find an expression as in (2–1) then $f \notin R$, and the algorithm stops.

Step 3. Otherwise, set $g := c \cdot f_1^{i_1} \cdots f_r^{i_r}$, and replace f by $f - g$. Repeat the previous steps until the algorithm stops or f is a constant in k.

In Section 5 we will also see how this algorithm can be used to produce an algorithm to compute SAGBI bases which is similar to Buchberger's algorithm for computing Gröbner bases.

As for the geometric interpretation, let $I \subset k[x_1, \ldots, x_n]$ be the ideal defining a variety X, and $>$ a term ordering in $k[x_1, \ldots, x_n]$. The question is:

How close are X and $\mathrm{Spec}\,(k[x_1, \ldots, x_n]/\mathrm{in}_> I)$?

The general theory of Gröbner basis says that one can construct a flat family of varieties over a one-dimensional scheme $\mathrm{Spec}\,(k[t])$, whose general fiber is isomorphic to X, and whose special fiber at $t = 0$ is $\mathrm{Spec}\,(k[x_1, \ldots, x_n]/\mathrm{in}_> I)$. In this sense we say that the original variety X deforms into $\mathrm{Spec}\,(k[x_1, \ldots, x_n]/\mathrm{in}_> I)$.

Now let Y be a variety parametrized by equations $f_1, \ldots, f_s \in k[t_1, \ldots, t_m]$, and let $>$ be a term ordering in $k[t_1, \ldots, t_m]$. If $\{g_1, \ldots, g_r\}$ is a canonical subalgebra basis of $k[f_1, \ldots, f_s]$, we will see in Section 6 that one can construct a one-parameter flat family of varieties, whose general fiber is isomorphic to Y, and whose special fiber is a toric variety; the generators of the algebras degenerate into monomials and the relations between them into binomials.

Perhaps the main difference between Gröbner bases and canonical subalgebra bases is that while the first are always finite, the second may fail to be so. This point is discussed in the next section.

3. When Are SAGBI Bases Finite?

Canonical subalgebra bases are not always finite; finiteness may even depend on the term ordering $>$ chosen on $k[x_1, \ldots, x_n]$. We examine some examples.

If $R \subset k[x, y]$ is generated by $\{x + y, xy, xy^2\}$, then R does not have a finitely generated canonical subalgebra basis, no matter what term ordering we fix in $k[x, y]$: If $x > y$, it can be shown that a SAGBI basis of R must contain the infinite set $S = \{x + y, xy^n : n > 0\}$. If $y > x$, note that R is also generated by $\{x + y, xy, x^2 y\}$. It can be shown that

$$S = \{x + y, yx^n : n > 0\}$$

should be contained in a SAGBI basis for R [Robbiano and Sweedler 1990, Example 1.20].

On the opposite extreme, the symmetric algebra $R \subset k[x_1, \ldots, x_n]$ always has a finitely generated canonical subalgebra basis B which does not depend on the order previously chosen: In this case, B is the set of elementary symmetric polynomials [Robbiano and Sweedler 1990, Theorem 1.14].

There are also examples of subalgebras that, depending on the order fixed, may or may not have a finite canonical subalgebra basis: Let $R \subset k[x, y]$ be the subalgebra generated by $\{x, xy - y^2, xy^2\}$. If we fix a term ordering on $k[x, y]$ such that $y > x$, then $B = \{x, xy - y^2, xy^2\}$ is indeed a canonical subalgebra basis for R, while if we fix a term ordering such that $x > y$ then it can be shown that $k[x, xy, xy^2, \ldots] \subset \text{in}_> R$, and therefore it cannot have a finite SAGBI basis [Robbiano and Sweedler 1990, Example 4.11]. For these and other examples, we refer the reader to [Göbel 2000; Göbel 1999b; Robbiano and Sweedler 1990].

In general, it is a hard problem to decide whether a given subalgebra does have a finite canonical subalgebra basis. Some conditions are as follows:

PROPOSITION 3.1 [Robbiano and Sweedler 1990, Proposition 4.7]. *Suppose that R is a subalgebra of $k[x_1, \ldots, x_n]$ and that C is a finitely generated subalgebra of $k[x_1, \ldots, x_n]$ containing $\text{in}_> R$. If C is integral over $\text{in}_> R$, then R has a finite SAGBI basis. In particular if $k[x_1, \ldots, x_n]$ is integral over $\text{in}_> R$, then R has a finite SAGBI basis.*

A corollary of the previous proposition is that when $n = 1$, things become less chaotic: Any subalgebra R of $k[x]$ has a finite subalgebra basis [Robbiano and Sweedler 1990, Corollay 4.8]. And when the number of generators is low, there are even easy criteria to decide if a given set of generators of a subalgebra of $k[x]$ is a canonical basis:

THEOREM 3.2 [Torstensson 2002, Theorems 10, 12]. *Let $f, g \in k[x]$ and consider the subalgebra $R \subset k[x]$ generated by them. Then:*

(i) *If f and g have relatively prime degrees, they form a canonical subalgebra basis for R.*

(ii) *If $\deg f$ divides $\deg g$, f and g form a canonical subalgebra basis for R if and only if g is a polynomial in f.*

For more along these lines see Propositions 6, 7, and Theorems 12 and 14 in [Torstensson 2002].

4. Finite SAGBI Bases

Let $R \subset k[x_1, \ldots, x_n]$ be a subalgebra generated by $B = \{f_1, \ldots, f_s\}$, let $>$ be a monomial ordering in $k[x_1, \ldots, x_n]$, and assume that R has a finite canonical subalgebra basis. The purpose of this section is to describe a criterion to decide whether B is a canonical basis for R. In Section 5 and 6 some consequences of this result will be discussed. The setup that follows can be found in [Sturmfels 1996, Chapter 11].

Consider the exact sequences

$$0 \longrightarrow I \longrightarrow k[t_1,\ldots,t_s] \longrightarrow k[f_1,\ldots,f_s] \longrightarrow 0$$
$$t_i \longmapsto f_i$$

$$0 \longrightarrow I_A \longrightarrow k[t_1,\ldots,t_s] \longrightarrow k[\mathrm{in}_> f_1,\ldots,\mathrm{in}_> f_s] \longrightarrow 0$$
$$t_i \longmapsto \mathrm{in}_> f_i$$

(4-1)

Since the kernel of the second map represents relations between monomials, the ideal I_A is generated by binomials — it is a toric ideal.

Let $\omega = (\omega_1,\ldots,\omega_n) \in \mathbb{R}^n$ be a weight vector which represents the term ordering $>$ for the polynomials $\{f_1,\ldots,f_s\}$.

Assume that $\mathrm{in}_> f_i = x_1^{a_{i1}} \cdots x_n^{a_{in}}$, for $i = 1,\ldots,s$. Then

$$A = \begin{pmatrix} a_{11} & \cdots & a_{s1} \\ \vdots & \ddots & \vdots \\ a_{1n} & \cdots & a_{sn} \end{pmatrix}$$

is an $n \times s$ matrix, and

$$A^T \omega = \begin{pmatrix} a_{11}\omega_1 + \ldots + a_{1n}\omega_n \\ \vdots \\ a_{s1}\omega_1 + \ldots + a_{sn}\omega_n \end{pmatrix}$$

is a vector in $\mathbb{R}^s$, which can be thought of as a weight vector defining an order in $k[t_1,\ldots,t_s]$. Therefore it can be used to form an initial ideal $\mathrm{in}_{A^T\omega} I$ of I. In general this will not be a monomial ideal since $A^T\omega$ may not be a generic vector, even if ω is (see Example 4.2 below).

The key point is that the comparison between I_A and $\mathrm{in}_{A^T\omega} I$ gives a criterion for deciding whether or not $\{f_1,\ldots,f_s\}$ is a canonical basis for the subalgebra that they generate.

In general $\mathrm{in}_{A^T\omega} I \subset I_A$ [Sturmfels 1996, Lemma 11.3], but if equality holds, then $\{f_1,\ldots,f_s\}$ is a canonical basis. More precisely:

THEOREM 4.1 [Sturmfels 1996, Theorem 11.4]. *The set* $\{f_1,\ldots,f_s\}$ *is a canonical basis if and only if* $\mathrm{in}_{A^T\omega} I = I_A$.

EXAMPLE 4.2. Let $R = k[x^2 + x^3, \, x + x^2] \subset k[x]$, and let us temporarily forget that we already know that $\{x^2 + x^3, \, x + x^2\}$ is a canonical subalgebra basis (since the degrees of the generators are coprime). With the notation above, we have $I = \langle t_2^3 - t_1^2 - t_1 t_2 \rangle$, and that $I_A = \langle t_2^3 - t_1^2 \rangle$. Let $A = (3, 2)$. Then $A^T\omega = \binom{3}{2}$ and $\mathrm{in}_{A^T\omega} I = \langle t_2^3 - t_1^2 \rangle = I_A$. Hence, by Theorem 4.1, $\{x^2 + x^3, \, x + x^2\}$ is a SAGBI basis for R.

5. An Algorithm to Compute SAGBI bases

Theorem 4.1 can be used to construct an algorithm for computing canonical subalgebra bases. With the same notation as in Section 4, first we state the following corollary to Theorem 4.1:

COROLLARY 5.1 [Sturmfels 1996, Corollary 11.5]. *Let $\{p_1, \ldots, p_t\}$ be generators for the toric ideal I_A. Then $\{f_1, \ldots, f_s\}$ is a canonical basis if and only if Algorithm 2.1 reduces $p_i(f_1, \ldots, f_s)$ to a constant for all $i = 1, \ldots, t$.*

Therefore, to apply the criterion of Theorem 4.1 there is no need to compute generators for I, since only the ones of I_A are used. Let us see how this works with an example:

EXAMPLE 5.2. Consider $R = [x^4 + x^3, x^2 + x] \subset k[x]$. By [Robbiano and Sweedler 1990, Corollay 4.8] there is a finite SAGBI basis for R. With the same notation as in Section 4 we have $A = (4, 2)$, $I = \langle t_1^2 - 2t_1 t_2^2 - t_1 t_2 + t_2^4 \rangle$, $A^T \omega = \binom{4}{2}$ and

$$\mathrm{in}_{A^T \omega} I = \langle t_1^2 - 2t_1 t_2^2 + t_2^4 \rangle \subsetneq I_A = \langle t_1 - t_2^2 \rangle.$$

Therefore, by Theorem 4.1, the set $\{x^4 + x^3, x^2 + x\}$ is not a SAGBI basis for R.

With the notation of Corollary 5.1, $p_1(t_1, t_2) = t_1 - t_2^2$ Algorithm 2.1 does not reduce $p_1(x^4 + x^3, x^2 + x)$ to a constant, since $p_1(x^4 + x^3, x^2 + x) = x^3 + x^2$. This is as expected. Hence we need to extend our generating set to

$$\{x^4 + x^3, x^2 + x, x^3 + x^2\}.$$

In this new setting, $I_A = \langle t_2^3 - t_3^2, t_1 - t_2^2 \rangle = \langle p_1(t_1, t_2, t_3), p_2(t_1, t_2, t_3) \rangle$, and it is easy to check that, in this case, Algorithm 2.1 reduces $p_i(t_1, t_2, t_3)$ to a constant for $i = 1, 2$, and therefore, $\{x^4 + x^3, x^2 + x, x^3 + x^2\}$ is a SAGBI basis for R.

REMARK 5.3. The algorithm that follows from Corollary 5.1 (used in Example 5.2) is similar to Buchberger's Algorithm to compute Gröbner bases of ideals, provided that we ahead of time know that there is a finite SAGBI basis.

6. Geometric Interpretation

As a final note we review the geometry behind the previous statements, specially diagram (4–1) and Theorem 4.1.

Assume that X is a variety parametrized by the equations $\{f_1, \ldots, f_s\}$. Using the same notation as in Section 4, if $\{f_1, \ldots, f_s\}$ is a canonical subalgebra bases then by Theorem 4.1, $I_A = \mathrm{in}_{A^T \omega} I$.

The following corollary to Theorem 4.1 relates any reduced Gröbner basis of I_A to a suitable reduced Gröbner basis of I:

COROLLARY 6.1 [Sturmfels 1996, Corollary 11.6]. *With the same notation as in Section 4, assume that* $\{f_1, \ldots, f_s\}$ *is a canonical subalgebra basis. Then every reduced Gröbner basis* $\mathcal{G}$ *of* I_A *lifts to a reduced Gröbner basis* $\mathcal{H}$ *of* I, *i.e. the elements of* $\mathcal{G}$ *are the initial forms with respect to* $A^T \omega$ *of the elements of* $\mathcal{H}$.

Now, the general theory of Gröbner bases tells us that we can construct a one-parameter flat family of varieties whose general fiber is isomorphic to X and whose special fiber is isomorphic to $\mathrm{Spec}\,(k[t_1, \ldots, t_s]/\mathrm{in}_{A^T\omega}I)$.

Therefore, Corollary 6.1, implies that we can construct a one-parameter flat family of varieties whose general fiber is isomorphic to X and whose special fiber is isomorphic to $\mathrm{Spec}\,(k[t_1, \ldots, t_s]/I_A)$: The parametric equations defining X degenerate into monomials and the relations among them into binomials. Therefore X degenerates to a toric variety.

Acknowledgments

I am indebted to Professors Sturmfels and Teissier for several useful conversations and suggestions.

References

[Conca et al. 1996] A. Conca, J. Herzog, and G. Valla, "Sagbi bases with applications to blow-up algebras", *J. Reine Angew. Math.* **474** (1996), 113–138.

[Göbel 1998] M. Göbel, "A constructive description of SAGBI bases for polynomial invariants of permutation groups", *J. Symbolic Comput.* **26**:3 (1998), 261–272.

[Göbel 1999a] M. Göbel, "The optimal lower bound for generators of invariant rings without finite SAGBI bases with respect to any admissible order", *Discrete Math. Theor. Comput. Sci.* **3**:2 (1999), 65–70.

[Göbel 1999b] M. Göbel, "The 'smallest' ring of polynomial invariants of a permutation group which has no finite SAGBI bases w.r.t. any admissible order", *Theoret. Comput. Sci.* **225**:1-2 (1999), 177–184.

[Göbel 1999c] M. Göbel, "Three remarks on SAGBI bases for polynomial invariants of permutation groups", pp. 190–201 in *Combinatorics, computation & logic '99* (Auckland, 1999), edited by C. S. Calude and M. J. Dinneen, Discrete Mathematics and Theoretical Computer Science **333**, Springer, Singapore, 1999.

[Göbel 2000] M. Göbel, "Rings of polynomial invariants of the alternating group have no finite SAGBI bases with respect to any admissible order", *Inform. Process. Lett.* **74**:1-2 (2000), 15–18.

[Göbel 2001] M. Göbel, "Visualizing properties of comprehensive SAGBI bases: two examples", *Appl. Algebra Engrg. Comm. Comput.* **12**:5 (2001), 429–435.

[Göbel 2002] M. Göbel, "Finite SAGBI bases for polynomial invariants of conjugates of alternating groups", *Math. Comp.* **71**:238 (2002), 761–765.

[Göbel and Maier 2000] M. Göbel and P. Maier, "Three remarks on comprehensive Gröbner and SAGBI bases", pp. 191–202 in *Computer algebra in scientific computing* (Samarkand, 2000), edited by V. G. Ganzha et al., Springer, Berlin, 2000.

[Goldin and Teissier 2000] R. Goldin and B. Teissier, "Resolving singularities of plane analytic branches with one toric morphism", pp. 315–340 in *Resolution of singularities* (Obergurgl, 1997), edited by H. Hauser et al., Progr. Math. **181**, Birkhäuser, Basel, 2000.

[Kapur and Madlener 1989] D. Kapur and K. Madlener, "A completion procedure for computing a canonical basis for a k-subalgebra", pp. 1–11 in *Computers and mathematics* (Cambridge, MA, 1989), edited by E. Kaltofen and S. M. Watt, Springer, New York, 1989.

[Miller 1996] J. L. Miller, "Analogs of Gröbner bases in polynomial rings over a ring", *J. Symbolic Comput.* **21**:2 (1996), 139–153.

[Miller 1998] J. L. Miller, "Effective algorithms for intrinsically computing SAGBI-Gröbner bases in a polynomial ring over a field", pp. 421–433 in *Gröbner bases and applications* (Linz, 1998), edited by B. Buchberger and F. Winkler, London Math. Soc. Lecture Note Ser. **251**, Cambridge Univ. Press, Cambridge, 1998.

[Nordbeck 2002] P. Nordbeck, "SAGBI bases under composition", *J. Symbolic Comput.* **33**:1 (2002), 67–76.

[Robbiano and Sweedler 1990] L. Robbiano and M. Sweedler, "Subalgebra bases", pp. 61–87 in *Commutative algebra* (Salvador, 1988), edited by W. Bruns and A. Simis, Lecture Notes in Math. **1430**, Springer, Berlin, 1990.

[Stillman and Tsai 1999] M. Stillman and H. Tsai, "Using SAGBI bases to compute invariants", *J. Pure Appl. Algebra* **139**:1-3 (1999), 285–302.

[Sturmfels 1996] B. Sturmfels, *Gröbner bases and convex polytopes*, University Lecture Series **8**, Amer. Math. Soc., Providence, 1996.

[Teissier 1975] B. Teissier, "Appendice", pp. 145–199 in *Le problème des modules pour les branches planes*, by O. Zariski, Centre de Mathématiques de l'École Polytechnique, Paris, 1975. Second edition, Hermann, 1986.

[Torstensson 2002] A. Torstensson, "Canonical bases for subalgebras on two generators in the univariate polynomial ring", *Beiträge Algebra Geom.* **43**:2 (2002), 565–577.

ANA BRAVO
DEPARTAMENTO DE MATEMÁTICAS
UNIVERSIDAD AUTÓNOMA DE MADRID
28049 MADRID
SPAIN
abravo@msri.org